Contributions to Management Science

The series *Contributions to Management Science* contains research publications in all fields of business and management science. These publications are primarily monographs and multiple author works containing new research results, and also feature selected conference-based publications are also considered. The focus of the series lies in presenting the development of latest theoretical and empirical research across different viewpoints.

This book series is indexed in Scopus.

Hasan Dinçer • Serhat Yüksel
Muhammet Deveci
Editors

Decision Making in Interdisciplinary Renewable Energy Projects

Navigating Energy Investments

Springer

Editors
Hasan Dinçer
Bostancı Mah.
Istanbul Medipol University
Kadıköy, Istanbul, Türkiye

Serhat Yüksel
Ortaköy Mah.
Istanbul Medipol University
Beşiktaş, Istanbul, Türkiye

Muhammet Deveci
Department of Industrial Engineering,
Turkish Naval Academy
National Defence University
Istanbul, Türkiye

ISSN 1431-1941 ISSN 2197-716X (electronic)
Contributions to Management Science
ISBN 978-3-031-51531-6 ISBN 978-3-031-51532-3 (eBook)
https://doi.org/10.1007/978-3-031-51532-3

This Springer imprint is published by the registered company Springer Nature Switzerland AG
The registered company address is: Gewerbestrasse 11, 6330 Cham, Switzerland

Contents

Chapter 1
An Analysis of Alternative Energy Sources and Applications in Maritime Transportation with a Strategic Management Approach

Filiz Mızrak

1.1 Introduction

Sea transportation is a very convenient transportation method in terms of cost and efficiency. In this respect, its place in the sustainable global economy is very valuable. In the maritime trade sector, the common expectation of the sector stakeholders is that the transportation business is cheaper, faster, and sustainable. Despite the rapid development of communication tools, this situation causes many serious problems in practice, such as marine pollution from ships, shipborne transport of living species that disrupt biodiversity, fluctuations in fuel prices, changes in international trade routes, cyberattacks, data theft, and geopolitical tensions (Xu et al., 2019).

Another of these problems is that although the maritime trade volume has increased approximately 2.5 times, the per capita income and population density have not increased at the same rate. As a result of this situation, the greenhouse gas released uncontrollably from the stacks of thousands of ships under different ownerships, registered to different flags and operating in international waters, causes global warming, and this situation cannot be adequately controlled (García-Olivares et al., 2018).

More than 80% of the world trade is carried out by sea. In general, fossil fuels are used in the machines ships. Looking at the statistical data, it has been revealed that the ships cause 3% of the global carbon dioxide (CO2) emissions, 15% of the nitrogen oxide (NOx) emissions, and 6% of the sulfur dioxide (SO2) emissions. Maritime industry representatives and researchers have started to work on the use of alternative energy systems on ships to fulfill the obligations of international conventions, to turn to cleaner energy sources against increasing environmental sensitivity

F. Mızrak (✉)
Logistics Management, Beykoz University, İstanbul, Turkey
e-mail: filizmizrak@beykoz.edu.tr

H. Dinçer et al. (eds.), *Decision Making in Interdisciplinary Renewable Energy Projects*, Contributions to Management Science,
https://doi.org/10.1007/978-3-031-51532-3_1

and to use cheaper energy sources due to rising energy costs (Eti et al., 2022). Researchers have focused on the use of alternative energy systems such as solar energy, wind energy, and onshore energy supply on ships to increase efficiency in the use of electrical energy (Van Fan et al., 2018).

In this context, minimizing the negative effects of ships on the environment emerges as an important field. The aim of the study is to shed light into the use of alternative energy systems in ship technology and contribute to investment decisions of maritime industry. In the second part of the study, the severity of the emission problem has been stressed with a detailed literature review. Furthermore, solar energy and wind energy have been suggested as alternative resources in the maritime industry with specific examples. Finally, onshore supply system has been explained benefiting from latest studies in the literature.

1.2 Exhaust Emissions from Ships

Compared to other types of transportation in the world, maritime transportation is the most preferred type of transportation because it is more economical and can carry a larger amount of cargo at once. About 93% of the shipping process on ships is provided by diesel machines. Around 67% of these machines are four-stroke and 26% are two-stroke diesel machines. In some studies, the diesel engine utilization rate of the entire maritime transport fleet in the world is stated as 99% (Jahangiri et al., 2018). The main reasons why diesel engines are preferred in ship propulsion systems can be listed as low-maintenance repair and spare parts costs and cheaper fuel used (Eti et al., 2023). The end products of combustion of fossil fuels used in main engines and generators constitute exhaust emissions. Therefore, almost all the exhaust emissions from ships originate from diesel engines used on ships (Xing et al., 2020).

Fossil fuels used in ships can be listed as heavy fuels, low-sulfur heavy fuel, marine distilled fuels, and recently ultralow-sulfur marine gas oil (MGO) in accordance with international rules. Generally, heavy fuel is used in ship diesel engines with low-speed high-fuel consumption capacity, and marine distilled fuel is generally used in medium- and high-speed ship diesel engines. Apart from these, the use of natural gas, which has fewer polluting properties, on ships has recently started to become widespread. To reduce sulfur oxide (SOX) emissions, ultralow sulfur-containing marine gas oil (MGO) has recently been used in ship diesel engines as required by international rules (Zhao et al., 2021).

With the expansion of the world trade volume, the number of ships used in maritime transportation has also increased. Naturally, there has been a significant increase in fossil fuel consumption and, accordingly, in exhaust emissions. Exhaust emissions from a ship's main engine most commonly include NOX, SOX, CO2, PM, HC, and CO gases. Therefore, it is crucial to look for alternative energy sources or improve application to take precautions against further pollution (Tichavska et al., 2019).

1.3 Solar Energy and Its Usage in Marine Vehicles

Solar energy is the radiant energy released from the conversion of hydrogen gas in the sun to helium and the fusion reactions that occur during this transformation. Energy in various wavelengths is emitted from the sun, which is a thermonuclear reactor, and only two billionth of the energy emitted from the entire surface of the sun comes to the earth. Radiation energy comes from the sun to the earth in 8 min (Yang et al., 2020a, 2020b). Not all the solar radiation reaches the earth's surface, 35% of it is reflected by the earth's atmosphere, 45% reaches the earth's surface by passing through the atmosphere, while 20% of the remaining radiation is kept in the atmosphere (Martínez et al., 2023). With the increase in energy prices and global climate change, the importance given to the use of renewable energy sources is increasing rapidly, and the most important of the known renewable energy sources has been solar energy (Yuan et al., 2018).

It is necessary to examine whether solar energy systems are applicable on ships; there are some existing factors that may affect this applicability, and the factors in question can be listed as follows (Nguyen & Dong, 2020):

- Availability of adequate deck area exposed to the sun
- Availability of a suitable grid-connected PV solar system
- Techno-economic selection of existing solar panels
- Scientific preparation of the system layout

The sun deck area plays an important role in the amount of solar energy absorbed by the solar panels; the deck area varies depending on the functions consisting of various factors such as ship type, ship characteristics, structure, and arrangement of onboard machinery. The more solar panels placed on the deck, the higher the amount of sunlight that these panels can collect, so that the collected energy can fill the batteries faster and provide the necessary energy to the system completely (Zhu et al., 2017). On the other hand, the amount of energy obtained is important in addition to the large deck area, the cloudiness of the air, and the low angle and low-intensity light level in the winter months. Charging time ranges from 4 to 16 h on a battery, depending on surface area and light conditions. Solar energy can be calculated based on geographic location, solar panel area, and panel efficiency (Wang et al., 2019).

The economic importance of the onboard solar energy system is to reduce fossil fuel consumption and make it a power source for electrical devices. Thus, the initial investment, maintenance and repair costs of fossil fuel powered auxiliary machines will be avoided. The application of this system to the ship is aimed at both cost reduction and minimization of emissions (Manickavasagam et al., 2019). By storing the energy obtained from solar panels in batteries, the energy required for lighting, air conditioning, and some light electric vehicles can be met with the energy coming from the solar panel while cruising or while the ship is in port, without the need for generators to operate. In this way, a more environmentally friendly ship design could be made by saving fuel (Krčum et al., 2018).

1.4 Wind Power and Its Usage in Marine Vehicles

With the increase in the world population and the progress of industrialization accompanying technological developments, the need for energy is increasing day by day. However, it is known that there are concerns about traditional-based energy sources, which are mainly used to meet this need, and it harms the environment (Dinçer et al., 2023a, 2023b, 2023c, 2023d, 2023e, 2023f). At this point, the contribution of renewable energy sources to electricity production is an important issue as it can be a solution. In addition, in recent years, it has been seen from industrial investments and applications as well as literature studies that there is an interest in the use of solar and wind energy-based power generation systems (Msigwa et al., 2022).

The investments made in the installed power-oriented wind energy in the last 10 years prove that there will be a serious interest in the future energy policy and that wind energy will play an even more effective role. Electricity generation from wind power not only provides economic benefits but also benefits the protection of the ecological environment. On the other hand, it is one of the effective factors in reducing or eliminating the countries' dependence on raw materials and energy from foreign countries (Carayannis et al., 2023). It is seen that the effective and efficient use of this type of energy to meet the growing energy demand is a very important issue and has been investigated in detail in the literature studies. It is also predicted that wind energy will play an even more active role in the future (Sadorsky, 2021).

The wind has a large size and scattered energy and is in constant motion on the Earth. Wind energy is an economical, clean, and renewable energy source that can contribute significantly to energy production. Today, the energy cost per kWh produced by wind energy is close to the cost of energy produced by coal. With a new technology to be developed in the field of wind energy, it is predicted that the amount of production will increase much more with the reduction of costs (Rehmatulla et al., 2017).

Wind turbines, which are used to convert the aerodynamic energy of the wind into mechanical energy, are the key devices in the energy conversion phase. The mechanical energy obtained by wind turbines is converted into electrical energy using an electrical generator. The electrical energy produced is transferred to the power unit in order to be converted into a form suitable for the consumer's demand. Wind energy is also called wind energy conversion system in the commercial and scientific world due to these conversion stages (Lundquist et al., 2019).

For centuries, world trade was conducted in the oceans only by ships powered by wind energy. But over time, fossil fuel-powered vehicles were invented, and wind and sun, known as renewable energy, were ignored. However, today it is clear that more use of fossil energy sources, which harm nature, will bring the end of life, so it is an alternative to fossil fuel vehicles that make our modern standard of living possible. What we are looking for is that we should turn to renewable energies. For this purpose, it is obvious that it is necessary to turn to renewable energy solutions such as wind and solar energy to power boats, freighters, and other ships (Gutierrez-Romero et al., 2019).

It should not be forgotten that this system is possible with globalization and economic growth. Looking at the history of ship designs, it is possible to design ships that can work with wind energy. For many years, scientists have worked on wind turbines to design various turbines, and in our age, these turbines can be turned into systems that power ships with a superior design that will further develop them (Sorensen et al., 2017).

Wind energy, one of the renewable energy types, is preferred in both conventional sails and propulsion systems in modern designs. These designs include the fletner rotors, the kite system, the drag sails, the soft sails, the wing sails, and the wind turbines. In propulsion systems installed through these devices, wind energy often does not serve as the main propulsion element of a typical ship. However, in bad weather conditions with high wind intensity, they can enable the ship to reach the desired speed by saving fuel without giving full sailing. A ship driven entirely by wind energy, on the other hand, can have a maximum carrying capacity of 3,000 to 10,000 tons in size, to be used in certain coastal voyages (Tillig & Ringsberg, 2020).

Toward the end of the twentieth century, the high increases in oil prices made wind energy a focus of great interest for commercial ships (Kumar et al., 2019a, 2019b). Some ships were built either new or converted. These include the 1600 DWT tanker Shin Aitoku Maru and the 26,000 dwt bulk carrier Usuki Pioneer. Japanese engineers designed and incorporated rigid sails into ships to reduce the fuel consumption of ships. Japan Machinery Development Association has proven that rigid sails reduce fuel consumption by designing Japanese ships such as Shin Aitoku Maru and Usuki Pioneer equipped with rigid sails. Thanks to these sails, it has been observed that the fuel consumption has decreased by about 10%–30%. Various rigid sails have been applied to a few small ships, but they have not been widely used on large ships due to the many engineering and operational difficulties (Prasanth et al., 2021).

1.5 Alternative Energy Applications on Ships

Electricity generation from solar energy is realized by PV panel, and electricity generation from wind energy is realized by wind turbine (Yüksel et al., 2023). Today, although the use of such renewable energy sources is encountered in ships with low-power needs, it is now seen that renewable energy sources are also used in ships engaged in commercial activities (Yuan et al., 2020). The ferry type ship "SolarSailor," built for commercial purposes by the Australian company OCIUS, is one of the first applications made in this context. The ship has a hybrid structure that can operate with solar, wind, battery, and LPG. The ship's energy needs are primarily met by solar, wind, and battery systems. If alternative energy sources are not sufficient, LPG fuel system is activated on board. The ship has the potential to reduce fuel emissions by 35% and emissions by 50% (Pan et al., 2021).

The cargo ship "M/V Auriga Leader," on the other hand, has managed to attract attention with its solar energy application. With 328 PV panels installed on the ship, an installed power capacity of approximately 40 kW was provided. With the solar energy system, approximately 10% of the electrical energy need on the ship can be met (Yang et al., 2020a, 2020b), The boat named "Turanor PlanetSolar," on the other hand, has the feature of being the first sea vehicle to tour the world, operating only with PV panel application. This project has been one of the important steps that excited the industry toward the use of renewable energy sources on ships, and the concept is a remarkable study showing that watercraft can be operated with zero fuel and zero emissions in the future (Nguyen et al., 2021).

In addition to the use of renewable energy sources such as solar and wind energy on ships, the application of energy supply from the shore, which is one of the alternative energy systems, has also become a subject of interest in the maritime sector. The energy supply system from the shore is expressed as the fact that the ships meet their electrical energy needs from the national grid instead of their own generators while they are at the port (Beriro et al., 2022).

The electrical energy taken from the national grid is transferred to the shore power supply panel located on the port side. Here, the electrical energy is adjusted to the appropriate voltage and frequency values for the ship to be connected, and the connection to the ship's main electrical panel is made. After this stage, the generator system on the ship is deactivated, and the required electrical energy needs are met from the national grid (Xu et al., 2023). According to a study, it has been estimated that if all ships in European ports use the shore power supply system by 2030, their annual electrical energy consumption will be 3543 GWh. This value corresponds to 0.1% of the 2012 electrical energy consumption of the countries in the European region. In addition, in the study, it was reported that the application of energy supply from the shore has the potential to reduce the CO2 emissions from ships in European ports by 800,000 tons (Kumar et al., 2019a, 2019b).

Onshore energy supply system is one of the solutions used to reduce the amount of exhaust emission in areas close to people's living spaces, especially in the port periods of ships. Under normal conditions, the required electrical power requirements are met by fossil fuel generators while the ships are at the berth. Exhaust emission resulting from the combustion of fossil fuels has many negative effects on the environment and constitutes an important expense item as fuel consumption among ship operating costs. As the first application area of the system, it is seen that there are large-capacity ports visited by large-sized passenger and RO-RO type ships making regular voyages (Han et al., 2023). In the following period, the application area of the system gradually expanded. With the use of this method, the concept of "green port" is realized by releasing less exhaust emissions to the environment, and the fuel consumption of the ships is reduced (Gutierrez-Romero et al., 2019).

Necessary infrastructure works have been completed in some ports around the world, especially for ships carrying out commercial activities, and the shore energy supply system has been started to be implemented (Kou et al., 2023a, 2023b). It is currently available in 22 ports in Germany, the United States, Belgium, Finland, the

Netherlands, Sweden, Canada, and Norway. In these ports, the electrical energy needs of Ro-Ro, ROPAX, cruise, and container-type commercial ships can be met (Kumar et al., 2019a, 2019b).

The Port of Los Angeles of the United States of America, on the other hand, took its place in the literature as the world's first port with a shore power supply system for container ships, which started operating in 2004. Afterward, the necessary infrastructure was established for the connection of cruise type passenger ships to the shore energy supply system. It has the feature of being the first port where two cruise ships can connect to the system at the same time. The port can meet the electrical energy needs of ships up to 40 MW power capacity (Wang et al., 2021).

Vancouver Port, which is one of the largest ports in Canada, has the feature of providing energy from the first coast of Canada and the third of the world, which was built for cruise-type ships and whose infrastructure works were completed in 2009. An average of 16 tons of fuel and 50.6 tons of CO2 emissions are saved from each passenger ship using the energy supply application from the shore. The port also supplies electricity from shore for container ships. An average of 30.7 tons of fuel and 94.6 tons of CO2 emissions are saved from each container ship using the onshore energy supply application (Kizielewicz & Skrzeszewska, 2021).

In addition to the high potential of the shore energy supply system to reduce the exhaust gas emissions, noise, and vibration on the ship, the environmental saving rates to be obtained from this application may vary and sometimes cause more emissions. The most important reason for this is the raw material used for the electrical energy produced for the national grid. For this reason, it is important to analyze the emission values released for unit electrical energy in ship generators and in the national grid. If the ports obtain their own electrical energy with renewable energy sources, such environmental problems will be solved and the energy supply system from the shore will offer a completely environmentally friendly solution (Krämer & Czermański, 2020).

Considering such applications, it is seen that the maritime industry has an important orientation and desire for the use of alternative energy sources on ships. In particular, the use of alternative energy sources in ship types that continue their commercial activities will significantly reduce the negative environmental impacts caused by ships and will contribute significantly to the sustainability of the maritime sector.

1.6 Conclusion

Energy is the most important input of development and is one of the indicators used to measure production, consumption, and development levels. As a result of the inadequacy of existing resources to meet the ever-increasing energy need, the efforts to find and develop alternative energy sources have accelerated because energy production and consumption with classical methods, especially fossil fuel-based energy production and consumption, causes irreparable damages in nature (Moiseev et al.,

2023). Issues such as protecting the livable environment in the world and reducing the damage caused by climate change impose responsibility on all humanity. As a requirement of this responsibility, issues such as making national and international legal regulations, taking environmental effects into consideration in energy production technologies and resource selection, and paying maximum attention to efficiency in energy use are gaining priority and weight (Yüksel & Dinçer, 2023).

Therefore, instead of fossil fuel-based energy production systems and traditional production technologies that cause environmental pollution and climate change on a global scale, it has become necessary to find energy sources with less environmental impacts and provide sustainability and renewability and to develop new technologies (Wan et al., 2023). The fact that fossil resources are a limited and irreplaceable energy source and the environmental pollution caused by production and consumption methods have made it necessary to research and develop renewable, unlimited, and environmentally compatible resources and technologies. Renewable energy is defined as "inexhaustible energy obtained from the continuous flow of energy in natural processes." Fossil fuels, which are widely used today, are energy sources that do not regenerate and disappear after use. On the other hand, natural resources such as hydroelectric, solar, wind, wave, and geothermal are renewable and clean energy resources. Solar and wind have gained importance due to their conversion to electricity faster than other types of renewable energy (Ai et al., 2023).

The power needed to operate the ship's auxiliary machinery, together with onboard uses, can be met with the power obtained from solar energy. Thus, both air pollution and fuel consumption can be significantly reduced (Niu et al., 2023). Although solar panels are seen as a suitable alternative energy system for boats in this study, it is seen that there are some difficulties to be overcome before integrating this system into a ship. The most important of these is the fact that open deck space is a major constraint. It is possible to use other renewable energy sources besides solar energy on ships, as wind energy was used in ancient times, it may be possible to propel ships with new generation sails with new technology. The design of the new-generation wind sails is not far away with today's technology.

It is thought that the invention of ships that can operate with hybrid energy systems will be revolutionary developments in the next 20 years. It has been shown that the proposed system will provide fuel savings economically over many years of operation. Moreover, using wind power alongside solar power to contribute to the total energy supply on a ship is a smart start for an environmentally friendly ship design. The results showed the possibility of achieving an overall reduction in the ship's emissions. It should not be forgotten that the amount of each unit of emission reduced is very important for a more livable environment and the Earth. With this thought, it should be aimed to reduce emissions in the whole transportation field by constantly working on new technologies.

Furthermore, it is seen that sustainability activities in the maritime sector are not limited to ships only, and green port concepts are tried to be developed on the port side. The application of energy supply from the shore is also the most important part of the green port concept (Aksoy et al., 2023). In the current situation, ships can meet their electrical energy needs from the national grid with the application of energy supply from the shore. By taking this concept one step further, green ship and green port concepts will complement each other and the environmental and economic disadvantages in the maritime sector will be minimized if the ports provide their own electrical energy with renewable energy sources independently of the national grid. Ship concepts operated entirely with alternative energy sources and port concepts that produce their own electrical energy will be a topic of interest in the upcoming period. In future studies, the feasibility of such systems will be evaluated separately from technical, environmental, and economic perspectives.

References

Ai, R., Zheng, Y., Yüksel, S., & Dinçer, H. (2023). Investigating the components of fintech ecosystem for distributed energy investments with an integrated quantum spherical decision support system. *Financial Innovation, 9*(1), 27.

Aksoy, T., Yüksel, S., Dinçer, H., Hacioğlu, Ü., Mikhaylov, A., & Muyeen, S. M. (2023). Assessing the electricity production capacities of emerging markets for the sustainable investments. *IEEE Access, 11*, 24574–24590.

Beriro, D., Nathanail, J., Salazar, J., Kingdon, A., Marchant, A., Richardson, S., et al. (2022). A decision support system to assess the feasibility of onshore renewable energy infrastructure. *Renewable and Sustainable Energy Reviews, 168*, 112771.

Carayannis, E., Kostis, P., Dinçer, H., & Yüksel, S. (2023). Quality function deployment-oriented strategic outlook to sustainable energy policies based on quintuple innovation helix. *Journal of the Knowledge Economy*, 1–19.

Dinçer, H., Yüksel, S., Hacioglu, U., Yilmaz, M. K., & Delen, D. (2023a). Development of a sustainable corporate social responsibility index for performance evaluation of the energy industry: A hybrid decision-making methodology. *Resources Policy, 85*, 103940.

Dinçer, H., Eti, S., Gökalp, Y., Yüksel, S., & Çelebi, B. (2023b). The role of small-scale wind turbines in increasing the competitiveness of hospitals. In *Handbook of research on quality and competitiveness in the healthcare services sector* (pp. 336–352). IGI Global.

Dinçer, H., Yüksel, S., Mikhaylov, A., Muyeen, S. M., Chang, T., Barykin, S., & Kalinina, O. (2023c). CO2 emissions integrated fuzzy model: A case of seven emerging economies. *Energy Reports, 9*, 5741–5751.

Dinçer, H., Yüksel, S., Aksoy, T., Hacıoğlu, Ü., Mikhaylov, A., & Pinter, G. (2023d). Analysis of solar module alternatives for efficiency-based energy investments with hybrid 2-tuple IVIF modeling. *Energy Reports, 10*, 61–71.

Dinçer, H., Eti, S., Yüksel, S., Özdemir, S., Yílmaz, A. E., & Ergün, E. (2023e). Integrating data mining and fuzzy decision-making techniques for analyzing the key minimizing factors of carbon emissions. *Journal of Intelligent & Fuzzy Systems, 45*, 1–17.

Dinçer, H., Eti, S., Aksoy, T., Yüksel, S., Hacioglu, U., Mikhaylov, A., & Muyeen, S. M. (2023f). Analysis of environmental impact for material production investments using a novel soft computing methodology. *IEEE Access, 11*, 37987–38001. https://doi.org/10.1109/ACCESS.2023.3266524

Eti, S., Dinçer, H., Gökalp, Y., Yüksel, S., & Kararoğlu, D. (2023). *Identifying key issues to handle the inflation problem in the healthcare industry caused by energy prices: An evaluation with decision-making models* (Managing inflation and supply chain disruptions in the global economy) (pp. 162–178).

Eti, S., Dinçer, H., Yüksel, S., & Gökalp, Y. (2022). Analysis of the suitability of the solar panels for hospitals: A new fuzzy decision-making model proposal with the T-spherical TOP-DEMATEL method. *Journal of Intelligent & Fuzzy Systems, Preprint*, 1–13.

García-Olivares, A., Solé, J., & Osychenko, O. (2018). Transportation in a 100% renewable energy system. *Energy Conversion and Management, 158*, 266–285.

Gutierrez-Romero, J. E., Esteve-Pérez, J., & Zamora, B. (2019). Implementing onshore power supply from renewable energy sources for requirements of ships at berth. *Applied Energy, 255*, 113883.

Han, Z., Gong, L., Chen, H., & Yüksel, S. (2023). The role of mineral and oil resources in achieving a clean environment in emerging economies: An asymmetric analysis. *Resources Policy, 86*, 104099.

Jahangiri, S., Nikolova, N., & Tenekedjiev, K. (2018). An improved emission inventory method for estimating engine exhaust emissions from ships. *Sustainable Environment Research, 28*(6), 374–381.

Kizielewicz, J., & Skrzeszewska, K. (2021). Identifying actions to prepare electricity infrastructure in seaports for future power supplying cruise ships with energy from land. *Energies, 14*(23), 8173.

Kou, G., Pamucar, D., Dinçer, H., & Yüksel, S. (2023a). From risks to rewards: A comprehensive guide to sustainable investment decisions in renewable energy using a hybrid facial expression-based fuzzy decision-making approach. *Applied Soft Computing, 142*, 110365.

Kou, G., Yuksel, S., & Dincer, H. (2023b). A facial expression and expert recommendation fuzzy decision-making approach for sustainable business investments within the metaverse world. *Applied Soft Computing*, 110849.

Krämer, I., & Czermański, E. (2020). Onshore power one option to reduce air emissions in ports. In *Sustainability management forum| nachhaltigkeits management forum* (Vol. 28, pp. 13–20). Springer.

Krčum, M., Gudelj, A., & Tomas, V. (2018). Optimal design of Ship's hybrid power system for efficient energy. *Transactions on Maritime Science, 7*(01), 23–32.

Kumar, J., Memon, A. A., Kumpulainen, L., Kauhaniemi, K., & Palizban, O. (2019b). Design and analysis of new harbour grid models to facilitate multiple scenarios of battery charging and onshore supply for modern vessels. *Energies, 12*(12), 2354.

Kumar, J., Kumpulainen, L., & Kauhaniemi, K. (2019a). Technical design aspects of harbour area grid for shore to ship power: State of the art and future solutions. *International Journal of Electrical Power & Energy Systems, 104*, 840–852.

Lundquist, J. K., DuVivier, K. K., Kaffine, D., & Tomaszewski, J. M. (2019). Costs and consequences of wind turbine wake effects arising from uncoordinated wind energy development. *Nature Energy, 4*(1), 26–34.

Manickavasagam, K., Thotakanama, N. K., & Puttaraj, V. (2019). Intelligent energy management system for renewable energy driven ship. *IET Electrical Systems in Transportation, 9*(1), 24–34.

Martínez, L., Dinçer, H., & Yüksel, S. (2023). A hybrid decision making approach for new service development process of renewable energy investment. *Applied Soft Computing, 133*, 109897.

Moiseev, N., Mikhaylov, A., Dinçer, H., & Yüksel, S. (2023). Market capitalization shock effects on open innovation models in e-commerce: Golden cut q-rung orthopair fuzzy multicriteria decision-making analysis. *Financial Innovation, 9*(1), 55.

Msigwa, G., Ighalo, J. O., & Yap, P. S. (2022). Considerations on environmental, economic, and energy impacts of wind energy generation: Projections towards sustainability initiatives. *Science of the Total Environment*, 157755.

Nguyen, X. P., & Dong, V. H. (2020, May). A study on traction control system for solar panel on vessels. In *AIP conference proceedings* (Vol. 2235, p. 020016). AIP Publishing LLC.

Nguyen, H. P., Hoang, A. T., Nizetic, S., Nguyen, X. P., Le, A. T., Luong, C. N., et al. (2021). The electric propulsion system as a green solution for management strategy of CO2 emission in ocean shipping: A comprehensive review. *International Transactions on Electrical Energy Systems, 31*(11), e12580.

Niu, X., Yüksel, S., & Dinçer, H. (2023). Emission strategy selection for the circular economy-based production investments with the enhanced decision support system. *Energy, 274*, 127446.

Pan, P., Sun, Y., Yuan, C., Yan, X., & Tang, X. (2021). Research progress on ship power systems integrated with new energy sources: A review. *Renewable and Sustainable Energy Reviews, 144*, 111048.

Prasanth, K., Prakash, M. S., & Sivaprasad, K. (2021). Optimization of wind sail using computational fluid dynamics simulation. *International Journal of Vehicle Structures & Systems, 13*(4), 477–481.

Rehmatulla, N., Parker, S., Smith, T., & Stulgis, V. (2017). Wind technologies: Opportunities and barriers to a low carbon shipping industry. *Marine Policy, 75*, 217–226.

Sadorsky, P. (2021). Wind energy for sustainable development: Driving factors and future outlook. *Journal of Cleaner Production, 289*, 125779.

Sorensen, A. J., Skjetne, R., Bo, T., Miyazaki, M. R., Johansen, T. A., Utne, I. B., & Pedersen, E. (2017). Toward safer, smarter, and greener ships: Using hybrid marine power plants. *IEEE Electrification Magazine, 5*(3), 68–73.

Tichavska, M., Tovar, B., Gritsenko, D., Johansson, L., & Jalkanen, J. P. (2019). Air emissions from ships in port: Does regulation make a difference? *Transport Policy, 75*, 128–140.

Tillig, F., & Ringsberg, J. W. (2020). Design, operation and analysis of wind-assisted cargo ships. *Ocean Engineering, 211*, 107603.

Van Fan, Y., Perry, S., Klemeš, J. J., & Lee, C. T. (2018). A review on air emissions assessment: Transportation. *Journal of Cleaner Production, 194*, 673–684.

Wan, Q., Miao, X., Wang, C., Dinçer, H., & Yüksel, S. (2023). A hybrid decision support system with golden cut and bipolar q-ROFSs for evaluating the risk-based strategic priorities of fintech lending for clean energy projects. *Financial Innovation, 9*(1), 1–25.

Wang, L., Liang, C., Shi, J., Molavi, A., Lim, G., & Zhang, Y. (2021). A bilevel hybrid economic approach for optimal deployment of onshore power supply in maritime ports. *Applied Energy, 292*, 116892.

Wang, H., Oguz, E., Jeong, B., & Zhou, P. (2019). Life cycle and economic assessment of a solar panel array applied to a short route ferry. *Journal of Cleaner Production, 219*, 471–484.

Xing, H., Spence, S., & Chen, H. (2020). A comprehensive review on countermeasures for CO2 emissions from ships. *Renewable and Sustainable Energy Reviews, 134*, 110222.

Xu, W., Xing, Y., Liu, J., Wu, H., Cui, Y., Li, D., et al. (2019). Efficient water transport and solar steam generation via radially, hierarchically structured aerogels. *ACS Nano, 13*(7), 7930–7938.

Xu, X., Yüksel, S., & Dinçer, H. (2023). An integrated decision-making approach with golden cut and bipolar q-ROFSs to renewable energy storage investments. *International Journal of Fuzzy Systems, 25*(1), 168–181.

Yang, R., Yuan, Y., Ying, R., Shen, B., & Long, T. (2020b). A novel energy management strategy for a ship's hybrid solar energy generation system using a particle swarm optimization algorithm. *Energies, 13*(6), 1380.

Yang, R., Jiang, L., Du, K., Zhang, Y., Wang, L., & Li, K. (2020a, November). Research and experimentation on energy management system for inland diesel-electric hybrid power ships. In *2020 IEEE 8th international conference on computer science and network technology (ICCSNT)* (pp. 102–106). IEEE.

Yuan, Y., Wang, J., Yan, X., Li, Q., & Long, T. (2018). A design and experimental investigation of a large-scale solar energy/diesel generator powered hybrid ship. *Energy, 165*, 965–978.

Yuan, Y., Wang, J., Yan, X., Shen, B., & Long, T. (2020). A review of multi-energy hybrid power system for ships. *Renewable and Sustainable Energy Reviews, 132*, 110081.

Yüksel, S., & Dinçer, H. (2023). Sustainability analysis of digital transformation and circular industrialization with quantum spherical fuzzy modeling and golden cuts. *Applied Soft Computing, 138*, 110192.

Yüksel, S., Dinçer, H., & Uluer, G. S. (2023). Socio-economic indicators of nuclear energy investments: A study on countries. In *Inclusive developments through socio-economic indicators: New theoretical and empirical insights* (pp. 33–44). Emerald Publishing Limited.

Zhao, J., Wei, Q., Wang, S., & Ren, X. (2021). Progress of ship exhaust gas control technology. *Science of the Total Environment, 799*, 149437.

Zhu, Y., Zhou, S., Feng, Y., Hu, Z., & Yuan, L. (2017). Influences of solar energy on the energy efficiency design index for new building ships. *International Journal of Hydrogen Energy, 42*(30), 19389–19394.

Chapter 2
Development and Implementation of an Enterprise's Energy Efficiency

Laura M. Baitenova, Lyailya M. Mutaliyeva, and Tarek Ismail Mohamed

2.1 Introduction

The concept of efficient use of electricity, proposed by the Energy Research Institute of the United States, assumes the definition of energy efficiency, taking into account the amount of initial resources, not the end use of the energy contained in them, and energy flows only to the end user when making decisions about the methods of its use. The developers of the concept propose to compare the expansion of the company's activities and measures to improve the energy efficiency of production (Eti et al., 2022). At the same time, the financing of activities related to energy saving is compared with the financing of other priority projects, and the final choice is made in favor of those projects that provide the maximum economic effect. In short, the maximum amount of capital expenditure on energy-saving measures is essentially determined by their economic efficiency. The concept of energy-saving policy considers energy conservation as an economic category. The authors note that energy-saving measures are constantly becoming more expensive due to the growing cost of primary energy resources. Hence, it is concluded that the implementation of these measures directly depends on the ratio of the growth rate of unit costs for the production and processing of primary and unit costs for energy conservation (Bhuiyan et al., 2021; Dong et al., 2021, Mikhaylov, 2021b; Barykin et al. 2022a, 2022b; Liu

L. M. Baitenova (✉)
Almaty University of Power Engineering and Telecommunications (AUPET) named after G. Daukeev, Almaty, Kazakhstan

L. M. Mutaliyeva
L.N. Gumilyov Eurasian National University, Nur-Sultan, Kazakhstan

T. I. Mohamed
College of Mass Communication, Ajman University, Ajman, United Arab Emirates
e-mail: t.abdellatif@ajman.ac.ae

H. Dinçer et al. (eds.), *Decision Making in Interdisciplinary Renewable Energy Projects*, Contributions to Management Science,
https://doi.org/10.1007/978-3-031-51532-3_2

et al., 2022a, 2022b; Bhuiyan et al., 2022a, 2022b; Danish et al., 2022a, 2022b; Saqib et al., 2021; Mukhametov et al., 2021, Candila et al., 2021; Mikhaylov & Grilli, 2022, Li et al., 2022).

The "Energy Strategy of Russia" until 2030 sets out new requirements in the context of a changed economic and energy policy. It notes that the planned pace of development of the country requires a reduction in the specific energy intensity of GDP by 2–2.1 times by 2030 (compared to 2005) and the specific electrical capacity—by 2.2 times. The need for continuous monitoring of the energy efficiency of industrial enterprises and PC production is caused by the fact that energy prices have increased significantly over the past two decades. In this regard, the most important place in the management activities of companies has become occupied by energy management, which provides for the management of energy resources to the same extent as the management of any other production resources. Statistics show that industrial companies, "armed" with energy management, achieve a significant reduction in energy costs by improving the energy efficiency of production (Nie et al., 2020; Mikhaylov et al., 2023a, 2023b; Mikhaylov, 2021a; Varyash et al., 2020; Zhao et al., 2021; An & Mikhaylov, 2020; Alwaelya et al., 2021; Yumashev & Mikhaylov, 2020; Yumashev et al., 2020; Mutalimov et al., 2021; An & Mikhaylov, 2021).

When implementing energy management, many companies use the well-known method of targeted energy monitoring, which is used at industrial enterprises in Western Europe and the United States as part of the overall management structure. According to experts, the introduction reduces energy consumption by 10%–20% without additional costs for technology modernization (Khan et al., 2022; Dinçer et al., 2022a, 2022b; Badr et al., 2022; Barykin et al., 2022a, 2022b; Mehta et al., 2022; Kalinina et al. 2022; Shaikh et al., 2022; Mikhaylov et al., 2022a, 2022b; Nyangarika et al., 2022).

2.2 Literature Review

Describing the term "energy efficiency," we note that today it is quite widely used, suggesting a reduction in energy costs, reduction of carbon dioxide emissions, improving the safety of energy supply, reducing overall costs in order to increase business competitiveness, etc. We add that the term "energy efficiency" is dual in nature, since, on the one hand, it is an indicator of efficiency On the one hand, it is a factor of development, since the amount of energy introduced into the system also increases with the increase in the volume of production (Denisova et al., 2019; Nyangarika et al., 2019a, 2019b; Huang et al., 2021a, 2021b; Mikhaylov, 2018, 2022; Mikhaylov et al., 2019; Conteh et al., 2021; Sediqi et al., 2022; Khan et al., 2021; Bhuiyan et al., 2022a, 2022b; Liu et al., 2021a, 2021b; Daniali et al., 2021; Moiseev et al., 2023a, 2023b).

In the context of this study, it should be noted that at the present stage of development, the vast majority of countries in the world face global challenges related to the need to reduce the energy intensity and environmental intensity of economic growth. This necessity is caused by frequent anomalous natural crisis phenomena (melting glaciers, rising or falling sea levels, changing climatic conditions, etc.) and forces us to look for an economic model of development that would establish a balance between the interests of nature and human needs (Mikhaylov et al., 2023a, 2023b; Wang et al. 2023; He et al., 2023; Irshad et al., 2023; Iqbal et al., 2023; Aksoy et al., 2023).

As such a model, the "green economy" model is considered, which assumes the production, distribution, and consumption of goods and services that should lead to an increase in human well-being in the long term, without exposing future generations to significant environmental crises or environmental scarcity (Dayong et al., 2020; Mikhaylov et al., 2018; Nyangarika et al., 2018; Danish et al., 2020, 2021; An et al., 2021; Uyeh et al., 2021; Tamashiro et al., 2021, 2023; Shaikh et al., 2021).

According to the UNEP definition, a green economy is "an economy that increases people's well-being and ensures social justice, while significantly reducing risks to the environment and its degradation." At the same time, one of the priority components of the green economy is a radical increase in energy efficiency. The transition to a "green economy" model should be accompanied by the introduction of new low-carbon technologies that take into account the reduction of greenhouse gas emissions (Eti et al., 2023). The United Nations Framework Convention on Climate Change (UNFCCC) and the Kyoto Protocol were among the first documents to officially limit greenhouse gas emissions. These documents served to create a new type of resource—quotas for greenhouse gas emissions. In Russia, the Kyoto Protocol to the UN Framework Convention on Climate Change was ratified on October 22, 2004, and entered into force (An et al., 2019a, 2019b, 2020a, 2020b, 2020c; 2022; Mikhaylov & Tarakanov, 2020; Moiseev et al., 2020, 2021; Gura et al., 2022; Dooyum et al., 2020; Mikhaylov et al., 2020; Mikhaylov, 2020a, 2020b, 2020c; Dinçer et al., 2023a, 2023b; Ivleva et al., 2023; Aissou et al., 2023; Mikhaylov, 2023).

Assessing the consequences of actions under the Kyoto Protocol, experts note positive aspects in terms of limiting the consumption of fuel and energy resources, which reduce the risks of depletion of mineral reserves, but, at the same time, express concerns about the possible containment of economic development. For the Russian industry, the entry into force of the Kyoto Protocol has a number of positive consequences, which include the following:

- In accelerating the process of technical and technological re-equipment
- In allocating rights to greenhouse gas emissions and then attracting investment for energy efficiency projects in exchange for transferring these reductions to investors
- In improving the energy efficiency of industrial complexes in general and individual technological processes in particular

2.3 The Creation of Concept of Energy Efficiency

In accordance with the green economy strategy, the priority task for the Russian economy is a structural and technological shift in the manufacturing industries associated with ferrous and nonferrous metallurgy. It should be noted that measures for the transition to a "green economy" related to the creation of energy-saving programs, the introduction of standards and certificates for equipment energy efficiency classes, and labeling of energy-efficient products are initiated "from above" and are still of a recommendatory nature. At the same time, it is already obvious that those production complexes that introduce low-carbon technologies and produce energy-efficient products are becoming more competitive in the market. The current concept of energy efficiency of production includes it in the energy policy, which is an integral part of the unified industrial policy, which encourages progressive structural changes in real production and, accordingly, has a direct impact on energy efficiency. Today, a different paradigm of industrial policy has developed as a result of the emergence of new industries that continue the development of the previous ones, but with different, more energy-efficient technologies, a different quantity and quality of labor resources, and labor in the cost of production takes an increasingly smaller share.

Increased attention to energy efficiency of production is explained by the significant benefits that this activity provides. This is, first of all, an increase in the profitability of production and the competitiveness of manufactured products. In addition, it is a solution to a complex of environmental and social problems that are very noticeable, especially in energy-intensive sectors of the economy (Martínez et al., 2023). And finally, it is the development of a socially oriented business within the framework of the concept of sustainable development. The efficiency strategy (ES) is a key functional strategy, which is a set of interrelated measures for choosing the technology and organizing the production of products that allow for sustainable energy-efficient consumption of energy resources. The energy strategy is an integral part of the company's development strategy and covers tasks related to improving the efficiency of energy use, as well as the development of enterprise energy supply systems (PCs) (Dinçer et al., 2023a, 2023b, 2023c, 2023d, 2023e, 2023f). From these positions, the energy efficiency of industrial enterprises (PCs) can be increased as a result of the implementation of three categories of ongoing projects (Carayannis et al., 2023):

- Investment projects aimed at upgrading production, increasing its productivity and improving product quality and accompanied by a reduction in the specific energy consumption per unit of production (Yüksel et al., 2023).
- Investment projects aimed at optimizing energy supply systems and energy-consuming equipment without changing the characteristics and productivity of the main production (Xu et al., 2023).
- Ongoing improvements in energy efficiency that provide acceptable operational efficiencies (Han et al., 2023).

Large enterprises and PCs develop strategies to improve production efficiency and competitiveness at four organizational levels. At the first level, a corporate strategy is developed for the entire company. At the second level, a business strategy is formed for each area of the company's activity. A functional strategy covering the activities of each functional unit is established at the third organizational level. Finally, at the fourth level, an operational strategy is developed (Kou et al., 2023a, 2023b).

The corporate strategy is largely decisive, since its development sets priorities for energy policy, which are specified at subsequent organizational levels. The business strategy of the enterprise (PC) is aimed at achieving optimal production productivity and, consequently, its optimal energy efficiency and energy saving (Moiseev et al., 2023a, 2023b).

The strategy of energy efficiency and energy saving, being functional, is the most important link in a business strategy, since it is included in almost all its components. Indeed, the financial component of the business strategy is related to the budgeting of the target energy program. The HR policy provides training of personnel to work on new equipment that increases energy efficiency and energy saving. The marketing strategy is aimed at increasing sales volumes in order to increase the profit volumes that form the budget of the energy program. The production strategy is aimed at a higher level of efficient use of equipment, which in turn increases the level of energy efficiency and energy saving (Yüksel & Dinçer, 2023).

Thus, the business strategy of an enterprise includes a whole range of measures to achieve competitive advantages, which are enhanced by the functional strategy of energy efficiency and energy saving and allow you to get a steady profit through the use of efficient technological processes that ensure a high level of energy efficiency.

It follows that the business strategy, in fact, forms a strategy of energy efficiency and energy saving, directly related to the key production ratio "income-costs-profit." And since fuel and energy resources are the defining constraints in the implementation of a business strategy, their effective use not only implements additional opportunities for competitive advantages but also provides them in the production process (Wan et al., 2023).

2.4 Conclusion

It should be noted that the development and implementation of an enterprise's energy efficiency and energy-saving strategy is significantly influenced by the systematization of factors for improving energy consumption efficiency. According to the authors, the works of many economists highlight the key factors for improving the energy efficiency of companies. Among them, rationing of fuel and energy consumption, price and tariff regulation, budget policy, and the monetary system are considered (Ai et al., 2023). Speaking about reserves for reducing energy losses, he highlights the quality of fuel and electricity, improvement of technological processes, modernization and renewal of equipment, introduction of energy-saving technologies, and a number of other factors that are essentially related to equipment modernization (Aksoy et al., 2023). It should be noted that in most of the mentioned works, there are no signs of systematization and, more often than usual, the most important market factors are not considered either (Niu et al., 2023).

References

Ai, R., Zheng, Y., Yüksel, S., & Dinçer, H. (2023). Investigating the components of fintech ecosystem for distributed energy investments with an integrated quantum spherical decision support system. *Financial Innovation, 9*(1), 27.

Aissou, A., Ahmed, H. S., Hussein, M. E., Hassan, M. R., Mikhaylov, A., & Mahmoud, E. (2023). Optimizing reliability of flow networks using RWGA subject to total capacity constraint. *Applied Mathematics and Information Sciences, 7*(4), 559–565. https://doi.org/10.18576/amis/170404

Aksoy, T., Yüksel, S., Dınçer, H., Hacioglu, U., Mikhaylov, A., & Muyeen, S. M. (2023). Assessing the electricity production capacities of emerging markets for the sustainable investment. *IEEE Access, 11*, 24574–24590.

Alwaelya, S. A., Yousif, N. B. A., & Mikhaylov, A. (2021). Emotional development in preschoolers and socialization. *Early Child Development and Care, 191*, 16.

An, J., & Mikhaylov, A. (2021). Analysis of energy projects financial efficiency and renewable energy generation in Russia. *Finance: Theory and Practice, 25*(5), 79–91.

An, J., & Mikhaylov, A. (2020). Russian energy projects in South Africa. *Journal of Energy in Southern Africa, 31*(3), 58–64.

An, J., Mikhaylov, A., & Richter, U. H. (2020a). Trade war effects: Evidence from sectors of energy and resources in Africa. *Heliyon, 6*, e05693. https://doi.org/10.1016/j.heliyon.2020.e05693

An, J., Mikhaylov, A., & Jung, S.-U. (2020b). The strategy of South Korea in the global oil market. *Energies, 13*(10), 2491.

An, J., Mikhaylov, A., & Kim, K. (2020c). Machine learning approach in heterogeneous group of algorithms for transport safety-critical system. *Applied Sciences, 10*(8), 2670.

An, J., Mikhaylov, A., & Jung, S.-U. (2021). A linear programming approach for Robust network revenue management in the airline industry. *Journal of Air Transport Management, 91*, 101979.

An, J., Mikhaylov, A., & Sokolinskaya, N. (2019a). Machine learning in economic planning: Ensembles of algorithms. *Journal of Physics: Conference Series, 1353*, 012126.

An, J., Mikhaylov, A., & Moiseev, N. (2019b). Oil price predictors: Machine learning approach. *International Journal of Energy Economics and Policy, 9*(5), 1–6.

An, J., Mikhaylov, A., Dinçer, H., & Yüksel, S. (2022). Economic modelling of electricity generation: Long short-term memory and Q-rung Orthopair fuzzy sets. *Heliyon, 8*, e10759.

Badr, A. M. N. A. A., Mohamed, T. I., Osman, N. A. Q., & Mikhaylov, A. (2022). A review of social media website users' interaction paths with governmental accounts during the COVID-19 pandemic. *Informatics, 9*, 50. https://doi.org/10.3390/informatics9030050

Barykin, S. E., Kapustina, I. V., Sergeev, S. M., Daniali, S. M., Kopteva, L. A., Semenova, G. N., Pryadko, I. P., Mikhaylov, A., Baboshkin, P., Datsyuk, P., & Senjyu, T. (2022a). Financial logistics models based on systematic approach improving management solutions. *F1000Research, 11*, 570. https://doi.org/10.12688/f1000research.111252.1

Barykin, S., Mikheev, A., Kiseleva, E., Putikhin, Y., Alekseeva, N., & Mikhaylov, A. (2022b). An empirical analysis of Russian regions' debt sustainability. *Economies, 10*(4), 296.

Bhuiyan, M. A., Dinçer, H., Yüksel, S., Mikhaylov, A., Danish, M. S. S., Pinter, G., Uyeh, D. D., & Stepanova, D. (2022a). Economic indicators and bioenergy supply in developed economies: QROF-DEMATEL and random forest models. *Energy Reports, 8*(2022), 561–570.

Bhuiyan, M. A., Zhang, Q., Khare, V., Mikhaylov, A., Pinter, G., & Huang, X. (2022b). Renewable energy consumption and economic growth nexus—a systematic literature review. *Frontiers in Environmental Science, 10*, 878394. https://doi.org/10.3389/fenvs.2022.878394

Bhuiyan, M. A., An, J., Mikhaylov, A., Moiseev, N., & Danish, M. S. S. (2021). Renewable energy deployment and COVID-19 measures for sustainable development. *Sustainability, 13*, 4418.

Candila, V., Maximov, D., Mikhaylov, A., Moiseev, N., Senjyu, T., & Tryndina, N. (2021). On the relationship between oil and exchange rates of oil-exporting and oil-importing countries: From the great recession period to the Covid-19 era. *Energies, 14*, 8065.

Carayannis, E., Kostis, P., Dinçer, H., & Yüksel, S. (2023). Quality function deployment-oriented strategic outlook to sustainable energy policies based on quintuple innovation helix. *Journal of the Knowledge Economy*, 1–19.

Conteh, F., Takahashi, H., Hemeida, A. M., Krishnan, N., Mikhaylov, A., & Senjyu, T. (2021). Hybrid grid-connected renewable power generation for sustainable electricity supply in Sierra Leone: Case study Lungi Town. *Sustainability, 13*(18).

Daniali, S. M., Barykin, S. E., Kapustina, I. V., Khortabi, F. M., Sergeev, S. M., Kalinina, O. V., Mikhaylov, A., Veynberg, R., Zasova, L., & Senjyu, T. (2021). Predicting volatility index according to technical index and economic indicators on the basis of deep learning algorithm. *Sustainability, 13*, 14011.

Danish, M. S. S., Bhattacharya, A., Stepanova, D., Mikhaylov, A., Grilli, M. L., Khosravy, M., & Senjyu, T. (2020). A systematic review of metal oxide applications for energy and environmental sustainability. *Metals, 10*(12), 1604.

Danish, M. S. S., Senjyu, T., Sabory, N. R., Khosravy, M., Grilli, M. L., Mikhaylov, A., & Majidi, H. (2021). A forefront framework for sustainable aquaponics modeling and design. *Sustainability, 13*, 9313.

Danish, M. S., Senjyu, T. S., Sabory, N. R., & Mikhaylov, A. (2022a). Lessons from COVID-19 conferring environmental re-engineering opportunity. In M. Danish & T. Senjyu (Eds.), *Eco-friendly and agile energy strategies and policy development* (pp. 99–118). IGI Global. https://doi.org/10.4018/978-1-7998-9502-2.ch005

Danish, M. S., Senjyu, T. S., Sabory, N. R., & Mikhaylov, A. (2022b). Interplay consequences of COVID-19 on global environmental sustainability. In M. Danish & T. Senjyu (Eds.), *Eco-friendly and agile energy strategies and policy development*. IGI Global. https://doi.org/10.4018/978-1-7998-9502-2.ch003

Dayong, N., Mikhaylov, A., Bratanovsky, S., Shaikh, Z. A., & Stepanova, D. (2020). Mathematical modeling of the technological processes of catering products production. *Journal of Food Process Engineering, 43*(2), 1–9.

Denisova, V., Mikhaylov, A., & Lopatin, E. (2019). Blockchain infrastructure and growth of global power consumption. *International Journal of Energy Economics and Policy, 9*(4), 22–29.

Dinçer, H., Yüksel, S., Mikhaylov, A., Barykin, S. E., Aksoy, T., & Hacioğlu, U. (2022a). Analysis of environmental priorities for green project investments using an integrated q-rung orthopair fuzzy modelling. *IEEE Access, 10*, 50996–51007. https://doi.org/10.1109/ACCESS.2022.3174058

Dinçer, H., Yüksel, S., Mikhaylov, A., Pinter, G., & Shaikh, Z. A. (2022b). Analysis of renewable-friendly smart grid technologies for the distributed energy investment projects using a hybrid picture fuzzy rough decision-making approach. *Energy Reports, 8*, 11466–11477. https://doi.org/10.1016/j.egyr.2022.08.275

Dinçer, H., Yüksel, S., Hacioglu, U., Yilmaz, M. K., & Delen, D. (2023a). Development of a sustainable corporate social responsibility index for performance evaluation of the energy industry: A hybrid decision-making methodology. *Resources Policy, 85*, 103940.

Dinçer, H., Eti, S., Gökalp, Y., Yüksel, S., & Çelebi, B. (2023b). The role of small-scale wind turbines in increasing the competitiveness of hospitals. In *Handbook of research on quality and competitiveness in the healthcare services sector* (pp. 336–352). IGI Global.

Dinçer, H., Yüksel, S., Mikhaylov, A., Muyeen, S. M., Chang, T., Barykin, S., & Kalinina, O. (2023c). CO2 emissions integrated fuzzy model: A case of seven emerging economies. *Energy Reports, 9*, 5741–5751.

Dinçer, H., Yüksel, S., Aksoy, T., Hacıoğlu, Ü., Mikhaylov, A., & Pinter, G. (2023d). Analysis of solar module alternatives for efficiency-based energy investments with hybrid 2-tuple IVIF modeling. *Energy Reports, 10*, 61–71.

Dinçer, H., Eti, S., Yüksel, S., Özdemir, S., Yı́lmaz, A. E., & Ergün, E. (2023e). Integrating data mining and fuzzy decision-making techniques for analyzing the key minimizing factors of carbon emissions. *Journal of Intelligent & Fuzzy Systems, 45*, 1–17.

Dinçer, H., Eti, S., Aksoy, T., Yüksel, S., Hacioglu, U., Mikhaylov, A., & Muyeen, S. M. (2023f). Analysis of environmental impact for material production investments using a novel soft computing methodology. *IEEE Access, 11*, 37987–38001. https://doi.org/10.1109/ACCESS.2023.3266524

Dong, B., Ikonnikova, I., Rogulin, R., Sakulyeva, T., & Mikhaylov, A. (2021). Environmental economic approach to optimization of transport communication in megacities. *Journal of Environmental Science and Health, Part A, 56*(6), 660–666.

Dooyum, U. D., Mikhaylov, A., & Varyash, I. (2020). Energy security concept in Russia and South Korea. *International Journal of Energy Economics and Policy, 10*(4), 102–107.

Eti, S., Dinçer, H., Gökalp, Y., Yüksel, S., & Kararoğlu, D. (2023). *Identifying key issues to handle the inflation problem in the healthcare industry caused by energy prices: An evaluation with decision-making models* (Managing inflation and supply chain disruptions in the global economy) (pp. 162–178).

Eti, S., Dinçer, H., Yüksel, S., & Gökalp, Y. (2022). Analysis of the suitability of the solar panels for hospitals: A new fuzzy decision-making model proposal with the T-spherical TOP-DEMATEL method. *Journal of Intelligent & Fuzzy Systems, Preprint*, 1–13.

Gura, D., Mikhaylov, A., Glushkov, S., Zaikov, M., & Shaikh, Z. A. (2022). Model for estimating power dissipation along the interconnect length in single on-chip topology. *Evolutionary Intelligence, 15*, 2369–2373. https://doi.org/10.1007/s12065-020-00407-7

Han, Z., Gong, L., Chen, H., & Yüksel, S. (2023). The role of mineral and oil resources in achieving a clean environment in emerging economies: An asymmetric analysis. *Resources Policy, 86*, 104099.

He, H., Huang, Y., Nakadomari, A., Masrur, H., Krishnan, N., Hemeida, A. M., Mikhaylov, A., Omine, E., & Senjyu, T. (2023). Potential and economic viability of green hydrogen production from seawater electrolysis using renewable energy in remote Japanese islands. *Renewable Energy, 202*, 1436–1447. https://doi.org/10.1016/j.renene.2022.12.046

Huang, Y., Masrur, H., Shigenobu, R., Hemeida, A. M., Mikhaylov, A., & Senjyu, T. (2021a). A comparative design of a campus microgrid considering a multi-scenario and multi-objective approach. *Energies, 14*, 2853.

Huang, Y., Yona, A., Takahashi, H., Hemeida, A. M., Mandal, P., Mikhaylov, A., Senjyu, T., & Lotfy, M. E. (2021b). Energy management system optimization of drug store operate electric vehicles charging station. *Sustainability, 14*, 1217836.

Iqbal, K. M. J., Khan, M. I., Mikhaylov, A., Shah, A. A., Yadykin, V., Filho, W. L., Tariq, M. A. U. R., & Ullah, W. (2023). Modeling principles, criteria and indicators to assess water sector governance for climate compatibility and sustainability. *Frontiers in Environmental Science, 11*, 989930.

Irshad, A. S., Ludin, G. A., Masrur, H., Ahmadi, M., Yona, A., Mikhaylov, A., Krishnan, N., & Senjyu, T. (2023). Optimization of grid-photovoltaic and battery hybrid system with most technically efficient PV technology after the performance analysis. *Renewable Energy, 207*, 714–730.

Ivleva, S., Hajimajeed, M., Kartashova, O., & Mikhaylov, A. (2023). A comparative study of regional variations in common visual impairments among school-age children. *Electronic Journal of General Medicine, 20*(5). https://doi.org/10.29333/ejgm/13264

Kalinina O.V., Barykin S.E., Sergeev S.M., Semenova G.N, Fatkullina A, Mikhaylov A, De La Poza E. (2022). Smart city perspectives in post-pandemic governance: Externalities reduction policy. F1000Research, 11, 1032. https://doi.org/10.12688/f1000research.123195.1.

Khan, A. A., Laghari, A. A., Gadekallu, T. R., Shaikh, Z. A., Javed, A. R., Rashid, M., Estrela, V. V., & Mikhaylov, A. (2022). A drone-based data management and optimization using metaheuristic algorithms and blockchain smart contracts in a secure fog environment. *Computers and Electrical Engineering, 102*, 108234. https://doi.org/10.1016/j.compeleceng.2022.108234

Khan, A. A., Shaikh, Z. A., Baitenova, L., Mutaliyeva, L., Moiseev, N., Mikhaylov, A., Laghari, A. A., Idris, S. A., & Alshazly, H. (2021). QoS-ledger: Smart contracts and metaheuristic for secure quality-of-service and cost-efficient scheduling of medical-data processing. *Electronics, 10*, 3083.

Kou, G., Pamucar, D., Dinçer, H., & Yüksel, S. (2023a). From risks to rewards: A comprehensive guide to sustainable investment decisions in renewable energy using a hybrid facial expression-based fuzzy decision-making approach. *Applied Soft Computing, 142*, 110365.

Kou, G., Yuksel, S., & Dincer, H. (2023b). A facial expression and expert recommendation fuzzy decision-making approach for sustainable business investments within the metaverse world. *Applied Soft Computing*, 110849.

Li, J., Dinçer, H., Yüksel, S., Mikhaylov, A., & Barykin, S. E. (2022). Bipolar q-ROF hybrid decision making model with golden cut for analyzing the levelized cost of renewable energy alternatives. *IEEE Access, 10*, 35900. https://doi.org/10.1109/ACCESS.2022.3168315

Liu, L., Kato, T., Mandal, P., Mikhaylov, A., Hemeida, A. M., & Senjyu, T. (2021a). Multi-area power system load frequency control by using model predictive control by introducing renewable energy sources. In *2021 International Conference on Science & Contemporary Technologies (ICSCT)* (pp. 1–6). https://doi.org/10.1109/ICSCT53883.2021.9642521

Liu, L., Kato, T., Mandal, P., Mikhaylov, A., Hemeida, A. M., Takahashi, H., & Senjyu, T. (2021b). Two cases studies of model predictive control approach for hybrid renewable energy systems. *AIMS Energy, 9*(6), 1241–1259.

Liu, Z.-J., Panfilova, E., Mikhaylov, A., & Kurilova, A. (2022a). COVID-19 crisis impact on the stability between parties in crowdfunding and crowdsourcing. *Wireless Personal Communications, 122*, 915–930.

Liu, Z., Panfilova, E., Mikhaylov, A., & Kurilova, A. (2022b). Assessing stability in the relationship between parties in crowdfunding and crowdsourcing projects during the COVID-19 crisis. *Journal of Global Information Management (JGIM), 30*(4), 1–18. https://doi.org/10.4018/JGIM.297905

Martínez, L., Dinçer, H., & Yüksel, S. (2023). A hybrid decision making approach for new service development process of renewable energy investment. *Applied Soft Computing, 133*, 109897.

Mehta, R., Verghese, J., Mahajan, S., Barykin, S., Bozhuk, S., Kozlova, N., Kapustina, I., Mikhaylov, A., Naumova, E., & Dedyukhina, N. (2022). Consumers' behavior in conversational commerce marketing based on messenger chatbots. *F1000Research, 11*(647). https://doi.org/10.12688/f1000research.122037.1

Mikhaylov, A., Dinçer, H., Yüksel, S., Pinter, G., & Shaikh, Z. A. (2023a). Bitcoin mempool growth and trading volumes: Integrated approach based on QROF multi-SWARA and aggregation operators. *Journal of Innovation & Knowledge, 8*(3), 100378.

Mikhaylov, A., Dinçer, H., & Yüksel, S. (2023b). Analysis of financial development and open innovation oriented fintech potential for emerging economies using an integrated decision-making approach of MF-X-DMA and golden cut bipolar q-ROFSs. *Financial Innovation, 9*(1), 10. https://doi.org/10.1186/s40854-022-00234-w

Mikhaylov, A. (2018). The yield of the carry trade strategy. *Finance: Theory and Practice, 22*(3), 52–63. https://doi.org/10.26794/2587-5671-2018-22-3-52-63

Mikhaylov, A. (2020a). Cryptocurrency market analysis from the open innovation perspective. *Journal of Open Innovation: Technology, Market, and Complexity, 6*(4), 197.

Mikhaylov, A. (2020b). Geothermal energy development in Iceland. *International Journal of Energy Economics and Policy, 10*(4), 31–35.

Mikhaylov, A. (2020c). Cryptocurrency market development: Hurst method. *Finance: Theory and Practice, 24*(3), 81–91.
Mikhaylov, A. (2021a). Development of Friedrich von Hayek's theory of private money and economic implications for digital currencies. *Terra Economicus, 19*(1), 53–62.
Mikhaylov, A. (2021b). Lichens as indicators of atmospheric pollution in urban ecosystems. *Israel Journal of Ecology & Evolution, 67*(1–2), 60–68. https://doi.org/10.1163/22244662-bja10016
Mikhaylov, A. (2022). Sustainable development and renewable energy: A new view to a global problem. *Energies, 15*, 1397. https://doi.org/10.3390/en15041397
Mikhaylov, A. (2023). Understanding the risks associated with wallets, depository services, trading, lending, and borrowing in the crypto space. *Journal of Infrastructure, Policy and Development, 6*, 2.
Mikhaylov, A., & Grilli, M. L. (2022). Machine learning methods and sustainable development: Metal oxides and multilayer metal-oxides. *Metals, 12*(4).
Mikhaylov, A., & Tarakanov, S. (2020). Development of Levenberg-Marquardt theoretical approach for electric network. *Journal of Physics: Conference Series, 1515*, 052006.
Mikhaylov, A., Yumashev, A., & Kolpak, E. (2022a). Quality of life, anxiety and depressive disorders in patients with extrasystolic arrhythmia. *Archives of Medical Science, 18*(2). https://doi.org/10.5114/aoms.2020.101359
Mikhaylov, A., Bhatti, I. M., Dinçer, H., & Yüksel, S. (2022b). Integrated decision recommendation system using iteration-enhanced collaborative filtering, golden cut bipolar for analyzing the risk-based oil market spillovers. *Computational Economics*. https://doi.org/10.1007/s10614-022-10341-8
Mikhaylov, A., Burakov, D., & Didenko, V. (2019). Relationship between oil price and macroeconomic indicators in Russia. *Finance: Theory and Practice, 23*(2), 105–116. https://doi.org/10.26794/2587-5671-2019-23-2-105-116
Mikhaylov, A., Moiseev, N., Aleshin, K., & Burkhardt, T. (2020). Global climate change and greenhouse effect. *Entrepreneurship and Sustainability Issues, 7*(4), 2897–2913. https://doi.org/10.9770/jesi.2020.7.4(21)
Mikhaylov, A., Sokolinskaya, N., & Nyangarika, A. (2018). Optimal carry trade strategy based on currencies of energy and developed economies. *Journal of Reviews on Global Economics, 7*, 582–592.
Moiseev, N., Mikhaylov, A., Dinçer, H., & Yüksel, S. (2023a). Effects of market capitalization shocks for open innovation models in e-commerce industry: Golden cut QROF MCDM analysis. *Financial Innovation, 9*(1), 32.
Moiseev, N., Mikhaylov, A., Dinçer, H., & Yüksel, S. (2023b). Market capitalization shock effects on open innovation models in e-commerce: Golden cut q-rung orthopair fuzzy multicriteria decision-making analysis. *Financial Innovation, 9*(1), 55.
Moiseev, N., Mikhaylov, A., Varyash, I., & Saqib, A. (2020). Investigating the relation of GDP per capita and corruption index. *Entrepreneurship and Sustainability Issues, 8*(1), 780–794.
Moiseev, N., Sorokin, A., Zvezdina, N., Mikhaylov, A., Khomyakova, L., & Danish, M. S. S. (2021). Credit risk theoretical model on the base of DCC-GARCH in time-varying parameters framework. *Mathematics, 9*, 2423. https://doi.org/10.3390/math9192423
Mukhametov, A., Bekhorashvili, A., Avdeenko, A., & Mikhaylov, A. (2021). The impact of growing legume plants under conditions of biologization and soil cultivation on Chernozem fertility and productivity of rotation crops. *Legume Research, 44*, 12.
Mutalimov, V., Kovaleva, I., Mikhaylov, A., & Stepanova, D. (2021). Assessing regional growth of small business in Russia. *Entrepreneurial Business and Economics Review, 9*(3), 119–133. https://doi.org/10.15678/EBER.2021.090308
Nie, D., Panfilova, E., Samusenkov, V., & Mikhaylov, A. (2020). E-learning financing models in Russia for sustainable development. *Sustainability, 12*(11), 4412.
Niu, X., Yüksel, S., & Dinçer, H. (2023). Emission strategy selection for the circular economy-based production investments with the enhanced decision support system. *Energy, 274*, 127446.

Nyangarika, A., Mikhaylov, A., Muyeen, S. M., Yadykin, V., Mottaeva, A., Pryadko, I., Barykin, S., Fomenko, N., Rykov, G., & Shvandar, K. (2022). Energy stability and decarbonization in developing countries: Random Forest approach for forecasting of crude oil trade flows and macro indicators. *Frontiers in Environmental Science, 10*, 1031343. https://doi.org/10.3389/fenvs.2022.1031343

Nyangarika, A., Mikhaylov, A., & Richter, U. (2019a). Influence oil price towards economic indicators in Russia. *International Journal of Energy Economics and Policy, 9*(1), 123–129.

Nyangarika, A., Mikhaylov, A., & Richter, U. (2019b). Oil price factors: Forecasting on the base of modified auto-regressive integrated moving average model. *International Journal of Energy Economics and Policy, 9*(1), 149–159.

Nyangarika, A., Mikhaylov, A., & Tang, B.-J. (2018). Correlation of oil prices and gross domestic product in oil producing countries. *International Journal of Energy Economics and Policy, 8*(5), 42–48.

Saqib, A., Chan, T.-H., Mikhaylov, A., & Lean, H. H. (2021). Are the responses of sectoral energy imports asymmetric to exchange rate volatilities in Pakistan? Evidence from recent foreign exchange regime. *Frontiers in Energy Research, 9*, 614463. https://doi.org/10.3389/fenrg.2021.614463

Sediqi, M. M., Nakadomari, A., Mikhaylov, A., Krishnan, N., Lotfy, M. E., Yona, A., & Senjyu, T. (2022). Impact of time-of-use demand response program on optimal operation of Afghanistan real power system. *Energies, 15*, 296.

Shaikh, Z. A., Kraikin, A., Mikhaylov, A., & Pinter, G. (2022). Forecasting stock prices of companies producing solar panels using machine learning methods. *Complexity, 2022*, 9186265. https://doi.org/10.1155/2022/9186265

Shaikh, Z. A., Moiseev, N., Mikhaylov, A., & Yüksel, S. (2021). Facile synthesis of copper oxide-cobalt oxide/nitrogen-doped carbon (Cu2O-Co3O4/CN) composite for efficient water splitting. *Applied Sciences, 11*, 9974. https://doi.org/10.3390/app11219974

Tamashiro, K., Omine, E., Krishnan, N., Mikhaylov, A., Hemeida, A. M., & Senjyu, T. (2023). Optimal components capacity based multi-objective optimization and optimal scheduling based MPC-optimization algorithm in smart apartment buildings. *Energy & Buildings, 278*, 112616. https://doi.org/10.1016/j.enbuild.2022.112616

Tamashiro, K., Alharbi, T., Mikhaylov, A., Hemeida, A. M., Krishnan, N., Lotfy, M. E., & Senjyu, T. (2021). Investigation of home energy management with advanced direct load control and optimal scheduling of controllable loads. *Energies, 14*, 7314. https://doi.org/10.3390/en14217314

Uyeh, D. D., Asem-Hiablie, S., Park, T., Kim, K.-M., Mikhaylov, A., Woo, S., & Ha, Y. (2021). Could Japonica rice be an alternative variety for increased global food security and climate change mitigation? *Food, 10*, 1869. https://doi.org/10.3390/foods10081869

Varyash, I., Mikhaylov, A., Moiseev, N., & Aleshin, K. (2020). Triple bottom line and corporate social responsibility performance indicators for Russian companies. *Entrepreneurship and Sustainability Issues, 8*(1), 313–331.

Wan, Q., Miao, X., Wang, C., Dinçer, H., & Yüksel, S. (2023). A hybrid decision support system with golden cut and bipolar q-ROFSs for evaluating the risk-based strategic priorities of fintech lending for clean energy projects. *Financial Innovation, 9*(1), 1–25.

Wang, Q., Guo, J., Li, R., Mikhaylov, A., & Moiseev, N. (2023). Does technical assistance alleviate energy poverty in sub-Saharan African countries? A new perspective on spatial spillover effects of technical assistance. *Energy Strategy Reviews, 45*, 101047. https://doi.org/10.1016/j.esr.2022.101047

Xu, X., Yüksel, S., & Dinçer, H. (2023). An integrated decision-making approach with golden cut and bipolar q-ROFSs to renewable energy storage investments. *International Journal of Fuzzy Systems, 25*(1), 168–181.

Yüksel, S., & Dinçer, H. (2023). Sustainability analysis of digital transformation and circular industrialization with quantum spherical fuzzy modeling and golden cuts. *Applied Soft Computing, 138*, 110192.

Yüksel, S., Dinçer, H., & Uluer, G. S. (2023). Socio-economic indicators of nuclear energy investments: A study on countries. In *Inclusive developments through socio-economic indicators: New theoretical and empirical insights* (pp. 33–44). Emerald Publishing Limited.

Yumashev, A., & Mikhaylov, A. (2020). Development of polymer film coatings with high adhesion to steel alloys and high wear resistance. *Polymer Composites, 41*(7), 2875–2880.

Yumashev, A., Ślusarczyk, B., Kondrashev, S., & Mikhaylov, A. (2020). Global indicators of sustainable development: Evaluation of the influence of the human development index on consumption and quality of energy. *Energies, 13*, 2768.

Zhao, Q., Cherkasov, S., Avdeenko, A., Kondratenko, L., & Mikhaylov, A. (2021). Integral estimate of the added value of agricultural crops in the synergetic agriculture on the example of Vicia faba (Vicia faba L.). *Polish Journal of Environmental Studies, 30*(5), 1–9. https://doi.org/10.15244/pjoes/131949

Chapter 3
Institutional Quality, Trade Openness, and Renewable Energy Consumption in the GCC Countries

Ibrahim Nandom Yakubu, Ayhan Kapusuzoglu, and Nildag Basak Ceylan

3.1 Introduction

Climate change is currently one of the most crucial concerns for policymakers and energy practitioners in recent decades. Conventional energy sources such as fossil oil and coal burning are among the widely cited primary factors promoting ecological deterioration. In response to global climate change challenges, governments and international organizations have strived to adopt policy frameworks that will aid in diversifying the energy portfolio (Bekun & Alola, 2022). Renewable energies are quickly gaining traction in the global energy grid and have been in the foreground in the climate change discourse due to their contribution to decreasing global warming. To ensure sustainable growth and the resilience of energy systems, policies encouraging the integration of renewable energies have been broadly pursued.

The GCC countries have the highest energy usage in the world, consuming about 6,260 MW per capita yearly, with natural gas and crude oil being the primary energy sources (Elrahmani et al., 2021). The rising domestic energy demand, as well as surges in environmental pollution resulting from industrial growth, has prompted the need for a cleaner energy source in the GCC states (Al-Sarihi & Mansouri, 2022). The adoption of technology that harnesses renewable sources of energy is an essential step to mitigating such environmental concerns. That is, the GCC states will be able to lower pollution by incorporating REN into the energy grid. In the GCC region, wind and solar systems presently account for 90% of REN investments, owing to their remarkable prospects and modern technological breakthroughs

I. N. Yakubu
School of Business, University for Development Studies, Tamale, Ghana

A. Kapusuzoglu (✉) · N. B. Ceylan
Faculty of Business, Ankara Yildirim Beyazit University, Ankara, Turkey
e-mail: akapusuzoglu@aybu.edu.tr; nbceylan@aybu.edu.tr

H. Dinçer et al. (eds.), *Decision Making in Interdisciplinary Renewable Energy Projects*, Contributions to Management Science,
https://doi.org/10.1007/978-3-031-51532-3_3

that have rendered them user-friendly and more cost-effective (Darwish, 2016). Solar energy is the most desirable REN source in the GCC, and its capacity is expected to increase to 65,490 MW by the year 2030 (IRENA, 2019).

Given the perceived role of REN in the reduction of environmental degradation, there have been growing attempts to unearth the factors that motivate the demand for renewable energies (Omri & Nguyen, 2014; Ergun et al., 2019; Amoah et al., 2020; Akarsu & Gümüşoğlu, 2019; Gershon & Emekalam, 2021; Bekun & Alola, 2022). Although numerous related researches sought to shed more light on the significant factors determining REN consumption, the effect of institutions on REN demand is still relatively new and understudied. Globally, institutions play a vital role in enhancing developmental goals including energy policy implementation. By investigating the role of governance quality, we offer a new dimension to the drivers of REN demand, particularly in the GCC nations. We also examine the effect of trade openness on the consumption of REN. For a successful energy transition, technological innovation is imperative. Trade openness will therefore aid in removing the constraints for the transfer of sophisticated technologies to facilitate the energy transition drive of the GCC states.

This study significantly contributes to the limited body of empirical research that has been done on the influence of institutional quality on REN demand. The remaining segments of the study are divided as follows: Section 3.2 delves into the literature on the drivers of REN. Section 3.3 outlines the research methodology. In Sect. 3.4, the results and discussions are presented, while Sect. 3.5 concludes the study.

3.2 Literature Review

3.2.1 Theoretical Background

A theoretical review of renewable energy consumption involves examining the existing literature on the topic to identify key concepts, theories, and research findings. The review can focus on various aspects of renewable energy consumption, such as the drivers of adoption, the barriers to adoption, and the policy measures that can promote renewable energy consumption. One key concept in the literature on renewable energy consumption is the "energy ladder," which refers to the idea that households and businesses will adopt renewable energy technologies as their income and energy needs increase. According to this theory, households and businesses will start with basic energy needs, such as lighting and heating, before moving on to more advanced energy needs, such as electricity for appliances and transportation. Another key concept in the literature on renewable energy consumption is the "rebound effect," which refers to the idea that increased energy efficiency can lead to increased energy consumption. According to this theory, individuals and businesses may use the savings from energy efficiency measures to purchase more energy-intensive goods and services, offsetting some of the energy savings. Barriers to the adoption of renewable energy include lack of access to finance, lack of information and awareness, and lack of proper regulations and standards. Policies that can promote renewable energy consumption include financial incentives, such as subsidies and tax credits, and regulations, such as renewable portfolio standards and building codes. Additionally, the literature also suggests that cultural and societal values play a crucial role in the adoption of renewable energy. For example, countries with a strong tradition of environmental conservation may have a higher rate of renewable energy adoption compared to countries where economic development is prioritized over environmental concerns. Overall, the literature on renewable energy consumption suggests that a combination of economic, social, and political factors influence the adoption of renewable energy. To effectively promote renewable energy consumption, it is important to consider these factors and design policies that address the specific barriers and drivers in a given context.

3.2.2 Empirical Literature

The demand for REN has risen dramatically in recent years, prompting research into the key factors influencing the usage of REN sources. Omri et al. (2015), for example, researched the drivers of REN usage across 64 nations. The author used data from 1990 to 2011 and used different panel methodologies. The study discovered that rising foreign trade, GDP, and carbon dioxide (CO_2) emissions primarily impact the adoption of REN. Attiaoui et al. (2017) studied the linkages between economic growth, CO_2 emission, and usage of REN using the pooled mean group estimation. The authors used data from African countries spanning from 1990 and 2011. The findings demonstrated that while CO_2 emissions harm REN consumption, the role of economic growth in boosting REN demand is negligible. Toumi and Toumi (2019) investigated the influence of economic freedom on African REN development. Using the Dynamic Ordinary Least Squares method, the study indicated that property rights and tax burden as indicators of economic freedom diminish demand for REN, whereas business and trade freedom spurs the use of renewables. Ergun et al. (2019) established that the rise in GDP per capita and the degree of human development decrease the use of REN. The authors, however, noted that FDI inflow positively affects REN demand. Uzar (2020a) explored the connection between income inequality and the usage of REN. The results indicated that as income inequality declines, the share of REN rises significantly. The author also demonstrated that reducing corruption promotes the use of REN. Utilizing data from Belt and Road Initiative countries, Khan et al. (2021) revealed that REN is adversely affected by FDI, technological innovation, and economic growth and the growth of the financial sector positively influences renewables. In a sample of OECD countries, Polcyn et al. (2021) found that while labor force decreases REN consumption, the influence of CO_2 emissions and GDP per capita on REN demand is positive and significant. Oluoch et al. (2021) revealed that education and GDP per capita directly impact REN usage in Sub-Saharan Africa. They also established that CO_2 emission and life expectancy reduce the share of renewables in the long-term. Chao et al. (2022) found that financial integration and ICT increase the share of REN consumption in China. Bunyaminu and Yakubu (2022) revealed that financial development and FDI diminish REN consumption in Africa while trade openness has no significant influence on REN consumption. Murshed (2018) noted that increased trade openness stimulates the use of REN in South Asia. Murshed (2020) also revealed that the boom in the trade of ICT goods boosts the consumption of REN in South Asia. Yakubu et al. (2022) reported similar results in the case of the BRICS economies.

While various factors have been identified as drivers of REN consumption, the influence of institutions on REN demand has been overlooked, presenting the prospects for further research. Among the few studies, Uzar (2020b) revealed that institutional quality influences REN consumption positively. Wu and Broadstock (2015) found a direct link between REN usage and institutional quality in the case of emerging economies. In the OECD countries, Wang et al. (2022) observed that quality institutions increase the demand for REN. Applying the ARDL approach, Saadaoui and Chtourou (2022) discovered that institutional quality boosts REN usage in Tunisia. Rahman and Sultana (2022) established that REN consumption is significantly motivated by institutional effectiveness in emerging market economies. Huang et al. (2022) evidenced that governance quality in the ASEAN region matters significantly for the consumption of REN. In Table 3.1, we provide a summary of the empirical findings on the drivers of REN consumption.

Table 3.1 Summary of the empirical literature on the drivers of REN consumption

Author(s)	Country/region	Period	Impact of variables on renewable energy
Omri et al. (2015)	Cross-country	1990–2011	CO_2 emissions (+) Trade (+) GDP (+)
Wu and Broadstock (2015)	Emerging economies	1990–2010	Institutional quality (+) Financial development (+)
Attiaoui et al. (2017)	Africa	1990–2011	CO_2 emissions (−) GDP (no relationship)
Lau et al. (2018)	Malaysia	1980–2015	FDI (+) Economic growth (+) Trade openness (−)
Toumi and Toumi (2019)	Africa	1996–2019	Property right (−) Tax burden (−) Business freedom (+) Trade freedom (+)
Ergun et al. (2019)	Africa	1990–2013	FDI (+) GDP per capita (−) Human development (−)
Uzar (2020a)	Developed and developing countries	2000–2015	Income equality (+) Corruption control (+) CO_2 emissions (+)

Table 3.1 (continued)

Author(s)	Country/region	Period	Impact of variables on renewable energy
Uzar (2020b)	Cross-country	1990–2015	Institutional quality (+) Economic growth (−) CO_2 emissions (+)
Muse (2021)	Nigeria	1990–2019	Oil prices (+) Population growth (+) Trade openness (−)
Khan et al. (2021)	Belt & Road Initiative countries	2000–2014	FDI (−) Economic growth (−) Technological innovation (−) Financial development (+)
Polcyn et al. (2021)	OECD countries	2000–2018	Total labor force (−) CO_2 emissions (+) GDP per capita (+)
Oluoch et al. (2021)	Sub-Saharan Africa	1998–2014	Education (+) GDP per capita (+) Life expectancy (−) CO_2 emissions (−)
Zhang et al. (2021)	BRIC economies	1997–2018	Financial development (+) FDI (+)
Bunyaminu and Yakubu (2022)	Africa	1990–2018	Financial development (−) FDI (−) Trade openness (no effect)
Chishti and Dogan (2022)	REN consuming economies	1990–2017	ICT (+) Economic growth (+) Globalization (+)
Khezri et al. (2022)	ASEAN countries	2000–2018	Energy consumption (+) Population growth (+)
Yakubu et al. (2022)	BRICS	2000–2017	ICT trade (+) FDI (no effect)
Wang et al. (2022)	OECD	1997–2019	Institutional quality (+) Per capita income (+) Political risk (−) Economic globalization (−)
Saadaoui and Chtourou (2022)	Tunisia	1984–2017	Financial development (−) Institutional quality (+) Economic growth (+)

(continued)

Table 3.1 (continued)

Author(s)	Country/region	Period	Impact of variables on renewable energy
Huang et al. (2022)	ASEAN countries	1980–2018	Governance quality (+) Urbanization (−) Trade openness (−) FDI (+)
Chu et al. (2023)	Energy consuming economies	1990–2017	Energy security risk (+/−) Economic complexity (+)
Iqbal et al. (2023)	Pakistan	1980–2019	Financial development (+) CO_2 emission (+) GDP (+)
Yank and Song (2023)	OECD economies	2000–2019	FDI (−) Urbanization (−) Trade (+)
Dogan et al. (2023)	EU countries	1995–2019	Oil prices (+) Environmental taxes (−) Economic growth (+) Energy taxes (−)
Kilinc-Ata and Dolmatov (2023)	BRICS and OECD economies	2000–2020	Research and development expenditure (+) Economic growth (+) CO_2 emissions (−) Electricity usage (−)
Foye (2023)	Nigeria	1990–2020	Inflation (+) Oil rent (−) Exchange rate (+)
Sachan et al. (2023)	BRICS	1990–2015	Human development (+) Gross fixed capital formation (−) CO_2 emissions (−)
Zhou et al. (2023)	United States	2005–2021	Climate policy uncertainty (+)
Akpanke et al. (2023)	West Africa	1990–2021	Public sector credit (+) Inflation (−) FDI (+)

3.3 Research Methodology

3.3.1 Data and Description of Variables

To realize the study's objective, we employ data from six countries in the GCC which covers from 1996 to 2020. Alphabetically, the sampled countries are Bahrain, Kuwait, Oman, Qatar, Saudi Arabia, and the United Arab Emirates. The data for the study are retrieved from the World Governance Indicators (WGI) and World Development Indicators (WDI) of the World Bank. The dependent variable is REN consumption. The main independent variables are institutional quality and trade openness. In measuring institutional quality, the study employs different sub-indices of governance and political institutions to generate an institutional quality index utilizing the principal component analysis (PCA). We also use FDI, urbanization, and population growth as control factors. The various definitions of the variables are presented in Table 3.2.

3.3.2 Model Specification

To examine the impact of institutional quality and trade openness on the consumption of renewable energy while adjusting for the impact of FDI, urbanization, and population growth, the empirical model below is specified.

Table 3.2 The definitions of the variables

Variables	Acronym	Definitions	Sources
Renewable energy consumption	REN	Share of renewable energy in total final energy consumption	WDI
Institutional quality	INSQ	The principal component of "control of corruption, government effectiveness, political stability, regulatory quality, rule of law, and voice and accountability"	WGI
Trade openness	TOP	Sum of exports and imports (% of GDP)	WDI
Foreign direct investment	FDI	Net inflows of FDI as a percentage of GDP)	WDI
Urbanization	URB	Urban population (% of total population)	WDI
Population growth	POP	Annual population growth rate	WDI

$$\ln \text{REN}_{it} = \alpha_0 + \beta_1 \ln \text{INSQ}_{it} + \beta_2 \ln \text{TOP}_{it} + \beta_3 \ln \text{FDI}_{it} + \beta_4 \text{lnURBit} + \beta_5 \ln \text{POPit} + \varepsilon_{it} \tag{3.1}$$

where the abbreviations REN, INSQ, TOP, FDI, URB, and POP are defined in Table 3.2. The subscripts i and t connote the respective countries and period, respectively, and the error term is symbolized by ε. The variables are transformed into natural logarithms.

3.3.3 *Analytical Technique*

To provide accurate results, the study conducts some preliminary tests prior to model estimation. Specifically, we perform the panel unit root and cointegration tests to check the stationarity status and the degree of association among the variables in the study. Following the preliminary tests, the study employs Pesaran et al.'s pooled mean group (PMG) under the panel ARDL. The PMG approach allows us to evaluate the long- and short-term effects of explanatory variables on renewable energy consumption.

The model for the PMG estimator is expressed as

$$Y_{it} = \sum_{j=1}^{p} \delta_i Y_{it-j} + \sum_{j=0}^{q} \lambda_{ij} X_{it-1} + \mu_i + \varepsilon_{it} \tag{3.2}$$

From Eq. (3.2), the dependent factor is denoted by Y. The coefficient matrix is defined by λ_{ij} and X_{it-1} is the vector of the explanatory factors. The unit-specific effects are shown by μ_i. The optimal lags are represented by letters p and q. Considering the error correction term (ECT), the following equation is specified based on Eq. (3.2):

$$\Delta Y_{it} = \vartheta_i \left\{ \text{Y}_{i(t-1)} - \emptyset_i X_{it} \right\} + \sum_{j=1}^{p-1} \pi_{ij} \Delta Y_{i(t-j)} + \sum_{j=0}^{q-1} \lambda_{ij} \Delta X_{i(t-1)} + \mu_i + \varepsilon_{it} \tag{3.3}$$

where the ECT is represented by $\{Y_i(t-1) - \phi_i X_{it}\}$ and π_{ij} and λ_{ij} are the short-term parameters. The vector representing the long-term estimates is ϕ_i.

3.4 Empirical Results

Table 3.3 summarizes the panel unit root test findings based on the Fisher-Phillips and Perron (1988) test. To avoid estimate biases, any variables with second-order integration must be excluded. The results demonstrate that only trade openness and urbanization exhibit stationarity at level. However, the test results demonstrate

Table 3.3 The results of panel unit root test—panel cointegration test—PMG estimate test

Panel unit root test results/Fisher-PP				
Level		Statistic	Prob.	
lnREN		4.601	0.799	
lnINSQ		4.475	0.923	
lnTOP		32.052	0.001	
lnFDI		18.186	0.110	
lnURB		61.435	0.000	
lnPOP		9.206	0.685	
First difference		**Statistic**	**Prob.**	
lnREN		36.031	0.000	
lnINSQ		33.799	0.000	
lnTOP		68.190	0.000	
lnFDI		114.129	0.000	
lnURB		20.168	0.064	
lnPOP		29.004	0.004	
Panel cointegration test results: Alternative hypothesis—common AR coefs. (within-dimension)				
	Statistic	**Prob.**	**Weighted statistic**	**Prob.**
Panel v	−1.539	0.938	−1.789	0.963
Panel rho	0.985	0.838	1.075	0.859
Panel PP	−2.216	0.013	−2.005	0.023
Panel ADF	−1.052	0.146	0.030	0.512
Panel cointegration test results: Alternative hypothesis—individual AR coefs. (between-dimension)				
	Statistic	Prob.		
Group rho	1.667	0.952		
Group PP	−2.293	0.011		
Group ADF	−1.556	0.060		
PMG estimate results/long-run estimates				
lnINSQ		−0.4320*** (0.0224)		
lnTOP		−4.6057*** (0.5034)		
lnFDI		−0.7996*** (0.0923)		
lnURB		−165.7930*** (7.0496)		
lnPOP		1.2741*** (0.0647)		
PMG Estimate Results/short-run estimates				
ECT		−2.5653 (2.5468)		
lnINSQ		−0.6868 (0.7380)		
lnTOP		4.2673 (4.3199)		
lnFDI		1.2786 (1.2363)		
lnURB		−638.8257** (284.1607)		
lnPOP		−13.8874 (13.8929)		
C		865.8228 (859.0665)		

Notes: Standard errors are in ( ); ** and *** are significance at 5% and 1%, respectively

stationarity for all the variables at first difference. Hence, it can be inferred that the variables are stationary without the need for additional differencing.

The panel cointegration method developed by Pedroni (1999) is used in the study to look at the long-term associations among the variables. The cointegration test results are displayed in Table 3.3. As indicated in the table, this technique specifies seven test statistics with estimates that are predicated on within and between dimensions. Under this test approach, the null hypothesis is that there is no cointegration among the series, as opposed to the alternative hypothesis of cointegration. Considering that some of the test statistics are statistically significant, we repudiate the null hypothesis and deduce that our series are cointegrated.

Given the cointegration of the examined variables, we estimate the long-run coefficients employing the pooled mean group (PMG) technique, and the findings are presented in Table 3.3. The regression results show that institutional quality exerts a negative significant effect on REN consumption in the long term. This suggests that institutional quality dampens REN demand in the GCC. To put it another way, while institutions play a vital role in enhancing developmental goals including energy policy implementation, the quality of institutions in the GCC does not auger well for increasing the demand for REN sources. This result stands in stark contrast to other research that evidenced a positive link between institutional quality and the use of REN (see Wang et al., 2022; Saadaoui & Chtourou, 2022; Huang et al., 2022). The short-run influence of institutions on renewables is also negative albeit insignificant.

The adoption of REN is negatively and significantly impacted by trade openness in the long run, suggesting that trade liberalization is unable to promote the transfer of renewable technologies or increase the demand for REN sources in the GCC region. The result concurs with those of Lau et al. (2018) and Huang et al. (2022). We find that trade openness in the short run positively affects REN though insignificantly.

FDI is found to decrease REN development significantly in the long term. This hints that the REN transition drive of the GCC through FDI has not been effective. Although the result is in contradiction to several earlier research, it is in alignment with Khan et al. (2021) and Bunyaminu and Yakubu (2022). We document a negative significant link between urbanization and REN consumption in both short- and long-run periods, suggesting that an increase in urban population reduces the demand for REN in the GCC. The study, however, finds that general population growth increases the usage of REN sources in the region. The implication is that as population grows, pressure on conventional energy sources like fossil fuels increases, driving up the demand for alternate effective energy sources like renewables.

3.5 Conclusion and Discussion

Recent years have seen resurgence in interest in the factors that influence the use of REN in an attempt to develop relevant policies that might aid in combating environmental deterioration. This study provides insight into how institutional quality and trade openness influence REN demand in the GCC countries. The study uses data from six countries in the GCC, spanning the years 1996 to 2020, and applies the pooled mean group method. We find that REN consumption in the long term is negatively influenced by institutional quality and trade openness. Controlling for other factors, we note that FDI and urbanization also lower the share of renewables in the GCC, while the influence of population growth on REN demand is positive in the long run.

In light of the findings, some key policy recommendations can be drawn. To begin with, countries in the GCC must strengthen their institutional frameworks to revert the decreasing effect that institutional quality has on renewable energy development. Strong institutions will pave the way for a better environment and prohibit initiatives with destructive environmental consequences. Given the detrimental effect of trade openness on renewables, policies should be enacted to ensure that trade in the region includes the transfer of renewable-related technologies. Similarly, for FDI to be a vehicle for encouraging the use of renewables, policymakers must guarantee that investment inflows are channeled toward green initiatives. Additionally, population growth has to go hand in hand with initiatives to enhance the use of REN in the GCC countries. Future research may examine the link between institutional quality and REN usage in each of the GCC nations to provide a clearer picture of how institutional quality influences renewable energy in these nations. A comparison with other regional blocs, such as the BRICS economies, would also provide a fascinating picture of the institutions-REN development link. Other panel approaches, in addition to the PMG, might be utilized to assure robust results.

References

Akarsu, G., & Gümüşoğlu, N. K. (2019). What are the main determinants of renewable energy consumption? A panel threshold regression approach. *Anadolu University Journal of Social Sciences, 19*(2), 1–22.

Akpanke, T. A., Deka, A., Ozdeser, H., & Seraj, M. (2023). Does foreign direct investment promote renewable energy use? An insight from west African countries. *Renewable Energy Focus, 44*, 124–131.

Al-Sarihi, A., & Mansouri, N. (2022). Renewable energy development in the Gulf Cooperation Council Countries: Status, barriers, and policy options. *Energies, 15*(5), 1923.

Amoah, A., Kwablah, E., Korle, K., & Offei, D. (2020). Renewable energy consumption in Africa: The role of economic well-being and economic freedom. *Energy, Sustainability and Society, 10*(1), 1–17.

Attiaoui, I., Toumi, H., Ammouri, B., & Gargouri, I. (2017). Causality links among renewable energy consumption, CO2 emissions, and economic growth in Africa: Evidence from a panel ARDL-PMG approach. *Environmental Science and Pollution Research, 24*(14), 13036–13048.

Bekun, F. V., & Alola, A. A. (2022). Determinants of renewable energy consumption in agrarian sub-Sahara African economies. *Energy, Ecology and Environment, 7*(3), 227–235.

Bunyaminu, A., & Yakubu, I. N. (2022). Green energy demand and financial development: Evidence from Africa. In *Finance for sustainability in a turbulent economy* (pp. 26–48). IGI Global.

Chao, T., Yunbao, X., Chengbo, D., Bo, L., & Ullah, S. (2022). Financial integration and renewable energy consumption in China: Do education and digital economy development matter? *Environmental Science and Pollution Research, 30*, 12944–12952.

Chishti, M. Z., & Dogan, E. (2022). Analyzing the determinants of renewable energy: The moderating role of technology and macroeconomic uncertainty. *Energy & Environment*. https://doi.org/10.1177/0958305X221137567

Chu, L. K., Ghosh, S., Doğan, B., Nguyen, N. H., & Shahbaz, M. (2023). Energy security as new determinant of renewable energy: The role of economic complexity in top energy users. *Energy, 263*, 125799.

Darwish, A. S. (2016). Sustainable green smart buildings: Future energy survivor. *ISESCO Journal of Science and Technology, 12*(21), 35–42.

Dogan, E., Hodžić, S., & Šikić, T. F. (2023). Do energy and environmental taxes stimulate or inhibit renewable energy deployment in the European Union? *Renewable Energy, 202*, 1138–1145.

Elrahmani, A., Hannun, J., Eljack, F., & Kazi, M. K. (2021). Status of renewable energy in the GCC region and future opportunities. *Current Opinion in Chemical Engineering, 31*, 100664.

Ergun, S. J., Owusu, P. A., & Rivas, M. F. (2019). Determinants of renewable energy consumption in Africa. *Environmental Science and Pollution Research, 26*(15), 15390–15405.

Foye, V. O. (2023). Macroeconomic determinants of renewable energy penetration: Evidence from Nigeria. *Total Environment Research Themes, 5*, 100022.

Gershon, O., & Emekalam, P. (2021). Determinants of renewable energy consumption in Nigeria: A Toda Yamamoto approach. In *IOP conference series: Earth and environmental science* (Vol. 665, p. 012005). IOP Publishing.

Huang, Y., Ahmad, M., & Ali, S. (2022). The impact of trade, environmental degradation and governance on renewable energy consumption: Evidence from selected ASEAN countries. *Renewable Energy, 197*, 1144–1150.

Iqbal, S., Wang, Y., Ali, S., Haider, M. A., & Amin, N. (2023). Shifting to a green economy: Asymmetric macroeconomic determinants of renewable energy production in Pakistan. *Renewable Energy, 202*, 234–241.

IRENA. (2019). Renewable energy market analysis: GCC 2019.

Khan, A., Chenggang, Y., Hussain, J., & Kui, Z. (2021). Impact of technological innovation, financial development and foreign direct investment on renewable energy, non-renewable energy and the environment in belt & road initiative countries. *Renewable Energy, 171*, 479–491.

Khezri, M., Karimi, M. S., Mamkhezri, J., Ghazal, R., & Blank, L. (2022). Assessing the impact of selected determinants on renewable energy sources in the electricity mix: The case of ASEAN countries. *Energies, 15*(13), 4604.

Kilinc-Ata, N., & Dolmatov, I. A. (2023). Which factors influence the decisions of renewable energy investors? Empirical evidence from OECD and BRICS countries. *Environmental Science and Pollution Research, 30*(1), 1720–1736.

Lau, L. S., Yii, K. J., Lee, C. Y., Chong, Y. L., & Lee, E. H. (2018). Investigating the determinants of renewable energy consumption in Malaysia: An ARDL approach. *International Journal of Business and Society, 19*(3), 886–903.

Murshed, M. (2018). Does improvement in trade openness facilitate renewable energy transition? Evidence from selected South Asian economies. *South Asia Economic Journal, 19*(2), 151–170.

Murshed, M. (2020). An empirical analysis of the non-linear impacts of ICT-trade openness on renewable energy transition, energy efficiency, clean cooking fuel access and environmental sustainability in South Asia. *Environmental Science and Pollution Research, 27*(29), 36254–36281.

Muse, B. O. (2021). Analysis of Co2 emission and economic growth as potential determinants of renewable energy demand in Nigeria: A nonlinear autoregressive distributed lags approach. *International Journal of Energy Economics and Policy, 11*(3), 510–516.

Oluoch, S., Lal, P., & Susaeta, A. (2021). Investigating factors affecting renewable energy consumption: A panel data analysis in sub-Saharan Africa. *Environmental Challenges, 4*, 100092.

Omri, A., & Nguyen, D. K. (2014). On the determinants of renewable energy consumption: International evidence. *Energy, 72*, 554–560.

Omri, A., Daly, S., & Nguyen, D. K. (2015). A robust analysis of the relationship between renewable energy consumption and its main drivers. *Applied Economics, 47*(28), 2913–2923.
Pedroni, P. (1999). Critical values for cointegration tests in heterogeneous panels with multiple regressors. *Oxford Bulletin of Economics and Statistics, 61*(S1), 653–670.
Phillips, P. C., & Perron, P. (1988). Testing for a unit root in time series regression. *Biometrika, 75*(2), 335–346.
Polcyn, J., Us, Y., Lyulyov, O., Pimonenko, T., & Kwilinski, A. (2021). Factors influencing the renewable energy consumption in selected European countries. *Energies, 15*(1), 108.
Rahman, M. M., & Sultana, N. (2022). Impacts of institutional quality, economic growth, and exports on renewable energy: Emerging countries perspective. *Renewable Energy, 189*, 938–951.
Saadaoui, H., & Chtourou, N. (2022). Do institutional quality, financial development, and economic growth improve renewable energy transition? Some evidence from Tunisia. *Journal of the Knowledge Economy*, 1–32.
Sachan, A., Sahu, U. K., Pradhan, A. K., & Thomas, R. (2023). Examining the drivers of renewable energy consumption: Evidence from BRICS nations. *Renewable Energy, 202*, 1402–1411.
Toumi, S., & Toumi, H. (2019). Asymmetric causality among renewable energy consumption, CO2 emissions, and economic growth in KSA: Evidence from a non-linear ARDL model. *Environmental Science and Pollution Research, 26*(16), 16145–16156.
Uzar, U. (2020a). Is income inequality a driver for renewable energy consumption? *Journal of Cleaner Production, 255*, 120287.
Uzar, U. (2020b). Political economy of renewable energy: Does institutional quality make a difference in renewable energy consumption? *Renewable Energy, 155*, 591–603.
Wang, E., Gozgor, G., Mahalik, M. K., Patel, G., & Hu, G. (2022). Effects of institutional quality and political risk on the renewable energy consumption in the OECD countries. *Resources Policy, 79*, 103041.
Wu, L., & Broadstock, D. C. (2015). Does economic, financial and institutional development matter for renewable energy consumption? Evidence from emerging economies. *International Journal of Economic Policy in Emerging Economies, 8*(1), 20–39.
Yakubu, I. N., Kapusuzoglu, A., & Ceylan, N. B. (2022). ICT trade and energy transition in the BRICS economies. In *Sustainability in energy business and finance* (pp. 13–24). Springer.
Yang, C., & Song, X. (2023). Assessing the determinants of renewable energy and energy efficiency on technological innovation: Role of human capital development and investment. *Environmental Science and Pollution Research*, 1–21.
Zhang, Y., Qamruzzaman, M., Karim, S., & Jahan, I. (2021). Nexus between economic policy uncertainty and renewable energy consumption in BRIC nations: The mediating role of foreign direct investment and financial development. *Energies, 14*(15), 4687.
Zhou, D., Siddik, A. B., Guo, L., & Li, H. (2023). Dynamic relationship among climate policy uncertainty, oil price and renewable energy consumption—findings from TVP-SV-VAR approach. *Renewable Energy, 204*, 722–732.

Chapter 4
Energy Efficiency Management in China

Alexey Mikhaylov

4.1 Introduction

Research economists both domestically and internationally use various criteria to evaluate developed strategies. By analyzing their work, it is possible to identify key evaluation criteria that include “compliance with the environment,” “competitive advantage,” and “efficiency.” The first criterion evaluates the adaptability of the strategy to the external environment while considering market threats and opportunities, industry development conditions, and competition severity (Aksoy et al., 2023). The “competitive advantage” criterion assesses the competitive capabilities of an enterprise by taking into account its strengths and weaknesses. The “efficiency” criterion confirms the positive dynamics of production and economic activity of the enterprise, strengthening its competitive position (Niu et al., 2023).

In the process of forecasting, a correlation and regression analysis of the dependence of PC energy efficiency indicators on indicators that characterize the conditions of economic development is performed (Ai et al., 2023). Next, the values of PC energy efficiency indicators are modeled as a result of the implementation of a portfolio of measures and energy-efficient PC development projects, which is the basis for building and implementing a competitive strategy for energy-efficient PC development. The final result for the third block is the forecast of PC competitiveness indicators achieved as a result of the energy efficiency development strategy (Mikhaylov et al., 2023b; Wang et al. 2023; He et al., 2023; Irshad et al., 2023; Iqbal et al., 2023; Aksoy et al., 2023).

Summing up the presented material, we note that energy consumption management in China of industrial enterprises and industrial complexes should be carried out in line with increasing energy efficiency and correspond to the energy-saving

A. Mikhaylov (✉)
Institute of China and Modern Asia of the Russian Academy of Sciences, Moscow, Russia

H. Dinçer et al. (eds.), *Decision Making in Interdisciplinary Renewable Energy Projects*, Contributions to Management Science,
https://doi.org/10.1007/978-3-031-51532-3_4

development of the industrial production economy within the framework of the new industrial policy (Wan et al., 2023). This should ensure sustainable socioeconomic development of industrial enterprises, industrial complexes, and the country as a whole (Ali et al., 2023; Dinçer et al., 2023a, 2023b; Ivleva et al., 2023; Aissou et al., 2023; Mikhaylov, 2023).

Energy comparison is the process of comparing different energy suppliers to find the one that offers the best value for money. Energy comparison involves comparing the cost of energy, the quality of service, and the terms and conditions of the energy supply contract. The goal of energy comparison is to find an energy supplier that offers the lowest cost of energy while providing high-quality service (Yüksel & Dinçer, 2023). Energy comparison works by using comparison websites that allow consumers to compare different energy suppliers. These websites use the consumer's energy usage data to calculate the estimated cost of energy from different suppliers. Consumers can then choose the supplier that offers the lowest cost of energy and the best terms and conditions. Energy comparison offers several benefits to consumers. One of the main benefits is cost savings (Moiseev et al., 2023a, 2023b). By comparing different energy suppliers, consumers can find the supplier that offers the lowest cost of energy. This can result in significant cost savings over time.

Another benefit of energy comparison is increased choice. By comparing different energy suppliers, consumers can choose the supplier that offers the best terms and conditions. This includes the length of the contract, payment options, and the quality of customer service. Energy comparison also promotes competition in the energy industry (Kou et al., 2023a, 2023b). By comparing different energy suppliers, consumers can encourage competition between suppliers. This can result in lower energy costs and improved service quality.

4.2 Literature Review

Staff training is an important factor that should not be overlooked when trying to save costs. As employees gain experience and qualifications, production costs can be reduced over time. It's common for the cost-saving benefits of staff training, advanced training, and learning new technologies to be underestimated when implementing energy efficiency and energy-saving strategies. Learning directly contributes to improving work organization, increasing efficiency, and creating high-tech products, which can lead to an increase in the volume of products produced. Additionally, optimizing costs for key resources like fuel and energy resources is crucial since they make up a significant portion of total production costs. Increasing the degree of utilization of production capacities can also help reduce energy intensity and distribute costs (Han et al., 2023). When developing an energy efficiency and energy-saving (ES) strategy, it's important to focus on energy-efficient technologies that can help reduce production costs and strengthen the company's competitive advantages (Nie et al. 2020; Mikhaylov et al., 2023a; Mikhaylov, 2021a; Varyash et al., 2020; Zhao et al., 2021; An & Mikhaylov, 2020; Alwaelya et al.,

2021; Yumashev & Mikhaylov, 2020; Yumashev et al., 2020; Mutalimov et al., 2021; An & Mikhaylov, 2021).

The economic indicators related to energy consumption and the organization of production within an enterprise are interdependent (An et al., 2019a, 2019b, 2020a, 2020b, 2020c, 2022; Mikhaylov & Tarakanov, 2020; Moiseev et al., 2020, 2021; Gura et al., 2022; Dooyum et al., 2020; Mikhaylov et al., 2020; Mikhaylov, 2020a, 2020b, 2020c). It is essential to consider energy economy from the perspective of enterprise functionality, where the interaction of system elements, maintaining integrity, and adaptation to changing energy consumption conditions play a critical role (Bhuiyan et al., 2021; Dong et al., 2021; Mikhaylov, 2021a, 2021b; Barykin et al., 2022a; Liu et al., 2022a, 2022b; Bhuiyan et al., 2022b; Danish et al., 2022a, 2022b; Saqib et al., 2021; Mukhametov et al., 2021, Candila et al., 2021; Mikhaylov and Grilli 2022; Li et al., 2022).

By adopting a systematic approach, the key factor in selecting energy efficiency measures is the savings achieved through reduced energy costs for the enterprise (Dayong et al., 2020; Mikhaylov et al., 2018; Nyangarika et al., 2018; Danish et al., 2020, 2021; An et al., 2021; Uyeh et al., 2021; Tamashiro et al., 2021, 2023; Shaikh et al., 2021).

This approach helps in identifying a list of energy efficiency factors that should be utilized in the development and implementation of energy programs to save fuel and energy (Denisova et al., 2019; Nyangarika et al., 2019a, 2019b; Huang et al., 2021a, 2021b; Mikhaylov, 2018, 2022; Mikhaylov et al., 2019; Conteh et al., 2021; Sediqi et al., 2022, Khan et al., 2021; Bhuiyan et al., 2022a; Liu et al., 2021a, 2021b; Daniali et al., 2021; Moiseev et al., 2023a, 2023b).

These factors include price levels and tariffs for fuel and energy, supply and demand for products and resources, access to the market for resources and new technologies, inflation and economic risks, efficiency in the use of production means, profitability of industrial production, efficiency of current costs, and the effective use of labor (Khan et al., 2022; Dinçer et al., 2022a, 2022b; Badr et al., 2022; Barykin et al., 2022b; Mehta et al., 2022; Kalinina et al. 2022; Shaikh et al., 2022; Mikhaylov et al., 2022a, 2022b; Nyangarika et al., 2022).

4.3 Results

The level of prices and tariffs for fuel and energy plays a vital role in the energy efficiency of production. This factor encourages the reduction of energy consumption and reflects the principle of respect for economic interests between energy companies and enterprises. Another crucial factor that affects energy efficiency and consumption of the enterprise is the demand for energy in the energy markets, which is associated with the level of prices and tariffs and the utilization of production capacities (Xu et al., 2023). The availability of resources and new technologies to the market, which determines the possibility of acquiring innovative technologies and the level of energy intensity of its products, is also crucial in the formation and implementation of the energy policy of an enterprise. When implementing investment projects of the company's energy program (EP), it is necessary to take into account inflation and economic risks. Additionally, the ratio of extensive and intensive factors of the use of fixed production assets is essential in determining the efficiency of the means of production involved and the value of the energy component in the cost of production (Yüksel et al., 2023).

To evaluate the profitability of industrial production, it's recommended to scrutinize the cost structure and identify significant cost items or groups to justify cost-saving directions, including energy costs. This analysis helps increase the efficiency of costs and the overall production (Carayannis et al., 2023). Labor efficiency, which reflects the effectiveness of labor at the lowest cost, is closely related to energy efficiency in the economic aspect of an enterprise. It's important to note that labor efficiency is not only about productivity but also labor savings and the quality of work that directly affects production efficiency, including the energy sector (Dinçer et al., 2023a, 2023b, 2023c, 2023d, 2023e, 2023f). In developing and implementing the energy strategy of an enterprise, several factors of energy efficiency in the production sector are equally important. These include the structure of industrial production, capacity utilization, physical and moral depreciation of fixed production assets, technological labor intensity, use of less energy-intensive materials, loss reduction, and level of electrification, among others. These factors are critical in forming and implementing energy programs for fuel and energy saving.

The way a company produces its products has a direct impact on their energy efficiency. This is due to the difference between energy-intensive and less energy-intensive production methods, which in turn affects the cost of production. Additionally, the scale of production also plays a role in determining the energy component of production costs. To ensure optimal energy efficiency, it is crucial for the management in China of the company to use simple and extended reproduction in a timely manner (Eti et al., 2023). The use of advanced technologies and the labor intensity of production also have a significant impact on the economic indicators of the company, such as productivity, production costs, profitability, and profit. The material intensity of production is also a key factor that affects energy consumption, and by reducing material consumption and using less energy-intensive materials, the company can increase its energy efficiency (Martínez et al., 2023).

The efficiency of energy consumption management in modern conditions requires the development of indicators and criteria for assessing the interaction of the state and the market with the industrial enterprise energy efficiency model. This approach combines the understanding of traditional energy efficiency and profitability of production and is aimed at sustainable satisfaction of energy needs and coordination of economic interests of market economy entities. The modern concept of energy consumption management in industrial production requires consideration of the system of factors of energy economy development. Many indicators of energy consumption are correlated, and, as a result, there are certain difficulties in attributing them to certain factors (Eti et al., 2022). In this regard, the authors suggest that the classification of factors should take into account certain proportions between the indicators that characterize the efficiency of using fuel and energy resources. This approach can be successfully used in the development of a program for improving the energy efficiency of industrial production, since it opens the way for the formation of a generalized criterion for energy efficiency and creates an opportunity for making management decisions to improve it.

In order to achieve better energy efficiency, it is important to reduce all types of production losses. This is directly linked to the level of organization and can be achieved through the implementation of lean production techniques. There are seven ways to reduce losses in lean manufacturing, including overproduction, transportation, waiting, excessive inventory, defects, excessive processing, and unnecessary movement. However, it is important to note that the majority of working time is spent on losses, with only 15% dedicated to creating value and generating profit. Lower costs of raw materials and materials can be achieved through the use of electrical technologies, which are more efficient than other energy carriers. Concentrating in industrial production through the consolidation of capacities is most effective when done through specialized production at large enterprises. This allows for more efficient use of high-performance equipment and leads to increased productivity. Increasing unit capacity in the electric power industry results in a decrease in unit cost and prime cost of output, including the energy component. Specialization also contributes to a decrease in the cost of production due to an increase in labor productivity and a decrease in conditionally fixed costs per unit of production. Additionally, specialization leads to an increase in the technical level of production and the quality of products, resulting in economies of scale and potential energy savings.

4.4 Conclusion

The integration of technology in production processes leads to an increase in the continuity of production, allowing for more efficient use of equipment. The use of raw materials, waste, and secondary resources results in a reduction in production costs and energy intensity. Industrial waste can be maximized by combining multistage processes with waste energy utilization to produce electricity, steam, and hot water. The technological combination of production is calculated by the ratio of processed raw materials and semifinished products to the total volume produced at the enterprise.

To achieve energy efficiency in the economic and industrial sphere of the enterprise, organizational factors like planning, accounting, and control should be considered when implementing energy-saving programs. Production organization forms and fuel and energy complex accounting and rationing should also be improved. Planning, accounting, and control are essential components of production management in China, while even advanced enterprises have room for improvement in production organization, which can reduce consumption by 5–10% with minimal costs.

References

Ai, R., Zheng, Y., Yüksel, S., & Dinçer, H. (2023). Investigating the components of fintech ecosystem for distributed energy investments with an integrated quantum spherical decision support system. *Financial Innovation, 9*(1), 27.

Aissou, A., Ahmed, H. S., Hussein, M. E., Hassan, M. R., Mikhaylov, A., & Mahmoud, E. (2023). Optimizing reliability of flow networks using RWGA subject to total capacity constraint. *Applied Mathematics and Information Sciences, 7*(4), 559–565. https://doi.org/10.18576/amis/170404

Aksoy, T., Yüksel, S., Dınçer, H., Hacioglu, U., Mikhaylov, A., & Muyeen, S. M. (2023). Assessing the electricity production capacities of emerging markets for the sustainable investment. *IEEE Access, 11*, 24574–24590.

Ali, K., Jianguo, D., & Kirikkaleli, D. (2023). How do energy resources and financial development cause environmental sustainability? *Energy Reports, 9*, 4036–4048.

Alwaelya, S. A., Yousif, N. B. A., & Mikhaylov, A. (2021). Emotional development in preschoolers and socialization. *Early Child Development and Care, 191*, 16.

An, J., & Mikhaylov, A. (2021). Analysis of energy projects financial efficiency and renewable energy generation in Russia. *Finance: Theory and Practice, 25*(5), 79–91.

An, J., & Mikhaylov, A. (2020). Russian energy projects in South Africa. *Journal of Energy in Southern Africa, 31*(3), 58–64.

An, J., Mikhaylov, A., & Jung, S.-U. (2020a). The strategy of South Korea in the global oil market. *Energies, 13*(10), 2491.

An, J., Mikhaylov, A., & Kim, K. (2020b). Machine learning approach in heterogeneous group of algorithms for transport safety-critical system. *Applied Sciences, 10*(8), 2670.

An, J., Mikhaylov, A., & Richter, U. H. (2020c). Trade war effects: Evidence from sectors of energy and resources in Africa. *Heliyon, 6*, e05693. https://doi.org/10.1016/j.heliyon.2020.e05693

An, J., Mikhaylov, A., & Jung, S.-U. (2021). A linear programming approach for Robust network revenue management in the airline industry. *Journal of Air Transport Management, 91*, 101979.

An, J., Mikhaylov, A., & Moiseev, N. (2019a). Oil price predictors: Machine learning approach. *International Journal of Energy Economics and Policy, 9*(5), 1–6.

An, J., Mikhaylov, A., & Sokolinskaya, N. (2019b). Machine learning in economic planning: Ensembles of algorithms. *Journal of Physics: Conference Series, 1353*, 012126.

An, J., Mikhaylov, A., Dinçer, H., & Yüksel, S. (2022). Economic modelling of electricity generation: Long short-term memory and Q-rung orthopair fuzzy sets. *Heliyon, 8*, e10759.

Badr, A. M. N. A. A., Mohamed, T. I., Osman, N. A. Q., & Mikhaylov, A. (2022). A review of social media website users' interaction paths with governmental accounts during the COVID-19 pandemic. *Informatics, 9*, 50. https://doi.org/10.3390/informatics9030050

Barykin, S., Mikheev, A., Kiseleva, E., Putikhin, Y., Alekseeva, N., & Mikhaylov, A. (2022a). An empirical analysis of Russian Regions' debt sustainability. *Economies, 10*(4), 296.

Barykin, S. E., Kapustina, I. V., Sergeev, S. M., Daniali, S. M., Kopteva, L. A., Semenova, G. N., Pryadko, I. P., Mikhaylov, A., Baboshkin, P., Datsyuk, P., & Senjyu, T. (2022b). Financial logistics models based on systematic approach improving management solutions. *F1000Research, 11*, 570. https://doi.org/10.12688/f1000research.111252.1

Bhuiyan, M. A., Dinçer, H., Yüksel, S., Mikhaylov, A., Danish, M. S. S., Pinter, G., Uyeh, D. D., & Stepanova, D. (2022a). Economic indicators and bioenergy supply in developed economies: QROF-DEMATEL and random forest models. *Energy Reports, 8*(2022), 561–570.

Bhuiyan, M. A., Zhang, Q., Khare, V., Mikhaylov, A., Pinter, G., & Huang, X. (2022b). Renewable energy consumption and economic growth nexus—a systematic literature review. *Frontiers in Environmental Science, 10*, 878394. https://doi.org/10.3389/fenvs.2022.878394

Bhuiyan, M. A., An, J., Mikhaylov, A., Moiseev, N., & Danish, M. S. S. (2021). Renewable energy deployment and COVID-19 measures for sustainable development. *Sustainability, 13*, 4418.

Candila, V., Maximov, D., Mikhaylov, A., Moiseev, N., Senjyu, T., & Tryndina, N. (2021). On the relationship between oil and exchange rates of oil-exporting and oil-importing countries: From the great recession period to the Covid-19 era. *Energies, 14*, 8065.

Carayannis, E., Kostis, P., Dinçer, H., & Yüksel, S. (2023). Quality function deployment-oriented strategic outlook to sustainable energy policies based on quintuple innovation helix. *Journal of the Knowledge Economy*, 1–19.

Conteh, F., Takahashi, H., Hemeida, A. M., Krishnan, N., Mikhaylov, A., & Senjyu, T. (2021). Hybrid grid-connected renewable power generation for sustainable electricity supply in Sierra Leone: Case study Lungi Town. *Sustainability, 13*(18).

Daniali, S. M., Barykin, S. E., Kapustina, I. V., Khortabi, F. M., Sergeev, S. M., Kalinina, O. V., Mikhaylov, A., Veynberg, R., Zasova, L., & Senjyu, T. (2021). Predicting volatility index according to technical index and economic indicators on the basis of deep learning algorithm. *Sustainability, 13*, 14011.

Danish, M. S. S., Bhattacharya, A., Stepanova, D., Mikhaylov, A., Grilli, M. L., Khosravy, M., & Senjyu, T. (2020). A systematic review of metal oxide applications for energy and environmental sustainability. *Metals, 10*(12), 1604.

Danish, M. S. S., Senjyu, T., Sabory, N. R., Khosravy, M., Grilli, M. L., Mikhaylov, A., & Majidi, H. (2021). A forefront framework for sustainable aquaponics modeling and design. *Sustainability, 13*, 9313.

Danish, M. S., Senjyu, T. S., Sabory, N. R., & Mikhaylov, A. (2022a). Lessons from COVID-19 conferring environmental re-engineering opportunity. In M. Danish & T. Senjyu (Eds.), *Eco-friendly and agile energy strategies and policy development* (pp. 99–118). IGI Global). https://doi.org/10.4018/978-1-7998-9502-2.ch005

Danish, M. S., Senjyu, T. S., Sabory, N. R., & Mikhaylov, A. (2022b). Interplay consequences of COVID-19 on global environmental sustainability. In M. Danish & T. Senjyu (Eds.), *Eco-friendly and agile energy strategies and policy development*. IGI Global. https://doi.org/10.4018/978-1-7998-9502-2.ch003

Dayong, N., Mikhaylov, A., Bratanovsky, S., Shaikh, Z. A., & Stepanova, D. (2020). Mathematical modeling of the technological processes of catering products production. *Journal of Food Process Engineering, 43*(2), 1–9.

Denisova, V., Mikhaylov, A., & Lopatin, E. (2019). Blockchain infrastructure and growth of global power consumption. *International Journal of Energy Economics and Policy, 9*(4), 22–29.

Dinçer, H., Yüksel, S., Mikhaylov, A., Pinter, G., & Shaikh, Z. A. (2022a). Analysis of renewable-friendly smart grid technologies for the distributed energy investment projects using a hybrid picture fuzzy rough decision-making approach. *Energy Reports, 8*, 11466–11477. https://doi.org/10.1016/j.egyr.2022.08.275

Dinçer, H., Yüksel, S., Mikhaylov, A., Barykin, S. E., Aksoy, T., & Hacioğlu, U. (2022b). Analysis of environmental priorities for green project investments using an integrated q-rung orthopair fuzzy modelling. *IEEE Access, 10*, 50996–51007. https://doi.org/10.1109/ACCESS.2022.3174058

Dinçer, H., Yüksel, S., Hacioglu, U., Yilmaz, M. K., & Delen, D. (2023a). Development of a sustainable corporate social responsibility index for performance evaluation of the energy industry: A hybrid decision-making methodology. *Resources Policy, 85*, 103940.

Dinçer, H., Eti, S., Gökalp, Y., Yüksel, S., & Çelebi, B. (2023b). The role of small-scale wind turbines in increasing the competitiveness of hospitals. In *Handbook of research on quality and competitiveness in the healthcare services sector* (pp. 336–352). IGI Global.

Dinçer, H., Yüksel, S., Mikhaylov, A., Muyeen, S. M., Chang, T., Barykin, S., & Kalinina, O. (2023c). CO2 emissions integrated fuzzy model: A case of seven emerging economies. *Energy Reports, 9*, 5741–5751.

Dinçer, H., Yüksel, S., Aksoy, T., Hacıoğlu, Ü., Mikhaylov, A., & Pinter, G. (2023d). Analysis of solar module alternatives for efficiency-based energy investments with hybrid 2-tuple IVIF modeling. *Energy Reports, 10*, 61–71.

Dinçer, H., Eti, S., Yüksel, S., Özdemir, S., Yflmaz, A. E., & Ergün, E. (2023e). Integrating data mining and fuzzy decision-making techniques for analyzing the key minimizing factors of carbon emissions. *Journal of Intelligent & Fuzzy Systems, 45*, 1–17.

Dinçer, H., Eti, S., Aksoy, T., Yüksel, S., Hacioglu, U., Mikhaylov, A., & Muyeen, S. M. (2023f). Analysis of environmental impact for material production investments using a novel soft computing methodology. *IEEE Access, 11*, 37987–38001. https://doi.org/10.1109/ACCESS.2023.3266524

Dong, B., Ikonnikova, I., Rogulin, R., Sakulyeva, T., & Mikhaylov, A. (2021). Environmental economic approach to optimization of transport communication in megacities. *Journal of Environmental Science and Health, Part A, 56*(6), 660–666.

Dooyum, U. D., Mikhaylov, A., & Varyash, I. (2020). Energy security concept in Russia and South Korea. *International Journal of Energy Economics and Policy, 10*(4), 102–107.

Eti, S., Dinçer, H., Gökalp, Y., Yüksel, S., & Kararoğlu, D. (2023). Identifying key issues to handle the inflation problem in the healthcare industry caused by energy prices: An evaluation with decision-making models. In *Managing inflation and supply chain disruptions in the global economy* (pp. 162–178).

Eti, S., Dinçer, H., Yüksel, S., & Gökalp, Y. (2022). Analysis of the suitability of the solar panels for hospitals: A new fuzzy decision-making model proposal with the T-spherical TOP-DEMATEL method. *Journal of Intelligent & Fuzzy Systems, Preprint*, 1–13.

Gura, D., Mikhaylov, A., Glushkov, S., Zaikov, M., & Shaikh, Z. A. (2022). Model for estimating power dissipation along the interconnect length in single on-chip topology. *Evolutionary Intelligence, 15*, 2369–2373. https://doi.org/10.1007/s12065-020-00407-7

Han, Z., Gong, L., Chen, H., & Yüksel, S. (2023). The role of mineral and oil resources in achieving a clean environment in emerging economies: An asymmetric analysis. *Resources Policy, 86*, 104099.

He, H., Huang, Y., Nakadomari, A., Masrur, H., Krishnan, N., Hemeida, A. M., Mikhaylov, A., Omine, E., & Senjyu, T. (2023). Potential and economic viability of green hydrogen production from seawater electrolysis using renewable energy in remote Japanese islands. *Renewable Energy, 202*, 1436–1447. https://doi.org/10.1016/j.renene.2022.12.046

Huang, Y., Masrur, H., Shigenobu, R., Hemeida, A. M., Mikhaylov, A., & Senjyu, T. (2021a). A comparative design of a campus microgrid considering a multi-scenario and multi-objective approach. *Energies, 14*, 2853.

Huang, Y., Yona, A., Takahashi, H., Hemeida, A. M., Mandal, P., Mikhaylov, A., Senjyu, T., & Lotfy, M. E. (2021b). Energy management system optimization of drug store operate electric vehicles charging station. *Sustainability, 14*, 1217836.

Iqbal, K. M. J., Khan, M. I., Mikhaylov, A., Shah, A. A., Yadykin, V., Filho, W. L., Tariq, M. A. U. R., & Ullah, W. (2023). Modeling principles, criteria and indicators to assess water sector governance for climate compatibility and sustainability. *Frontiers in Environmental Science, 11*, 989930.

Irshad, A. S., Ludin, G. A., Masrur, H., Ahmadi, M., Yona, A., Mikhaylov, A., Krishnan, N., & Senjyu, T. (2023). Optimization of grid-photovoltaic and battery hybrid system with most technically efficient PV technology after the performance analysis. *Renewable Energy, 207*, 714–730.

Ivleva, S., Hajimajeed, M., Kartashova, O., & Mikhaylov, A. (2023). A comparative study of regional variations in common visual impairments among school-age children. *Electronic Journal of General Medicine, 20*(5), em504. https://doi.org/10.29333/ejgm/13264

Kalinina, O. V., Barykin, S. E., Sergeev, S. M., Semenova, G. N., Fatkullina, A., Mikhaylov, A., & De La Poza, E. (2022). Smart city perspectives in post-pandemic governance: Externalities reduction policy. *F1000Research, 11*, 1032. https://doi.org/10.12688/f1000research.123195.1

Khan, A. A., Laghari, A. A., Gadekallu, T. R., Shaikh, Z. A., Javed, A. R., Rashid, M., Estrela, V. V., & Mikhaylov, A. (2022). A drone-based data management and optimization using metaheuristic algorithms and blockchain smart contracts in a secure fog environment. *Computers and Electrical Engineering, 102*, 108234. https://doi.org/10.1016/j.compeleceng.2022.108234

Khan, A. A., Shaikh, Z. A., Baitenova, L., Mutaliyeva, L., Moiseev, N., Mikhaylov, A., Laghari, A. A., Idris, S. A., & Alshazly, H. (2021). QoS-ledger: Smart contracts and metaheuristic for secure quality-of-service and cost-efficient scheduling of medical-data processing. *Electronics, 10*, 3083.

Kou, G., Pamucar, D., Dinçer, H., & Yüksel, S. (2023a). From risks to rewards: A comprehensive guide to sustainable investment decisions in renewable energy using a hybrid facial expression-based fuzzy decision-making approach. *Applied Soft Computing, 142*, 110365.

Kou, G., Yuksel, S., & Dincer, H. (2023b). A facial expression and expert recommendation fuzzy decision-making approach for sustainable business investments within the metaverse world. *Applied Soft Computing*, 110849.

Li, J., Dinçer, H., Yüksel, S., Mikhaylov, A., & Barykin, S. E. (2022). Bipolar q-ROF hybrid decision making model with golden cut for analyzing the levelized cost of renewable energy alternatives. *IEEE Access, 10*, 35900–35900. https://doi.org/10.1109/ACCESS.2022.3168315

Liu, L., Kato, T., Mandal, P., Mikhaylov, A., Hemeida, A. M., & Senjyu, T. (2021a). Multi-area power system load frequency control by using model predictive control by introducing renewable energy sources. In *2021 international conference on science & contemporary technologies (ICSCT)* (pp. 1–6). https://doi.org/10.1109/ICSCT53883.2021.9642521

Liu, L., Kato, T., Mandal, P., Mikhaylov, A., Hemeida, A. M., Takahashi, H., & Senjyu, T. (2021b). Two cases studies of model predictive control approach for hybrid renewable energy systems. *AIMS Energy, 9*(6), 1241–1259.

Liu, Z.-J., Panfilova, E., Mikhaylov, A., & Kurilova, A. (2022a). COVID-19 crisis impact on the stability between parties in crowdfunding and crowdsourcing. *Wireless Personal Communications, 122*, 915–930.

Liu, Z., Panfilova, E., Mikhaylov, A., & Kurilova, A. (2022b). Assessing stability in the relationship between parties in crowdfunding and crowdsourcing projects during the COVID-19 crisis. *Journal of Global Information Management (JGIM), 30*(4), 1–18. https://doi.org/10.4018/JGIM.297905

Martínez, L., Dinçer, H., & Yüksel, S. (2023). A hybrid decision making approach for new service development process of renewable energy investment. *Applied Soft Computing, 133*, 109897.

Mehta, R., Verghese, J., Mahajan, S., Barykin, S., Bozhuk, S., Kozlova, N., Kapustina, I., Mikhaylov, A., Naumova, E., & Dedyukhina, N. (2022). Consumers' behavior in conversational commerce marketing based on messenger chatbots. *F1000Research, 11*(647). https://doi.org/10.12688/f1000research.122037.1

Mikhaylov, A., Dinçer, H., & Yüksel, S. (2023a). Analysis of financial development and open innovation oriented fintech potential for emerging economies using an integrated decision-making approach of MF-X-DMA and golden cut bipolar q-ROFSs. *Financial Innovation, 9*(1), 10. https://doi.org/10.1186/s40854-022-00234-w

Mikhaylov, A., Dinçer, H., Yüksel, S., Pinter, G., & Shaikh, Z. A. (2023b). Bitcoin mempool growth and trading volumes: Integrated approach based on QROF multi-SWARA and aggregation operators. *Journal of Innovation & Knowledge, 8*(3), 100378.

Mikhaylov, A. (2018). The yield of the carry trade strategy. *Finance: Theory and Practice, 22*(3), 52–63. https://doi.org/10.26794/2587-5671-2018-22-3-52-63

Mikhaylov, A. (2020a). Cryptocurrency market analysis from the open innovation perspective. *Journal of Open Innovation: Technology, Market, and Complexity, 6*(4), 197.

Mikhaylov, A. (2020b). Geothermal energy development in Iceland. *International Journal of Energy Economics and Policy, 10*(4), 31–35.

Mikhaylov, A. (2020c). Cryptocurrency market development: Hurst method. *Finance: Theory and Practice, 24*(3), 81–91.

Mikhaylov, A. (2021a). Development of Friedrich von Hayek's theory of private money and economic implications for digital currencies. *Terra Economicus, 19*(1), 53–62.

Mikhaylov, A. (2021b). Lichens as indicators of atmospheric pollution in urban ecosystems. *Israel Journal of Ecology & Evolution, 67*(1–2), 60–68. https://doi.org/10.1163/22244662-bja10016

Mikhaylov, A. (2022). Sustainable development and renewable energy: A new view to a global problem. *Energies, 15*, 1397. https://doi.org/10.3390/en15041397

Mikhaylov, A. (2023). Understanding the risks associated with wallets, depository services, trading, lending, and borrowing in the crypto space. *Journal of Infrastructure, Policy and Development, 6*, 2.

Mikhaylov, A., & Grilli, M. L. (2022). Machine learning methods and sustainable development: Metal oxides and multilayer metal-oxides. *Metals, 12*(4).

Mikhaylov, A., & Tarakanov, S. (2020). Development of Levenberg-Marquardt theoretical approach for electric network. *Journal of Physics: Conference Series, 1515*, 052006.

Mikhaylov, A., Bhatti, I. M., Dinçer, H., & Yüksel, S. (2022a). Integrated decision recommendation system using iteration-enhanced collaborative filtering, golden cut bipolar for analyzing the risk-based oil market spillovers. *Computational Economics*. https://doi.org/10.1007/s10614-022-10341-8

Mikhaylov, A., Yumashev, A., & Kolpak, E. (2022b). Quality of life, anxiety and depressive disorders in patients with extrasystolic arrhythmia. *Archives of Medical Science, 18*(2). https://doi.org/10.5114/aoms.2020.101359

Mikhaylov, A., Burakov, D., & Didenko, V. (2019). Relationship between oil price and macroeconomic indicators in Russia. *Finance: Theory and Practice, 23*(2), 105–116. https://doi.org/10.26794/2587-5671-2019-23-2-105-116

Mikhaylov, A., Moiseev, N., Aleshin, K., & Burkhardt, T. (2020). Global climate change and greenhouse effect. *Entrepreneurship and Sustainability Issues, 7*(4), 2897–2913. https://doi.org/10.9770/jesi.2020.7.4(21)

Mikhaylov, A., Sokolinskaya, N., & Nyangarika, A. (2018). Optimal carry trade strategy based on currencies of energy and developed economies. *Journal of Reviews on Global Economics, 7*, 582–592.

Moiseev, N., Mikhaylov, A., Dinçer, H., & Yüksel, S. (2023a). Effects of market capitalization shocks for open innovation models in e-commerce industry: Golden cut QROF MCDM analysis. *Financial Innovation, 9*(1), 32.

Moiseev, N., Mikhaylov, A., Dinçer, H., & Yüksel, S. (2023b). Market capitalization shock effects on open innovation models in e-commerce: Golden cut q-rung orthopair fuzzy multicriteria decision-making analysis. *Financial Innovation, 9*(1), 55.

Moiseev, N., Mikhaylov, A., Varyash, I., & Saqib, A. (2020). Investigating the relation of GDP per capita and corruption index. *Entrepreneurship and Sustainability Issues, 8*(1), 780–794.

Moiseev, N., Sorokin, A., Zvezdina, N., Mikhaylov, A., Khomyakova, L., & Danish, M. S. S. (2021). Credit risk theoretical model on the base of DCC-GARCH in time-varying parameters framework. *Mathematics, 9*, 2423. https://doi.org/10.3390/math9192423

Mukhametov, A., Bekhorashvili, A., Avdeenko, A., & Mikhaylov, A. (2021). The impact of growing legume plants under conditions of biologization and soil cultivation on Chernozem fertility and productivity of rotation crops. *Legume Research, 44*, 12.

Mutalimov, V., Kovaleva, I., Mikhaylov, A., & Stepanova, D. (2021). Assessing regional growth of small business in Russia. *Entrepreneurial Business and Economics Review, 9*(3), 119–133. https://doi.org/10.15678/EBER.2021.090308

Nie, D., Panfilova, E., Samusenkov, V., & Mikhaylov, A. (2020). E-learning financing models in Russia for sustainable development. *Sustainability, 12*(11), 4412.

Niu, X., Yüksel, S., & Dinçer, H. (2023). Emission strategy selection for the circular economy-based production investments with the enhanced decision support system. *Energy, 274*, 127446.

Nyangarika, A., Mikhaylov, A., Muyeen, S. M., Yadykin, V., Mottaeva, A., Pryadko, I., Barykin, S., Fomenko, N., Rykov, G., & Shvandar, K. (2022). Energy stability and decarbonization in developing countries: Random Forest approach for forecasting of crude oil trade flows and macro indicators. *Frontiers in Environmental Science, 10*, 1031343. https://doi.org/10.3389/fenvs.2022.1031343

Nyangarika, A., Mikhaylov, A., & Richter, U. (2019a). Influence oil price towards economic indicators in Russia. *International Journal of Energy Economics and Policy, 9*(1), 123–129.

Nyangarika, A., Mikhaylov, A., & Richter, U. (2019b). Oil price factors: Forecasting on the base of modified auto-regressive integrated moving average model. *International Journal of Energy Economics and Policy, 9*(1), 149–159.

Nyangarika, A., Mikhaylov, A., & Tang, B.-J. (2018). Correlation of oil prices and gross domestic product in oil producing countries. *International Journal of Energy Economics and Policy, 8*(5), 42–48.

Saqib, A., Chan, T.-H., Mikhaylov, A., & Lean, H. H. (2021). Are the responses of sectoral energy imports asymmetric to exchange rate volatilities in Pakistan? Evidence from recent foreign exchange regime. *Frontiers in Energy Research, 9*, 614463. https://doi.org/10.3389/fenrg.2021.614463

Sediqi, M. M., Nakadomari, A., Mikhaylov, A., Krishnan, N., Lotfy, M. E., Yona, A., & Senjyu, T. (2022). Impact of time-of-use demand response program on optimal operation of Afghanistan real power system. *Energies, 15*, 296.

Shaikh, Z. A., Kraikin, A., Mikhaylov, A., & Pinter, G. (2022). Forecasting stock prices of companies producing solar panels using machine learning methods. *Complexity, 2022*, 9186265. https://doi.org/10.1155/2022/9186265

Shaikh, Z. A., Moiseev, N., Mikhaylov, A., & Yüksel, S. (2021). Facile synthesis of copper oxide-cobalt oxide/nitrogen-doped carbon (Cu2O-Co3O4/CN) composite for efficient water splitting. *Applied Sciences, 11*, 9974. https://doi.org/10.3390/app11219974

Tamashiro, K., Omine, E., Krishnan, N., Mikhaylov, A., Hemeida, A. M., & Senjyu, T. (2023). Optimal components capacity based multi-objective optimization and optimal scheduling based MPC-optimization algorithm in smart apartment buildings. *Energy & Buildings, 278*, 112616. https://doi.org/10.1016/j.enbuild.2022.112616

Tamashiro, K., Alharbi, T., Mikhaylov, A., Hemeida, A. M., Krishnan, N., Lotfy, M. E., & Senjyu, T. (2021). Investigation of home energy management with advanced direct load control and optimal scheduling of controllable loads. *Energies, 14*, 7314. https://doi.org/10.3390/en14217314

Uyeh, D. D., Asem-Hiablie, S., Park, T., Kim, K.-M., Mikhaylov, A., Woo, S., & Ha, Y. (2021). Could Japonica rice be an alternative variety for increased global food security and climate change mitigation? *Food, 10*, 1869. https://doi.org/10.3390/foods10081869

Varyash, I., Mikhaylov, A., Moiseev, N., & Aleshin, K. (2020). Triple bottom line and corporate social responsibility performance indicators for Russian companies. *Entrepreneurship and Sustainability Issues, 8*(1), 313–331.

Wan, Q., Miao, X., Wang, C., Dinçer, H., & Yüksel, S. (2023). A hybrid decision support system with golden cut and bipolar q-ROFSs for evaluating the risk-based strategic priorities of fintech lending for clean energy projects. *Financial Innovation, 9*(1), 1–25.

Wang, Q., Guo, J., Li, R., Mikhaylov, A., & Moiseev, N. (2023). Does technical assistance alleviate energy poverty in sub-Saharan African countries? A new perspective on spatial spillover effects of technical assistance. *Energy Strategy Reviews, 45*, 101047. https://doi.org/10.1016/j.esr.2022.101047

Xu, X., Yüksel, S., & Dinçer, H. (2023). An integrated decision-making approach with golden cut and bipolar q-ROFSs to renewable energy storage investments. *International Journal of Fuzzy Systems, 25*(1), 168–181.

Yüksel, S., & Dinçer, H. (2023). Sustainability analysis of digital transformation and circular industrialization with quantum spherical fuzzy modeling and golden cuts. *Applied Soft Computing, 138*, 110192.

Yüksel, S., Dinçer, H., & Uluer, G. S. (2023). Socio-economic indicators of nuclear energy investments: A study on countries. In *Inclusive developments through socio-economic indicators: New theoretical and empirical insights* (pp. 33–44). Emerald Publishing Limited.

Yumashev, A., & Mikhaylov, A. (2020). Development of polymer film coatings with high adhesion to steel alloys and high wear resistance. *Polymer Composites, 41*(7), 2875–2880.

Yumashev, A., Ślusarczyk, B., Kondrashev, S., & Mikhaylov, A. (2020). Global indicators of sustainable development: Evaluation of the influence of the human development index on consumption and quality of energy. *Energies, 13*, 2768.

Zhao, Q., Cherkasov, S., Avdeenko, A., Kondratenko, L., & Mikhaylov, A. (2021). Integral estimate of the added value of agricultural crops in the synergetic agriculture on the example of Vicia faba (Vicia faba L.). *Polish Journal of Environmental Studies, 30*(5), 1–9. https://doi.org/10.15244/pjoes/131949

Chapter 5
Energy Crisis: An Actual Collapse? Energy Crisis and Apocalypse Societies in Popular Films

Başak Gezmen and Hasan Gürkan

5.1 Introduction

Today, the issues of energy crisis, sustainability, and efficient energy use are constantly on the agenda, and all states are working on these issues (Department for Business, Energy, and Industrial Strategy, 2021; Liddell & Morris, 2010; Middlemiss et al., 2019). Worldwide, natural gas dependency and supply-demand-oriented problems in oil production, competitive markets, high energy prices, increased electricity production costs, etc. The global energy crisis, which is experienced due to the problems caused by renewable energy, is predicted to be solved within a few years. The energy crisis in the world and the issue of energy efficiency, which has an essential place in public opinion, are frequently on the agenda of many developed countries. The importance of energy efficiency as a policy goal is increasing daily; there are many reports and books on this subject, and there is increasing policy interest and research on energy efficiency (Gezmen, 2022). In this context, media messages are essential in raising awareness of the energy crisis and solution-oriented approaches. Considering the influence of the media, and the speed and effectiveness of the sharing of messages in the digital world formed by the development of new communication technologies, the importance of the media in this regard emerges. Within the scope of this subject, there are many public service announcements, advertisements, news, social media content, and TV and radio

B. Gezmen (✉)
The School of Communication, İstanbul Medipol University, İstanbul, Turkey
e-mail: bgezmen@medipol.edu.tr

H. Gürkan
İstinye University Philology and Communication Department, Girona University, Girona, Turkey

H. Dinçer et al. (eds.), *Decision Making in Interdisciplinary Renewable Energy Projects*, Contributions to Management Science,
https://doi.org/10.1007/978-3-031-51532-3_5

programs made for awareness purposes in the media. Many movies reveal the importance and effectiveness of the subject.

The study deals with energy, the energy crisis, and technology. Energy-themed films of 2015 and later on the subject were selected. The study analyzed the movies April and the Extraordinary World and The Cloverfield Paradox. In both films, the energy crisis experienced worldwide is the subject, and the solution-oriented works, interstate rivalries, and wars that take place as predicted by the capitalist system, sustainable life, and access to new energy sources are highlighted.

5.2 Energy and Energy Crisis

Energy is an essential resource for the infrastructure of industrialization and an indispensable element of daily life. Thus, energy may affect a country socially. Moreover, energy is vital to economic development as it is the industry's raw material (Martínez et al., 2023). Therefore, governments have to provide the energy they need to develop. These needs must be met for countries to achieve industrial and economic growth. Otherwise, ensuring sustainability in the country's economy can be challenging (Eti et al., 2022).

Moreover, with the help of energy, people can meet their daily needs, examples of which are energy, warming, and enlightenment (Dinçer et al., 2023a, 2023b, 2023c, 2023d, 2023e, 2023f). In this sense, providing energy to meet people's needs is essential because they are imperative (Liu et al., 2021). In case of failure to ensure the sustainability of energy resources, states may face an energy crisis. This is a global problem (Carayannis et al., 2023).

Today, the issue of the energy crisis has become the most crucial issue of the whole world (Yüksel et al., 2023). Energy also plays a vital role in interstate policies. Regionalization, supported by energy as the primary governance object, is characterized by various actors, including governments and international energy companies, investors, consumers, and regulators (Goldthau et al., 2020). The average energy price is essential for energy security, but what matters most is its volatility (standard deviation per period). High energy price volatility increases the risk of any energy investment (Xu et al., 2023). In addition, energy prices affect the entire economy, as many studies reveal (Papanikos, 2017).

The issue of energy security is a topic that is discussed all over the world (Han et al., 2023). Although it plays an increasingly important role in Euro-Atlantic agendas, concerns over European countries' dependence on Russian gas have overshadowed energy security issues in the Southern and Eastern Mediterranean regions (Kou et al., 2023a, 2023b). Reliable energy exports from the Southern and Eastern Mediterranean are vital for Europe's energy security and the resilience of many states and societies (Franza et al., 2020).

Many studies have been initiated, and projects have been put forward, both in the world and in the states, on the issue of the energy crisis (Moiseev et al., 2023).

Awareness areas on sustainability, increasing energy efficiency, and consumption of energy resources have begun to be created (Eti et al., 2023).

Especially from the mid-1970s to the present, the energy crisis has become a frequent topic in literature, films, and TV series that tell of the imminent radical break-in and that the procedural stability of the hybrid network society is subject to repeated interruptions (Galloway & Thacker, 2007). After presenting the event from a general perspective as a rationale and category of investigation of pop-culture narratives, the blackout phenomenon should be profiled accordingly as an epistemic event oscillating between material reality and fiction. This is followed by a discursive differentiation which, on the one hand, reads the imagination of unconsciousness as a biopolitically impregnated negotiation of human-machine relations and, on the other hand, interprets it as a fear-fantasy of social unrest. In this context, it is possible to encounter narratives about energy and energy crises in visual cultural products (films, series, TV programs, etc.). The energy issue comes up, especially in dystopian and science fiction-themed narratives.

Science fiction genre, like other genres, is a genre based on the early ages of literature (Bleiler, 1990; Luckhurst, 2005; Rieder, 2017). Jean Baudrillard (1991) relates the emergence and function of science fiction to the political problems and social movements that began in the nineteenth century. According to Baudrillard, the classical science fiction narrative is based on a worldview dominated by colonialism.

Especially science fiction cinema is in a close relationship with the dominant ideological structure. For this reason, science fiction cinema is the watchdog of the political and cultural conditions of its age. Science fiction, which has developed and changed over the ages, is based on both political changes and periodic changes in the individual's perspective on nature and technology. Science supports science fiction as much as science fiction supports science (Barnett et al., 2006). In the light of scientific developments, the genre of science fiction also creates new fields for itself. The genre of science fiction not only criticizes development and the future but also develops new ideas and technologies for the future. Sociologists and media theorists (Kitchin & Kneale, 2001; Lempert, 2014; Collins, 2003; Ronay, 2011) have also been involved in discussions on this issue. Cinema and media theorist Vivian Sobhack (1997) draws attention to the connection between science fiction cinema and imagination and states that fictional worlds contribute to science by influencing the imagination.

The genre of science fiction, which has grown over the last 150 years, has become the most popular genre not only for writers or filmmakers but also for scientists. Science fiction examples, which have gained a lot of space in the visual media, have turned into a branch of art in themselves. Science fiction, which conveys the connection with reality with developments that have not yet been experienced, establishes a tight bond with the audience and readers. It is possible to define dystopia, which is often referred to as science fiction, as follows: In dystopian societies, individuals have generally accepted the social norms and system, and most societies do not even notice the disorder of the system. Dystopian narratives often take their subject from individuals who oppose the system by expressing that this system is wrong. Dystopian narratives do not have a happy ending. At best, the main character of the story and his loved ones are saved, but there is no social change. The purpose of dystopias is to show societies the dangers they may face in the future. This danger sometimes appears with mechanization and sometimes as purely ideological and sociological (Yürür, 2017; Westfahl, 2005).

Beyond these, two other important factors affecting the transformation of paradise into a hell on earth are bureaucracy and technology. While bureaucracy is used to keep individuals under control and pressure, technology provides this control and pressure with the necessary method and technique. In this fictionalized world, it is constantly documented how individuals will live, how they will spend their free time, who they should associate with, and what they should think. This requires a lot of bureaucracy. Technology also undertakes the task of providing production and control devices, developing the necessary methods for the mechanical understanding and analysis of the individual. Thus, in dystopias, society functions like a two-sided mechanism consisting of bureaucracy and technology. In this mechanism, while the social organization becomes mechanized, the individual becomes objectified and isolated (Bezel, 2001: 8–10).

Most of the works on the concept of dystopia deal with the concept of dictatorship and sovereign power. The dictator, who makes the society obey him, holds the administration completely and forces the individual to the system. This administration determines the social roles and identities of individuals and divides the society into social classes. All tools and control mechanisms, including science and technology, belong to the state, that is, to the sovereign power. By creating fictional realities, individuals are instilled with the idea that the most ideal and fair system is the functioning one. Thus, individuals continue to live in the wheel of the system, in an acceptance where there is no awareness, and they cannot oppose the system.

Terri Chung lists the characteristic features of dystopia as follows:

- In dystopias, individuals worship a concept or figure.
- Knowledge, free thought, and freedom are restricted.
- The government uses propaganda as a tool to keep the society under control.
- A pessimistic, oppressed society is envisioned.
- Individuals are supervised by surveillance mechanisms at all times.
- The world is portrayed as an unsafe place by the government and is banned.
- The survival of the state overshadows the rights and freedoms of the individual.

Among the most popular films presenting a dystopian society fiction are V for Vendetta, The Maze Runner series, and the Nineteen Eighty-Four films written by George Orwell in 1948 and brought to the big screen by Michael Radford in 1984.

5.3 Methodology

For this study, first of all, movie searches were carried out on the Internet Movie Database, which is a movie database (IMDb), with the keyword "energy crisis" (see Gürkan, 2012). It is possible that researchers who have studied and researched this subject before have not provided much helpful information on the theme of the anonymous energy crisis. For this reason, the cast list on IMDb was searched for this study. Besides, the films with the themes of energy crisis, energy, and alternative energy, which are not represented as very important in the film, and documentary or fictional series were eliminated. In addition, due to the constant representation of the energy crisis in the media in recent years, energy-themed films after 2015 were included in the study. The films included in the study in this context are (i) April and the Extraordinary World (2015) and (ii) The Cloverfield Paradox (2018).

In this context, text analysis method, which is a kind of qualitative analysis, was used for this study. Text analysis helps us understand meanings and representations in a text (Gürkan, 2017). Therefore, in this study, the use of text analysis will be helpful in obtaining an in-depth analysis of the energy crisis in the series. According to McKee (2003), understanding the context is the most important aspect of text analysis. This includes understanding the entire text, its genre, and the broader public context in which the text circulates (McKee, 2003). At this point, an analysis is made on the patterns of energy and energy crisis, as the interpretation of the series without taking into account the background elements may cause misunderstanding. At this point, this study seeks answers to the following questions:

RQ1: How do energy-themed films deal with the energy issue?

RQ2: How do science, scientists, and technology do in the absence of energy?

The answers to these research questions are sought through the visual narrative examples (i) April and the Extraordinary World (2015) and (ii) The Cloverfield Paradox (2018) films. Visual culture is seen as a source of entertainment that relieves people of the inevitable stress of working in a complex society. Since the authors of this study see this cultural consumption as something belonging to the private domain of social actors, in this study, the product of visual culture is placed at the center of the study. Visual media has become the dominant ideological device today, replacing literate culture. At this point, these three films show exactly how a power that watches and punishes works, but this is a power that also seduces. Therefore, in this study, by placing a visual culture product at the center, the effect of the libidinal power of globalization, not the disciplinary power of modernity, is emphasized, and it is emphasized that (visual) media reproduce and strengthen systems. The use of text analysis in the context of this research facilitates the discussion of the reasons behind the behavior, gender, and beliefs of the characters in the selected films. The coding used to understand the selected sequence was developed based on the research questions and identified themes.

The following stages were taken into account in the analysis of the data (sequence):

(a) Each film was watched two times by the researchers. In order to answer the research questions, the main issue "energy crisis" and its relation to other dynamics were taken into consideration. The first viewing is the stage in which the films are watched by the ordinary viewer. In the second viewing, the importance of the scenes, shootings, characters, and dialogs of the films was determined under the collected and noted themes. Besides, themes (energy, energy crisis) were evaluated and notes were taken.
(b) The findings in this section reveal (i) how the concept of energy and energy crisis are handled in the narratives, (ii) how energy crisis opens a new order, (iii) how these concepts come to the fore, (iv) and how the struggle for a cure occurs, especially among the world's population.
(c) Text analysis was applied based on Mullet's (2018) general analytical framework: (1) selection of discourse, (2) locating data sources, (3) analysis of each text, (4) coding the text and identifying overarching themes, (5) analysis of external relations in texts, (6) analysis of internal relations in texts, and (7) interpretation of data.

    Analytical framework for analysis (adapted from Mullet, 2018)

    1. Choice of discourse: The description of energy crisis in the selected films has been chosen as the subject of the research. Thus, it is tried to determine what people do when they face with an energy crisis situation and what their reactions are.
    2. Finding data source: April and the Extraordinary World and The Cloverfield Paradox films were selected. This series is important in the sense that it represents this energy crisis issue.
    3. Analysis of the text: These two films, released in 2015 and 2018, based on the energy crisis issue to the subject of this research, are examined.
    4. Coding the text and identifying overarching themes: In these two films, which was selected using the qualitative coding method, the concepts of energy and energy crisis are presented, and how people face this important issue and how they handle are described.
    5. Analysis of external relations in the text: In the event that all energy sources are destroyed and consumed, the main reason what people do for the cures and how their basic needs will be met, is that they do not know exactly what to do.
    6. Analysis of internal relations in the text: the depiction of the dominant and the nondominant, racial-sexual-religious characteristics.
    7. Interpretation of data: Data provides insight into which messages are prioritized. The plot and characters in the films and their descriptions and life practices will shed light on how a cure is built for energy crisis.

(d) The following findings are reached to examine the connections between the plot in the films, the progression of events in a cause-effect relationship, and the energy and the energy crisis in the dystopian universe:

    - Energy crisis and dystopian world

- Energy crisis and science's search for a cure

To better understand the films, it would be meaningful to mention their plots.

April and the Extraordinary World (2015) is set where Napoleon III was killed in a building explosion, and his son's cunning diplomacy averted the Franco-Prussian War. But, over the next 70 years, scientists worldwide would disappear, causing technological progress to come to a standstill and pollution to spiral out of control.

The Cloverfield Paradox (2018) follows the crew of an earth-orbiting space station and their testing of a dangerous device that can provide unlimited power. At the same time, it exposes them to a dark alternate reality at the risk of creating a terrifying paradox.

5.4 Findings

Dystopia can be defined as a way of life of a possible future or an alternative present that can be evaluated negatively for humanity. This alternative lifestyle is the subject of a broad spectrum, from written narratives to visual narratives. Furthermore, dystopian narratives depict a dark future or an uncertain time (Gürkan & Ergene, 2022). In this context, it is possible to encounter unusual situations in dystopian narratives. The issues, such as how individuals and societies are affected by the concepts of sovereign power, social balances, management, the resources of the world and how these resources are consumed, and the situations that may arise in the absence of especially energy, are placed in the center of these two films' narratives.

In both films, in the absence of energy, individuals somehow establish the hegemonic balance in different systems and societies (dystopic). At this point, individuals (especially scientists) realize the construction of a new world order by promising to ensure the continuity of the values and traditions of society and to seek alternative resources when the world's resources are exhausted. In this context, mentioning ten different utopia/dystopia genres of Balasopoulos (2006) seems essential. These are (i) satirical anti-utopias, (ii) dogmatic fictional anti-utopias, (iii) nonfiction dogmatic anti-utopias, (iv) preventive anti-utopias, (v) critical anti-utopias, (vi) dystopias of tragic failure, (vii) dystopias of authoritarian oppression, (viii) dystopias of catastrophic probability, (ix) nihilistic dystopias, and (x) critical dystopias. According to Gürkan and Ergene (2022), this classification depends on the interpretation of the texts they apply. It aptly illustrates how vital this typology of Balasopoulos is to the question of definition in any discussion of the subject.

In this context, the dystopian world can be described as follows in the energy-themed film examined in this context: As Westfahl (2005) states, in both films, individuals in the society have accepted the social system and norms, and individuals do not even realize the disorders/defects in the system. In these two film narratives, three factors effectively transform the concepts of heaven and hell, good and evil: energy resources and technology and science. The lack of energy resources and the search for new energy are factors of control, pressure, and fear for individuals. At the same time, technology and science enable the necessary techniques and methods to be providers of this pressure, fear, and control mechanism. In these two narratives, the construction of society is built in the triangle of energy resources, technology, and science. Individuals become isolated and objectified in this social mechanism, while social organization mechanizes.

For example, in the movie, April and the Extraordinary World, how all the elements of the capitalist system in every aspect of our life practices provide movement toward the system is conveyed through lizards. Lizards, created to build an invincible army, point to the necessity of working toward a sustainable world and not destroying energy resources. The film aims to prolong the human life with endurance serums. Lizards aim to build a sustainable society/world by destroying this serum. In this context, an alternative reality is created in the movie April and the Extraordinary World. It is portrayed as what would be like in a world where everything that has no electricity, no atoms, and no automobiles operates with the power of steam—bridge construction between England and Russia, etc. The scenes about the absence of scientists, magicians, and circus acrobats trying to make unscientific potions in circus entertainment centers and scenes showing charlatans are the visual elements depicting this dystopian world in cinematographic terms.

In The Cloverfield Paradox, the fact that the world is facing a significant energy crisis is shown with the scenes of power cuts. It is supported by news from news sources that everyday lives are under threat. In this way, the dystopian atmosphere of the film is built with the cinematographic elements used and transferred to the audience. By emphasizing the system and hierarchy in the film, these two (system and hierarchy) elements that form the basis of dystopian narratives are constantly repeated. In the film, it is pointed out that the wars between states have increased due to the energy crisis. Germany is preparing for war as a significant power; hunger is increasing in Russia; hierarchy is emphasized by using discourses such as creating order; and discourses are made about the continuity of the system and order in the world.

Considering that science offers solutions to the difficulties of daily life and helps humanity solve its great mysteries, science and scientific perspective are placed at the center of the narrative in both films. Science is a branch of knowledge based on observation and experimentation applied regularly and systematically to gain knowledge about the physical world and its phenomena. In this context, in the films April and the Extraordinary World and The Cloverfield Paradox, science is used to develop some laws and hypotheses in the despotic world experiencing the energy crisis. It is depicted that scientists act according to the targeted purpose. As Hunt (1991) stated, when it is remembered that the primary purpose of science is to develop laws and theories in order to explain, predict, understand, and control phenomena, in the narratives of both films, some goals are determined after the scientific research is carried out by scientists, and scientists are guided toward these goals. It is depicted that they are on a quest.

In April and the Extraordinary World, the mines run out, and coal is replaced by charcoal. The effect of the technological regression and progress underlying these discourses is the ineffective use of energy resources and the adverse effects of this on human life. The importance of science and technology is being repeated with the energy crises that started with the disappearance of forests, the conflicts between the states due to the use of coal, and the wars that occurred as a result. In this context, in many scenes of the film, the need for scientists is emphasized as the glorification of science and the establishment and assignment of special teams in search of lost scientists. In addition, in a world stuck in the era of steam vehicles in 1831, a different perspective on technology is introduced in the film. A society that produces science for war has been built, and at this point, a comparative reading with today's world is tried to be realized. Regardless of the period we are in, it is pointed out that methane can replace steam. There are no anti-technology discourses in the scenes that emphasize the ways of using technology. However, inferences can be made that correct results can be achieved for effective and successful use—symbols of the totalitarian regime, imperial security, etc. Although expressions are not frequently used, the film's central theme is the "negative" results of science serving systemic interests.

The movie The Cloverfield Paradox states that free energy production resources are needed because the world's energy resources will be completely depleted in 5 years. Financial and energy resources have been used unnecessarily in international initiatives, away from their practical use. For this reason, astronauts build spacecraft to reach new and endless energy sources outside the world. Thus, scientists describe the search for new remedies for humanity in the light of science in this way. These scientists are testing the Shepard Separation Accelerator (SPA) device, which can provide unlimited energy and resolve the current energy crisis, potentially resulting in a dark alternate reality. The film has scenes emphasizing the importance of the discovery of infinite and unlimited energy. There are also segments in the film that produce rhetoric that efforts to solve the energy crisis may spread demons. Thus, at a time when energy resources are depleted, science and religion, seeking a cure, come face to face. The film opens with a radio announcement: "Government sources tell us that the world's energy resources will be completely depleted in 5 years." This discourse shows the seriousness of the energy problem in the world. Scenes of prayer and supplication support glorifying religious motifs to provide energy for god support, and scenes of supplication to prevent wars.

In this context, it is also important to mention how the concept of energy creates itself and supports the messages with the cinematographic images used in two different films about energy and the energy crisis. Films make their messages cinematic through images and generate new ideas. Thoughts/ideas overflow through cinematographic images in movies. As a result of this overflow, thoughts/ideas on the conceptual plane are stretched and reach an unthinkable dimension (Bordwell, 1985; Bordwell, 1979; Gürkan, 2015). While the atmosphere created in both films depicts a dystopian world, light, space, sound, and plot are essential in describing how the world entered a new order in a period when it was deprived of energy resources. Disappearing scientists, hegemonic new management approaches, dark interiors and exteriors, and diegetic and non-diegetic sounds (Sounds produced naturally or accidentally by the environment and character are defined. Diegetic sounds are naturally produced sounds used for dramatic effect loaded with meanings that are not accidental—creating atmosphere, creating ambiance. On the other hand, non-diegetic sounds and soundtracks are used to set the movie's tone, manipulate emotions, add drama, express uncertainty, or provide an element of surprise. Sound for dramatic effects is added and is assumed not to be heard by the characters in the movie; a dystopian world based on energy is built and supported by the use of light.

5.5 Conclusion

An energy crisis is a situation of a sudden shortage (or a drastic increase in price) in the supply of energy resources (Ai et al., 2023). It usually refers to a decrease in oil, electricity, or other natural resources (Niu et al., 2023). The crisis often spreads to the rest of the economy, causing a recession. In particular, electricity generation costs rise, which increases production costs. In addition, from the consumer's point of view, the increase in the price of automotive fuel (gasoline or diesel) causes the consumer to reduce their expenditures on other items and decrease their self-confidence (Aksoy et al., 2023).

The energy crisis, which has been on the agenda for the last 20 years, has become one of the main issues of the written and visual media (Yüksel & Dinçer, 2023). In this context, the subject of popular narratives is that films based on energy and energy crisis are produced and meet with the audience (Wan et al., 2023). Especially in 2000 and after, energy and energy crises are depicted and discussed in many TV series, fictional films, and documentaries.

This study investigates how the energy issue and the search for science, technology, and remedies that come up with this issue are depicted in popular narratives made in and after 2015. In this context, the films April and the Extraordinary World and The Cloverfield Paradox were included in the study. The study analyzes how energy-themed films deal with the energy issue and how science, scientists, and technology act in the absence of energy. The fact that energy-themed films are based on dystopian narratives and placing these subjects at the center of their narratives show how individuals and societies are affected by the concepts of sovereign power, social balances, governing management, the resources of the world, how these resources are consumed, and the situations that may arise in the absence of energy. However, these films emphasize the importance of science, scientists, and technology and show that advancing in the footsteps of science and technology can offer a way out/emancipation when the world is experiencing energy supply problems.

References

Ai, R., Zheng, Y., Yüksel, S., & Dinçer, H. (2023). Investigating the components of fintech ecosystem for distributed energy investments with an integrated quantum spherical decision support system. *Financial Innovation, 9*(1), 27.

Aksoy, T., Yüksel, S., Dinçer, H., Hacioğlu, Ü., Mikhaylov, A., & Muyeen, S. M. (2023). Assessing the electricity production capacities of emerging markets for the sustainable investments. *IEEE Access, 11*, 24574–24590.

Balasopoulos, A. (2006). Anti-Utopia and Dystopia: Rethinking the generic field. In *Utopia Project Archive, 2006–2010* (pp. 59–67). School of Fine Arts Publications.

Barnett, M., Wagner, H., Gatling, A., et al. (2006). The impact of science fiction film on student understanding of science. *Journal of Science Education Technology, 15*, 179–119.

Baudrillard, J. (1991). Simulacra and science fiction (Simulacres et science-fiction). *Science Fiction Studies, 18*(3), 309–313.

Bezel, N. (2001). *Yeryüzü Cennetlerinin Sonu*. Güldikeni Yayınları.

Bleiler, E. F. (1990). *Science-fiction, the early years: A full description of more than 3,000 science-fiction stories from earliest times to the appearance of the genre magazines in 1930: With author, title, and motif indexes*. Kent State University Press.

Bordwell, D. (1979). The art cinema as a mode of film practice. *Film Criticism*.

Bordwell, D. (1985). *Narration in the fiction film*. The University of Wisconsin Press.

Carayannis, E., Kostis, P., Dinçer, H., & Yüksel, S. (2023). Quality function deployment-oriented strategic outlook to sustainable energy policies based on quintuple innovation helix. *Journal of the Knowledge Economy*, 1–19.

Collins, S. (2003). Sail on! Sail on!: Anthropology, science fiction, and the enticing future. *Science Fiction Studies, 30*(2), 180–198.

Department for Business, Energy, and Industrial Strategy. (2021). Quarterly energy prices statistical release 23rd December 2021, BEIS.

Dinçer, H., Yüksel, S., Hacioglu, U., Yilmaz, M. K., & Delen, D. (2023a). Development of a sustainable corporate social responsibility index for performance evaluation of the energy industry: A hybrid decision-making methodology. *Resources Policy, 85*, 103940.

Dinçer, H., Eti, S., Gökalp, Y., Yüksel, S., & Çelebi, B. (2023b). The role of small-scale wind turbines in increasing the competitiveness of hospitals. In *Handbook of research on quality and competitiveness in the healthcare services sector* (pp. 336–352). IGI Global.

Dinçer, H., Yüksel, S., Mikhaylov, A., Muyeen, S. M., Chang, T., Barykin, S., & Kalinina, O. (2023c). CO2 emissions integrated fuzzy model: A case of seven emerging economies. *Energy Reports, 9*, 5741–5751.

Dinçer, H., Yüksel, S., Aksoy, T., Hacıoğlu, Ü., Mikhaylov, A., & Pinter, G. (2023d). Analysis of solar module alternatives for efficiency-based energy investments with hybrid 2-tuple IVIF modeling. *Energy Reports, 10*, 61–71.

Dinçer, H., Eti, S., Yüksel, S., Özdemir, S., Yı́lmaz, A. E., & Ergün, E. (2023e). Integrating data mining and fuzzy decision-making techniques for analyzing the key minimizing factors of carbon emissions. *Journal of Intelligent & Fuzzy Systems, 45*, 1–17.

Dinçer, H., Eti, S., Aksoy, T., Yüksel, S., Hacioglu, U., Mikhaylov, A., & Muyeen, S. M. (2023f). Analysis of environmental impact for material production investments using a novel soft computing methodology. *IEEE Access, 11*, 37987–38001. https://doi.org/10.1109/ACCESS.2023.3266524

Eti, S., Dinçer, H., Gökalp, Y., Yüksel, S., & Kararoğlu, D. (2023). Identifying key issues to handle the inflation problem in the healthcare industry caused by energy prices: An evaluation with decision-making models. In *Managing inflation and supply chain disruptions in the global economy* (pp. 162–178).

Eti, S., Dinçer, H., Yüksel, S., & Gökalp, Y. (2022). Analysis of the suitability of the solar panels for hospitals: A new fuzzy decision-making model proposal with the T-spherical TOP-DEMATEL method. *Journal of Intelligent & Fuzzy Systems, Preprint*, 1–13.

Franza, L., van der Linde, C., & Stapersma, P. (2020). European energy security and the resilience of southern Mediterranean countries. In E. Cusumano & S. Hofmaier (Eds.), *Projecting resilience across the Mediterranean*. Palgrave Macmillan. https://doi.org/10.1007/978-3-030-23641-0_7

Galloway, A. R., & Thacker, E. (2007). *The exploit. A theory of networks*. University of Minnesota Press.

Gezmen, B. (2022). "The Role of Environmental Journalism in Raising Awareness on Energy Efficiency", contributions to economics. In H. Dinçer & S. Yüksel (Eds.), *Circular economy and the energy market* (pp. 91–104). Springer.

Goldthau, A., Richert, J., & Stetter, S. (2020). Leviathan awakens. Gas finds, energy governance, and the emergence of the eastern Mediterranean as a geopolitical region. In *The review of policy research*. https://doi.org/10.1111/ropr.12387

Gürkan, H. (2012). Halkla ilişkiler sektörünün sinemada temsili: "Again you?","Sex and the City" ve "Thank you for smoking" filmlerinde halkla ilişkiler mesleği ve halkla ilişkiler uzmanının temsili. *Gümüşhane Üniversitesi İletişim Fakültesi Elektronik Dergisi, 1*(3), 135–158.

Gürkan, H. (2015). *Karşı Sinema*. Es Yayınları.

Gürkan, H. (2017). The portrayal of journalists in Turkish cinema: A study about journalism ethics through cinema. *Medijske studije, 8*(16), 41–59.

Gürkan, H., & Ergene, İ. (2022). The free individual in a tale of democracy: Hegemony and dystopia in visual narratives. KOME. *International Journal of Pure Communication Inquiry, 10*(2), 25–36. https://doi.org/10.17646/KOME.75672.90

Han, Z., Gong, L., Chen, H., & Yüksel, S. (2023). The role of mineral and oil resources in achieving a clean environment in emerging economies: An asymmetric analysis. *Resources Policy, 86*, 104099.

Hunt, S. D. (1991). *Modern marketing theory: Critical issues in the philosophy of marketing science*. South-Western Publishing.

Kitchin, R., & Kneale, J. (2001). Science fiction or future fact? Exploring imaginative geographies of the new millennium. *Progress in Human Geography, 25*(1), 19–35.

Kou, G., Pamucar, D., Dinçer, H., & Yüksel, S. (2023a). From risks to rewards: A comprehensive guide to sustainable investment decisions in renewable energy using a hybrid facial expression-based fuzzy decision-making approach. *Applied Soft Computing, 142*, 110365.

Kou, G., Yuksel, S., & Dincer, H. (2023b). A facial expression and expert recommendation fuzzy decision-making approach for sustainable business investments within the metaverse world. *Applied Soft Computing*, 110849.

Lempert, W. (2014). Decolonizing encounters of the third kind: Alternative futuring in native science fiction film. *Visual Anthropology Review, 30*(2), 164–176.

Liddell, C., & Morris, C. (2010). Fuel poverty and human health: A review of recent evidence. *Energy Policy, 38*, 2987–2997.

Liu, Y., Gong, X., Yüksel, S., Dinçer, H., & Ve Aydın, R. (2021). A multidimensional outlook to energy investments for the countries with continental shelf in east mediterranean region with hybrid decision making model based on IVIF logic. *Energy Reports, 7*, 158–173. https://doi.org/10.1016/j.egyr.2020.11.196

Luckhurst, R. (2005). Science fiction. *Polity*.

Martínez, L., Dinçer, H., & Yüksel, S. (2023). A hybrid decision making approach for new service development process of renewable energy investment. *Applied Soft Computing, 133*, 109897.

McKee, A. (2003). *Textual analysis: A Beginner's guide*. Sage Publications.

Middlemiss, L., Ambrosio-Albalá, P., Emmel, N., Gillard, R., Gilbertson, J., Hargreaves, T., Mullen, C., Ryan, T., Snell, C., & Tod, A. (2019). Energy poverty and social relations: A capabilities approach. *Energy Research & Social Science, 55*, 227–235.

Moiseev, N., Mikhaylov, A., Dinçer, H., & Yüksel, S. (2023). Market capitalization shock effects on open innovation models in e-commerce: Golden cut q-rung orthopair fuzzy multicriteria decision-making analysis. *Financial Innovation, 9*(1), 55.

Mullet, D. R. (2018). A general critical discourse analysis framework for educational research. *Journal of Advanced Academics, 29*(2), 116–142.

Niu, X., Yüksel, S., & Dinçer, H. (2023). Emission strategy selection for the circular economy-based production investments with the enhanced decision support system. *Energy, 274*, 127446.
Papanikos, G. T. (2017). Energy security, the European energy union and the Mediterranean countries. *Athens Journal of Mediterranean Studies, 3*(4), 341–354.
Rieder, J. (2017). *Science fiction and the mass cultural genre system*. Wesleyan University Press.
Ronay, I. S. (2011). *The seven beauties of science fiction*. Wesleyan University Press.
Sobhack, V. (1997). *Screening space: The American science fiction film*. Rutgers University Press.
Wan, Q., Miao, X., Wang, C., Dinçer, H., & Yüksel, S. (2023). A hybrid decision support system with golden cut and bipolar q-ROFSs for evaluating the risk-based strategic priorities of fintech lending for clean energy projects. *Financial Innovation, 9*(1), 1–25.
Westfahl, G. (2005). *The greenwood encyclopedia of science fiction and fantasy*. Greenwood Press.
Xu, X., Yüksel, S., & Dinçer, H. (2023). An integrated decision-making approach with golden cut and bipolar q-ROFSs to renewable energy storage investments. *International Journal of Fuzzy Systems, 25*(1), 168–181.
Yüksel, S., & Dinçer, H. (2023). Sustainability analysis of digital transformation and circular industrialization with quantum spherical fuzzy modeling and golden cuts. *Applied Soft Computing, 138*, 110192.
Yüksel, S., Dinçer, H., & Uluer, G. S. (2023). Socio-economic indicators of nuclear energy investments: A study on countries. In *Inclusive developments through socio-economic indicators: New theoretical and empirical insights* (pp. 33–44). Emerald Publishing Limited.
Yürür, F. (2017). *Kıyametin Sineması-Post Apokaliptik Filmler*. Cinius Yayınları.

Chapter 6
Evaluation of Psychosocial Risk Factors Encountered for Energy Sector Employees by MAIRCA Method

Serap Tepe, Serkan Eti, Haruna M. Moda, and Zeyneb Kahraman

6.1 Introduction

Energy is a crucial element in improving the quality of life and promoting economic development for nations (Sindhu et al., 2016). The increasing global population, driven by factors such as urbanization and technological innovations, has led to a rise in production and consumption demands, consequently resulting in a rapid growth of global energy demand (Artan, 2020; Yıldız & Ulucak, 2022). Particularly, due to global population growth and industrialization, energy poverty has reached unprecedented levels in developing countries (Gray, 2017). While the demand for energy is increasing, the share of electricity in final energy consumption is also on the rise. Conferring to all scenarios presented in the International Energy Agency's (IEA) World Energy Outlook, electricity demand has increased by approximately 50% in just 20 years. Looking ahead, electricity is expected to play a leading role

S. Tepe
The Department of Occupational Health and Safety, The University of Health Sciences, İstanbul, Turkey
e-mail: serap.tepe@sbu.edu.tr

S. Eti (✉)
The Department of Computer Programming, İstanbul Medipol University, İstanbul, Turkey
e-mail: seti@medipol.edu.tr

H. M. Moda
The Department of Occupational Health and Safety, University of Doha for Science and Technology, Doha, Qatar
e-mail: haruna.moda@udst.edu.qa

Z. Kahraman
Graduate School of Natural and Applied Sciences, Kastamonu University, Kastamonu, Turkey

H. Dinçer et al. (eds.), *Decision Making in Interdisciplinary Renewable Energy Projects*, Contributions to Management Science,
https://doi.org/10.1007/978-3-031-51532-3_6

not only in heating, cooling, and transportation but also in various sectors that are digitally integrated, such as communication, finance, and healthcare (International Energy Agency, 2020, 2022). As a result, the growing production and consumption demands have led to significant growth in global production sectors to meet societal needs (Yıldız & Ulucak, 2022).

The energy sector encompasses a vast and comprehensive network of interconnected companies directly or indirectly involved in energy production and distribution. It plays a crucial role in strengthening the economy and facilitating the production and transportation of goods (Chen, 2022). Increasing energy consumption, coupled with environmental and socioeconomic impacts, as well as technological advancements such as digitalization, is forcing the energy system to undergo a transformation (Acar et al., 2021; International Energy Agency, 2020). Alongside this change and transformation, the necessity of ensuring safe, healthy, and dignified working conditions emerges, highlighting the importance of prioritizing the health and safety of workers in energy production facilities.

The necessity of human factors in production has been proven through conducted study (Diao & Ghorbani, 2018). In a production facility, the most crucial element of production is the human factor, that is, the worker or the personnel. Employees in these facilities are exposed to various factors due to job-specific reasons (Koruca et al., 2015). Perhaps one of the most significant factors among them is the psychosocial elements that leave both psychological and physical damages. The effective management of risk factors in a business is associated with having a positive/negative psychosocial work environment. In this sense, in a positive psychosocial environment, meaningful work, utilizing skills appropriately, and engaging the employee in the work are essential. Conversely, in a negative psychosocial work environment, factors such as meaningless work, inadequate utilization of skills, and high levels of uncertainty are present. The negative response of individuals to excessive pressure or other types of demands likely to arise from such working conditions is associated with exposure to psychosocial hazards in work environments (Leka et al., 2017). Despite the existence of various psychosocial hazards in the work environment, accordingly, Howard et al. (2022) study affirmed within the 27 EU countries (ESENER3) that stress associated with dealing with difficult customers and time pressure is considered among top occupational hazards related to the sector (Table 6.1).

Table 6.1 The status of psychosocial risk factors identified in establishments in 27 EU countries in ESENER3 (Howard et al., 2022)

Risk factors	EU average (%)
Having to deal with difficult customers	59
Time pressure	45
Long or irregular working hours	21
Poor communication or cooperation within the organization	18
Job insecurity	11

The energy sector poses high risks, and therefore, it is crucial to comprehensively evaluate and carefully manage anything that hinders an individual's ability to make sound operational decisions (Prostect, 2019). Considering the indispensability of energy production on a global scale, the psychosocial well-being of employees directly involved in production is a matter that deserves significant attention. Moreover, despite many adverse conditions, psychosocial risks often emerge as unnoticed and insufficiently highlighted risks. Therefore, this study aims to identify the psychosocial issues that employees in energy production facilities may encounter and focus on developing strategies to prevent these risk factors.

6.2 Literature Review

Due to decades of poor management practices in the energy sector, factors such as increased workload, long working hours, and workplace bullying have led to deteriorating working conditions for employees (Prostect, 2019). Particularly in electricity production facilities, the execution of production and the performance of necessary maintenance/repair activities are complex, specialized, and labor-intensive operations (Özcan et al., 2017). While there are various production facilities operating in this field, each production facility has its unique working principles and job-specific risks accordingly. A literature review conducted on energy production facilities, which encompass a wide range, has provided the following headings as the general psychosocial working conditions in energy production facilities.

With the advancements in technology, changes in facility structures, and an increase in the number of employees, the risks in terms of occupational health and safety (OHS) have also escalated. Particularly in relatively new sectors such as wind energy facilities and emerging sectors like floating solar energy, it is likely that most employees will lack experience and proper training. The term "workforce" primarily refers to the local community residing in the vicinity of the power plant. According to the European Agency for Safety and Health at Work (EU-OSHA) report, skill shortages are widespread in renewable energy industries due to rapid innovations, and hazardous tasks are often performed by less skilled workers. Similarly, employees working in small-scale solar energy operations or offices lack sufficient knowledge and skills to address different risks. This mismatch in skills and the constantly evolving workplace arrangements give rise to various OHS issues (European Agency for Safety and Health at Work [EU-OSHA], 2013a, 2013b). In particular, inexperienced workers who are involved in the processes without adequate training pose skill gaps and put their health and safety at risk (EU-OSHA, 2022). A study highlights the inadequacy of education and training, defining it as a barrier in the renewable energy industry, resulting in a shortage of qualified personnel in many countries (Lucas et al., 2018). Furthermore, due to the complexity of the tasks involved, training workers for facilities like nuclear power plants proves to be challenging (Li et al., 2021). Nuclear power plants, when evaluated in terms of hazards and risks, pose significant threats (Kahraman & Y. Özdemir, 2022). In the mentioned nuclear energy facility, both the high risks involved and the complexity of the work make it increasingly difficult to directly find qualified personnel in this field. Additionally, personnel shortages can occur in all types of energy production facilities due to varying conditions that require different skills and experiences.

The subject of psychosocial risks and work organization is important in small-scale solar energy facilities, where a diverse workforce with various characteristics, skills, and needs can be employed, including subcontracted workers, migrant workers, undocumented workers, as well as new and unskilled workers (EU-OSHA, 2013a). This is because most workers in these facilities are immigrants with a weak occupational safety and health (OSH) culture and low skills (EU-OSHA, 2013b). Due to the introduction of new technologies or work processes that require new combinations of skills, the "old" OSH knowledge cannot be easily transferred to the new field, resulting in limited knowledge about OSH (Sen et al., 2021; EU-OSHA, 2022). Furthermore, the long-term effects on occupational health and worker well-being are unknown (Cooper et al., 2014; Karanikas et al., 2021). However, the standards and regulations in the field of occupational safety and health are inadequate. Earlier study with regard floating solar energy sector asserts the lack of training programs and OSH standards specific to the field. To which, the absence of OSH standards and field-specific training modules exposes untrained workers to complex and unfamiliar working conditions (Sen et al., 2021) and the creation of related job hardship for workers (Li et al., 2021). On the other hand, with the growth of the green economy, low-skilled workers may be compelled to accept worse working conditions, leading to a stronger polarization in terms of skills in the workforce (EU-OSHA, 2022). It is evident that many unskilled workers working under these challenging conditions in order to avoid job loss will be exposed to increasingly negative psychosocial conditions due to the inadequacy of regulations in the field of occupational safety and health. The lack of sufficient knowledge among workers with regard new technologies and systems implies an increased responsibility in terms of job skills and a constant need for learning in order to adapt to the pace of change taking place.

In complex large-scale systems with stringent requirements for human safety, such as nuclear energy, numerous unfavorable working conditions exist, including limited workspace, extensive teamwork, demanding tasks, tight deadlines, and unknown risks (Zhang, 2004). Continuous operation monitoring is necessary in such facilities, necessitating the rotational, fixed, or modified shift work of employees (Stockwell, 2020). In comparison with modern power plants, older technology-based facilities (e.g., hydroelectric) naturally employ a larger number of shift workers. These plants are more susceptible to operator errors due to the presence of shift workers (Özcan et al., 2017). Alvesson and Einola (2018) emphasize that the number of hours individuals are required to work has a pervasive effect not only on their physical health but also on their mental well-being. Relatedly, Li et al. (2021) highlight the work difficulties faced by nuclear power plant workers in their study, including heavy mental workload (e.g., coping with excessive information and decision-making), the necessity of completing multiple tasks, the demanding nature of the job, prolonged project durations, or time constraints, all of which impose high cognitive demands on workers. In furtherance, Shanfa (2008) reports high effort and low reward among workers in thermal power plants as another factor impacting employee well-being within the sector. In a study conducted by Badre (2021) among 120 employees, both senior and non-senior, in thermal power plants, it was found that senior employees experienced higher workloads. Moreover, the primary determinant of psychological well-being in both employee groups was identified as experiencing higher pressure inherent in the nature of the job. Aydın (2020) indicates, based on risk assessment in a wind energy plant, that during the planning of work programs, the processes of turbine control, repairs due to unexpected malfunctions, and regular maintenance activities continue as the turbines are in operation, with a limited number of personnel. Attention is also drawn to the presence of a small number of workers for shifts and maintenance tasks.

One of the factors that create difficulties in terms of job performance in organizations is the lack of clarity regarding tasks, roles, and responsibilities (Li et al., 2021). Findings from a study have shown that employees' occupational stress is primarily caused by role inadequacy and role ambiguity, which are significantly associated with job satisfaction (Hoboubi et al., 2017). Therefore, uncertainties about employees' tasks and responsibilities are important psychosocial conditions in the energy sector, as in many other sectors. Workers in unfavorable conditions experience stress due to both task ambiguity and unmet needs (such as social support). Li et al. (2021) have demonstrated that certain occupational stress factors, such as the emotional needs of workers and their families, which are often overlooked, can be more pronounced in nuclear power plants compared to other sectors. Ceylan et al. (2015), in their study involving employees from various departments (e.g., production) of Turkey Petroleum Corporation and Turkish Petroleum International Company, found a positive relationship between job satisfaction, managerial support, and staff empowerment.

Due to a competitive and unhealthy work environment created by cultural factors, some employees may exhibit a tendency to engage in excessively aggressive behavior that can be characterized as bullying (Rahmadani, 2016). For instance, increased aggression can arise as a result of exposure to noise (Organization, 2019). The findings of a study conducted by Taoussi et al. (2022) identified nonauditory symptoms such as irritability, headaches, and a tendency to raise one's voice during conversations in 68.5% of employees, which were associated with noise. Another study involving 40 blue-collar workers in the energy sector revealed that some employees engage in mocking or exploiting the weaknesses of their colleagues, driven by factors such as stress, jealousy, and a desire to appear powerful. Furthermore, some employees exploit their co-workers' vulnerabilities or phobias for different purposes. Based on the collected data, it is considered disruptive behavior in the workplace for employees within a hierarchical structure to inflict psychological harm on each other through the use of derogatory language, insults, or threats, which disrupts workplace communication and harmony. One of the most commonly encountered disruptive behaviors in the workplace is the psychological disregard of individuals, known as psychological ostracism (Tünay Doğan, 2019).

The concern of not being able to make ends meet, commonly referred to as "end-of-the-month anxiety," causes stress among employees. In a study evaluating salary conditions, it was found that employees earning below a certain threshold had lower job satisfaction and experienced higher levels of job stress compared to those earning higher salaries. In this regard, one of the most significant factors that enhance job satisfaction is a higher salary (Zincirkıran et al., 2015). Similarly, in a study on job insecurity in a private energy company in Turkey, it was noted that employees expressed concerns about salary and promotion, emphasizing these subcategories. Moreover, it was found that individuals who were parents focused more on salary and promotion opportunities, while those without parenting responsibilities emphasized the working conditions and personal development opportunities provided by the job (Boz-Semerci, 2018). Furthermore, a study conducted among nuclear power plant employees indicated that comparing career development with peers, economic problems arising from education and family life, distance from the workplace, and lack of support from family members in career advancement were identified as stressors (Li et al., 2021). Badre (2021) found that for senior employees, the factor that exerted significant pressure and stress on them was the need to perform for career progression.

It is widely recognized that the work environment can significantly impact the health of employees in psychosocial terms. Employees face psychosocial challenges due to difficulties in accessing work areas, harsh weather conditions, and the remote locations of power plants (Aydın, 2020). Wind energy turbines are typically found in isolated areas that require extensive travel (Webster et al., 2013), while workers at nuclear power plants cannot stay in one place for an extended period. Additionally, employees face challenges in their work due to factors such as the complexity of tools and equipment, temperature, humidity, noise, and the ambiguous identification of facilities (Li et al., 2021). In solar energy, workers are often found in outdoor settings with high solar radiation and extremely hot weather conditions (Samaniego-Rascón et al., 2019; SolarPower Europe, 2019). A study conducted on solar energy workers in Mexico found that working conditions exceeded the recommended exposure limits due to heat. The study also revealed that acclimatized workers had a higher work capacity compared to non-acclimatized workers (Samaniego-Rascón et al., 2019). Occupational risks identified in these energy plants include electric shocks, slipping, falling from heights, straining, fire hazards, mud, severe weather, mechanical injuries, and wildlife encounters (SolarPower Europe, 2019). Floating solar energy systems entail adverse environmental conditions such as heat, ultraviolet radiation, humidity, rain, snow, noise, and dust (Sen et al., 2021). In thermal power plants, 15% of managers reported inadequate facilities and resources, and 55% of them indicated a hot working environment as a significant concern related to environmental factors. Other significant problems include excessive noise and a dusty environment (Hole & Pande, 2009). In coal-fired thermal power plants, which account for approximately 38% of global electricity production, workers are exposed to energy sources with low levels of automation. Moreover, these facilities involve ergonomic issues, nonionizing radiation, electrical hazards, fire and explosion risks, as well as airborne pollutants and other similar substances (Ersoy et al., 2022).

Studies examining the wind energy sector specifically highlight problems such as weather conditions, working in confined spaces, electric shocks and working at heights, challenging climatic conditions, and inadequate occupational training (Aydın, 2020). A study investigated the effects of noise and electromagnetic fields on the psychological well-being of personnel at an electricity plant in Shiraz. The results indicated that a significant number of individuals (78.2%) were exposed to electromagnetic fields and noise, and there were suspicions of some form of mental disorder among these individuals (Zamanian et al., 2010).

6.3 Psychosocial Risks Encountered in Energy Production Facilities

Employees in the energy sector face some unique challenges related to mental health and well-being (Canadian Mental Health Association, 2022). A study conducted on power plant workers found that 79.9% had good mental health, while 14.2%, 5%, and 0.9% experienced mild, moderate, and severe mental disorders, respectively (Wahyuningtias et al., 2023). Mental health issues in the workplace not only affect individuals but also have serious consequences for organizational productivity. In Western countries, one-third to half of all long-term illnesses and disabilities among the working-age population are attributed to mental health problems in organizations (Seitz, 2018). Additionally, 69% of individuals with moderate mental health issues report experiencing difficulties in job performance, compared to 26% among those without mental health problems (Alvesson & Einola, 2018). Consequently, work-related psychological disorders rank among the most significant problems for employees (Zamanian et al., 2010). The challenges faced in the work environment are long term and can cause psychological and physiological stress for employees, as well as an increased risk of accidents (Li et al., 2021). Psychosocial risks can lead to destruction and losses, even in cases where they are not the main factor in work accidents (Tepe et al., 2022). Research conducted on energy production facilities has identified various psychosocial risks associated with unfavorable working conditions, including stress, fatigue, aggressive behavior, bullying, depression, burnout, work-life imbalance, job insecurity, work accidents, and absenteeism, which manifest as problems in work life (Tepe et al., 2022). A study conducted on thermal power plants revealed that different types of thermal power plants face different problems in terms of efficiency, facilities, environment, occupational health and safety (OHS), and ergonomics. In other words, companies with higher health, facility, and environmental issues tend to experience more performance-related problems such as lower efficiency and higher absenteeism (Hole & Pande, 2009).

Stress is not only becoming an increasing concern for organizations (Badre, 2021), but it is also one of the most significant psychosocial risks commonly encountered in negative psychosocial work environments. Research indicates that employees in the energy sector experience higher levels of job-related stress and general stress and have a higher incidence of long-term health issues compared to the general population (Canadian Mental Health Association, 2022). Stress is an adaptive response to an external situation that results in physical, psychological, and behavioral deviations. The physiological and emotional reactions that arise from the imbalance between job demands and the ability to meet those demands are referred to as occupational stress (Narban et al., 2016). Recent findings from a study conducted in thermal power plants indicate that employees experience stress due to factors such as personal responsibilities, organizational roles and positions, expectations from higher authorities, shift scheduling, career advancement, and commuting to work. Job stress manifests in various ways and affects mental and physical health in different ways (Badre, 2021). Further to this, Khan (2020) emphasizes that employees' productivity is directly influenced by the work environment and highlights the tendency to underestimate psychological stress. For instance, exposure to noise can lead to communication disruptions (Bal, 2018), increasing the effort required for communication and resulting in stress, fatigue, decreased productivity, and an increased risk of accidents (Le et al., 2017). In a study conducted in the Netherlands by Van der Klauw et al. (2016), it was found that 33.0% of work accidents leading to disability were associated with psychological harm such as trauma or stress. On the other hand, accidents also give rise to psychological effects, creating a vicious cycle. A study conducted on 1495 individuals 2–3 months after a nuclear plant accident demonstrated the persistence of psychological impact following the accident. According to the results of the study investigating the impact of the accident on the employees of Daiichi and Daini nuclear plants, general psychological distress and post-traumatic stress disorder were found to be prevalent among nuclear plant workers, with Daiichi workers (n = 885) being more frequently exposed to stress factors related to the disaster (n = 412; 47%) (Shigemura et al., 2012). This association is attributed to the greater exposure of Daiichi workers to stress factors. Additionally, under arid and moderate-mild working conditions, heat stress levels exceed exposure limits (Samaniego-Rascón et al., 2019). This situation can exacerbate existing health conditions and result in risks of heat exposure, such as decreased mental and physical work capacity, fatigue, and cramps (Di Corleto & Jodie, 2018). Therefore, heat stress is a critical occupational risk factor for solar energy workers. In a report, the International Labour Organization (ILO) estimates that by 2030, heat stress will cause a loss equivalent to 2.2% of total global working hours (International Labour Organization [ILO], 2019). Even in acclimatized workers, when work intensity is excessively high, productivity in these environments can only reach 50% (Samaniego-Rascón et al., 2019).

Shift work leads to both physiological and psychological effects, particularly chronic fatigue, sleep disorders, and increased job dissatisfaction during night shifts (Stockwell, 2020). Some effects of sleep deprivation due to increased working hours include (a) increased likelihood of accidents, (b) higher rates of absenteeism, (c) decreased job performance, and (d) reduced quality of life and increased healthcare utilization (Blagoeu et al., 2018). Furthermore, fatigue is one of the most common physiological factors associated with a decrease in functional ability and performance in an organ, muscle, or the entire organism due to workload. While the level of fatigue varies from person to person, engaging in the same or similar tasks at a specific pace repetitively every day and year after year leads to physiological and psychological fatigue, depending on an individual's physical and psychological makeup (Koruca et al., 2015). It is evident that fatigue has consequences for both job performance and safety, with numerous examples of major disasters caused by fatigue. One such example is the partial reactor meltdown that occurred at the Three Mile Island Nuclear Power Plant in March 1979, partially attributed to human factors. Shift workers failed to notice a coolant leakage at the end of their shifts, and other workers had to analyze the problem in the reactor control room for several hours due to inadequate or misleading information from the instruments. This situation highlights the tendency for performance to decline toward the end of a shift, particularly during long working hours and inadequate shift schedules, indicating a clear connection between shift work and performance decline (Stockwell, 2020).

The need for work-life balance arises as a response to the mismatch between the demands of one's personal role and the demands of one's productive role, leading to work-family conflict (Pradhan et al., 2016). In the Nigerian oil and gas industry, 80% of workers who are employed in rotation struggle to achieve a balance between their work and personal lives (Kumar & Janakiran, 2017). It is noted that blue-collar workers experience more work-life imbalance compared to non-blue-collar workers, which is attributed to the difficulty blue-collar workers face in integrating their work and family responsibilities (Fernandez et al., 2017). Work-life conflict affects organizations due to high turnover and absenteeism rates, low performance, and job dissatisfaction (Jolaolu, 2020; Visser et al., 2016). Harding et al. (2018) argue that the literature consistently demonstrates that employees experiencing high work-family conflict are subject to a wide range of negative outcomes, including decreased satisfaction in both work and family domains, as well as compromised physical health and emotional well-being. In a study investigating the strategies required to maintain work-life balance among workers in the petroleum and gas energy sector, field notes, document analysis, and data obtained from employees with over 5 years of experience revealed that the key factors for sustaining work-life balance are family support and synergy (collaboration) (Jolaolu, 2020).

One of the significant problems encountered by employees working in energy production facilities is job insecurity. Concerns regarding the loss of the job itself and the difficulties of finding new employment in the future are considered the quantitative dimension of job insecurity, while the losses related to wages, promotions, working conditions, individual skills development, and social opportunities are evaluated as the qualitative dimension of job insecurity (Boz-Semerci, 2018). Global economic uncertainty, the rapid pace of technological change, and the development of new occupational sectors have historically led to workforce restructuring, increased occupational demands (e.g., retraining, learning new skills, changing existing job tasks), and changes in policies and regulations (Strietska-Ilina et al., 2011). This situation has an even more exacerbating effect on job insecurity, particularly in energy production industries that are directly affected by newly established facilities or technological advancements. A study conducted in the United

States found a significantly higher likelihood of reporting job insecurity among green-collar workers in the solar energy industry, similar to work-life imbalance (Fernandez et al., 2017). Additionally, job insecurity arises in nuclear power plants when the reactor's lifespan expires, leading to the end of employment in these energy facilities. As a result of job insecurity, Boz-Semerci's (2018) study indicates that the outputs reported by participants were stress (24%), job satisfaction (22%), commitment to work (18%), organizational commitment (16%), absenteeism (10%), and turnover (10%).

Exposure to unfavorable working conditions not only affects an individual's work capacity but also leads to various mental problems. In the short term, this can manifest as anxiety, irritability, and mood disorders, while in the long term, it can result in serious issues such as burnout, depression, or even suicide. Studies have demonstrated the emergence of various mental problems as a consequence of exposure to psychosocial risk factors. Research highlights the limited knowledge regarding the determinants of cognitive workload in mentally demanding professions, thereby maintaining the uncertainty surrounding the relative importance of personal and work-related factors in such occupations (Jolaolu, 2020). Furthermore, excessive mental load can cause information processing delays or cognitive errors, leading to occupational stress, while physically demanding cognitive tasks can impair mental processing and affect mental workload (DiDomenico & Nussbaum, 2011; Elfering et al., 2011). High levels of occupational stress are associated with an increased risk of depression and an increase in negative behaviors such as anger, anxiety, and irritability (Mosadeghrad, 2014). On the other hand, exposure to noise can give rise to various mental problems such as poor mood, attention deficits, lack of concentration, restlessness, communication difficulties, weakness, fatigue, decreased productivity, increased aggressive behaviors, feelings of helplessness, and confusion associated with job conditions (Bal, 2018; Organization, 2019; Shan & Neis, 2020). Additionally, rising temperatures can lead to psychosocial problems for workers, including difficulties in decision-making, communication, social adaptation, decreased job performance, aggression, and hostility (Tepe et al., 2022).

One of the psychological effects that can arise due to shift work is depression. Recent research has demonstrated that shift workers, especially women, are at a higher risk of poor mental health, including depressive symptoms (Torquati et al., 2019). Another study indicated that the intensity of noise is associated with the emergence of depression symptoms and negatively affects the emotional well-being and quality of life of employees (Kuru, 2017). Depression can directly contribute to inadequate sleep, negative eating behaviors, and other disorders (Stockwell, 2020).

Workplace bullying, which is associated with employees' mental health, originates from various factors such as work environment, individual characteristics, and measures taken against bullying. It is emphasized that most companies lack regulations regarding bullying, and even if they exist, they may not have the capacity to stop such behavior (Rahmadani, 2016). A study examining the prevalence, distress level, and life satisfaction among workers in seven power plants found that the majority of workers did not experience negative actions related to workplace bullying (84%), but some employees still experienced bullying (11.4%) or became targets of workplace bullying (Wahyuningtias et al., 2023). Workplace bullying is associated with job burnout, which can lead to increased fatal accidents and company losses (Wahyuningtias et al., 2023). Employees who experience bullying are more likely to suffer from job burnout due to work-related stress and stress arising from interpersonal relationships (Young-Ran & Jeong-Won, 2016). Additionally, workplace bullying has a negative relationship with employees' life satisfaction (Gabriele et al., 2015).

6.4 Method

Multi-criteria decision-making (MCDM) is a crucial area in decision science that aims to handle complex problems with multiple criteria. Numerous MCDM methods have been developed over the years to help decision-makers tackle various challenges effectively. One such method is the multi-attribute ideal-real comparative analysis (MAIRCA) method. This method is a multi-criteria decision-making approach used to solve problems with multiple conflicting attributes. Developed by researchers as an extension of the analytic hierarchy process (AHP), MAIRCA integrates elements of ideal point analysis and reference point theory to facilitate complex decision-making processes. The key principles of the MAIRCA method involve the establishment of an ideal solution and a real solution for each criterion under evaluation. The ideal solution represents the most favorable state for a given criterion, while the real solution represents the actual performance. By comparing these solutions, decision-makers gain insights into the gap between the ideal and real performance, aiding in prioritizing and selecting the most appropriate alternative.

6.4.1 Steps in MAIRCA Method

The MAIRCA method follows a structured approach to assist decision-makers in reaching a well-informed choice. The main steps involved in the MAIRCA technique are as follows:

Step 1. Problem Identification and Criteria Definition: Identify the problem and determine the criteria relevant to the decision context. These criteria should be measurable, mutually exclusive, and collectively exhaustive.

Step 2. Data Collection and Evaluation: Gather data on the performance of each alternative concerning the established criteria. Evaluate the real performance of each alternative based on the collected data.

Step 3. Ideal Solution Determination: Identify the ideal performance for each criterion, which represents the most favorable state. This can be obtained through benchmarking, expert judgment, or historical data.

Step 4. Comparative Analysis: Calculate the discrepancies between the real and ideal performance for each criterion. This analysis helps in determining the overall performance of each alternative concerning all criteria.

Step 5. Weighted Aggregation: Apply weights to each criterion to represent their relative importance. Aggregate the discrepancies from Step 4 using these weights to obtain a comprehensive performance score for each alternative.

Step 6. Ranking and Decision-making: Rank the alternatives based on their aggregated performance scores. The alternative with the highest score indicates the most suitable option (Donyatalab et al., 2020).

The MAIRCA method finds applications in various fields, including project selection, supplier evaluation, financial investment, environmental impact assessment, and healthcare management. Its ability to accommodate multiple criteria and handle conflicting attributes makes it a valuable tool for real-world decision-making challenges. MAIRCA addresses complex decision problems with multiple criteria and provides a systematic framework for evaluating alternatives. The method considers both ideal and real performance, facilitating a comprehensive analysis. MAIRCA enables decision-makers to incorporate their preferences through weighted aggregation. The accuracy of the results heavily depends on the quality of data and the selection of appropriate benchmarks for the ideal performance. The method can be time-consuming, especially when dealing with a large number of alternatives and criteria. MAIRCA method is a powerful tool in the domain of multi-criteria decision-making. Its ability to handle complex problems with conflicting attributes and provide a structured approach to decision-making makes it valuable for various applications across diverse industries. Although MAIRCA has some limitations, its advantages outweigh them, making it a valuable addition to the toolbox of decision science.

6.4.2 Findings

In the preceding sections, psychosocial risk factors in the energy sector are introduced. This section of the study entails the analysis of expert opinions through the MAIRCA method, and the assessments obtained from the experts have been consolidated. The decision matrix, as presented in Table 6.2, is constructed based on the average of these expert evaluations.

After the decision matrix is obtained, the theoretical evaluation matrix and the actual evaluation matrices are obtained, respectively. Then, the gap matrix is obtained by taking the difference between the two matrices. The total gap score (Q) is obtained by obtaining the row sums of the gap matrix. The Q scores of each renewable energy type are given in Table 6.3.

When Table 6.3 is examined, the riskiest sector in terms of psychosocial risks for the renewable energy sector employees is determined as biomass energy, and the wind energy sector has a lower psychosocial risk level compared to other sectors.

Table 6.2 The decision matrix

	Shortage of qualified personnel	The inadequacy of standards	Job programs and work pace	Duties and expectations in the organization	Interpersonal relationships	Career development and compensation	Environmental factors	Job stress	Fatigue and sleep deprivation	Work-life imbalance	Job insecurity	Mental issues	Depression	Bullying
Wind energy	7.6	8	8	4.4	7.6	8.2	8.8	8.2	7.8	8	8.4	7.6	8	5
Solar energy	7.2	7.4	7.6	5	6.4	7.2	8.4	7	7.2	7.4	7.8	6.4	7.2	5
Hydroelectric energy	7.4	5.8	6.8	5,4	6.8	6.8	8	6.6	6.6	7	7.8	5.8	6.6	5
Biomass energy	7	5.4	6	3.6	5.8	6.6	7.6	5.8	5.6	6.2	6.6	5	5.6	3.2
Geothermal energy	7	5.4	6.4	4	5.6	6.2	7.8	5.2	6	5.6	6.4	5.4	5.2	3.4

Table 6.3 Q score and ranking

Renewable energy sources	Q	Rank
Wind energy	0.192	5
Solar energy	0.133	4
Hydroelectric energy	0.107	3
Biomass energy	0.014	1
Geothermal energy	0.015	2

6.5 Conclusion

Biomass energy, or the process of turning organic materials into energy, has drawn a lot of interest since it can help reduce greenhouse gas emissions and promote the use of renewable energy sources. However, this industry is not without its challenges. Working in the biomass energy sector exposes employees to various physical hazards, such as handling machinery and dealing with combustible materials. Moreover, throughout the energy conversion process, workers may be exposed to hazardous fumes and volatile compounds. Furthermore, long working hours and the demands of meeting production targets can lead to increased stress levels among workers. Biomass energy is a relatively young industry that can be influenced by fluctuating government policies and economic conditions. Because of the uncertainty surrounding the long-term profitability of biomass energy projects, workers in this industry may face increased job instability, which can have a detrimental effect on their mental health. Due to the evolving nature of biomass energy technology, workers might encounter a lack of adequate training and awareness about potential risks. This can lead to a heightened sense of vulnerability and anxiety among employees, as they may feel ill-equipped to handle challenging situations safely. The biomass energy sector sometimes faces opposition from environmental activists and communities concerned about potential environmental impacts. Workers in this industry may experience social stigma and isolation, leading to feelings of job dissatisfaction and reduced self-esteem. Compared to more established energy sectors, biomass energy might offer limited career development opportunities for its employees. The absence of clear career progression paths could cause frustration and demotivation among workers. Employees in the biomass energy sector face a range of psychosocial risks that make their working conditions potentially more challenging than those in other energy sectors. The physical hazards, uncertainty surrounding job security, lack of training, social stigma, and limited career growth prospects can collectively impact the mental well-being and overall job satisfaction of workers. Recognizing and addressing these psychosocial risks is crucial for fostering a healthier and more resilient workforce in the biomass energy industry. Properly managed, the sector can continue to contribute to renewable energy goals while prioritizing the well-being of its employees.

Today, wind energy stands out as an important component of environmentally friendly and sustainable energy production. The wind energy sector is mostly technology-oriented and has a modern structure. Therefore, while employees have the opportunity to quickly adapt and develop new technologies, they experience less stress compared to more traditional energy sectors. In addition, the environmentally friendly nature of wind energy projects makes the contribution of employees to their jobs more meaningful and satisfying. Wind power plants are generally installed in large areas far from city centers. This allows employees to be in a more peaceful working environment away from heavy city traffic and crowds. This situation increases the quality of life of the employees and reduces the psychosocial risks. Moreover, the wind energy sector is an attractive career path for young and dynamic professionals. Those who want to work in this sector are generally environmentally conscious, innovative, and passionate about their work. Because motivation and job satisfaction levels are high, psychosocial problems are less common. Those working in the wind energy sector are at less psychosocial risk compared to other energy sectors, thanks to factors such as environmental awareness, technological developments, and a more peaceful working environment. This situation increases the interest and participation of both employees and employers in the sector and provides a healthier working environment in the field of sustainable energy.

In conclusion, there are many psychosocial risk factors waiting for energy sector employees, and various improvements can be made by working on these factors. Improvements to be made will increase employee well-being and motivation.

References

Acar, A., Aksoy, H., Demir, B., Düşmez Tek, E., Yaylıoğlu, İ., Aydoğmuş, K., Deniz Kavas, M., & Aksoy, Y. N. (2021). Türkiye'de enerji sektörünün dijitalleşmesi kapsamında iş modellerinin, teknolojilerin ve gerekli mevzuat altyapısının değerlendirilmesi. SHURA Enerji Dönüşümü Merkezi, Sabancı Üniversitesi ISBN 978-625-7329-61-3.

Alvesson, M., & Einola, K. (2018). Excessive work regimes and functional stupidity. *German Journal of Human Resource Management, 32*, 283–296. https://doi.org/10.1177/2397002218791410

Artan, A. (2020). Güneş Enerjili Elektrik Sistemlerinin Kurulum Maaliyet Analizi: Adıyaman Örneği, Yüksek Lisans Tezi, Amasya Üniversitesi Fen Bilimleri Enstitüsü, Amasya.

Aydın, M. G. 2020). Rüzgâr enerji santrallerinde işçi sağlığı ve iş güvenliği açısından riskler ve bir örnek alandaki risk değerlendirmesi. Yüksek Lisans Tezi, İstanbul Yeni Yüzyıl Üniversitesi, Sağlık Bilimleri Enstitüsü, İstanbul.

Badre, P. (2021). A study on occupational stress among employees working in select thermal power plants of Nagpur District. *The Journal of Oriental Research Madras, XCII-IV*, 83–96.

Bal, F. (2018). Gürültünün stres üzerindeki etkisinin incelenmesi. *Uluslararası Hakemli Psikiyatri ve Psikoloji Araştırmaları Dergisi, 13*, 1–28.

Blagoeu, B., Muhr, S. L., Ortlieb, R., & Schreyogg, G. (2018). Organizational working time regimes. Drivers consequences and attempts to change pattern of excessive working hours. *German Journal of Human Resource Management, 32*, 155–167. https://doi.org/10.1177/2397002218791408

Boz-Semerci, A. (2018). Çalışanların İş Güvencesizliği Algısı: Ebeveyn Olan ve Olmayan Çalışanlar Arasındaki Farklılıklar. *Sosyoekonomi, 26*(37), 81–102.

Canadian Mental Health Association. (2022). Improving the mental health of energy sector workers. https://cmha.ca/news/improving-the-mental-health-of-energy-sector-workers/, Erişim Tarihi. March 6, 2023.

Ceylan, A. K., Mat Çelik, G., & Emhan, A. (2015). Personel Güçlendirmesi ve Yönetici Desteğinin İş Memnuniyeti Üzerindeki Etkisi: Enerji Sektöründe Bir Uygulama. *İşletme Araştırmaları Dergisi, 7/1*, 168–185.

Chen, J. (2022). Energy sector: Understanding what types of companies comprise it. https://www--investopedia-com.translate.goog/terms/e/energy_sector.asp?_x_tr_sl=en&_x_tr_tl=tr&_x_tr_hl=tr&_x_tr_pto=sc#citation-8, Erişim Tarihi. February 22, 2023.

Cooper, K., Kirkpatrick, P., & Stewart, A. (2014). Health effects associated with working in the wind power generation industry: a comprehensive systematic review. *JBI Evidence Synthesis*, (11), 12. https://doi.org/10.11124/jbisrir-2014-1710

Diao, H., & Ghorbani, M. (2018). Production risk caused by human factors: A multiple case study of thermal power plants. *Frontiers of Business Research in China, 12*, 15. https://doi.org/10.1186/s11782-018-0035-9

Di Corleto, R., Jodie, B. (2018). The thermal environment. In: Reed, S., Pisianello, D., Benke, G. (Eds.), Principles of Occupational Health and hygiene: An introduction. 444–445.

DiDomenico, A., & Nussbaum, M. A. (2011). Effects of different physical workload parameters on mental workload and performance. *International Journal of Industrial Ergonomics, 41*(3), 255–260.

Donyatalab, Y., Seyfi-Shishavan, S. A., Farrokhizadeh, E., Kutlu Gündoğdu, F., & Kahraman, C. (2020). Spherical fuzzy linear assignment method for multiple criteria group decision-making problems. *Informatica, 31*(4), 707–722. https://doi.org/10.15388/20-INFOR433

Elfering, A., Grebner, S., & Dudan, A. (2011). Job characteristics in nursing and cognitive failure at work. *Safety and Health at Work, 2*(2), 194–200.

Ersoy, N., Bingöl, N., & Ekmekci, İ. (2022). Kömürlü Termik Santrallerdeki İSG Çalışmalarının Değerlendirilmesi. *OHS Academy, 5*(2), 103–130. https://doi.org/10.38213/ohsacademy.1135505

European Agency for Safety and Health at Work [EU-OSHA]. (2013a). E-Fact 69. Hazard Identification Cheklist: OSH Risks Associated with small-scale solar Energy applications. https://osha.europa.eu/en/publications/e-fact-69-hazard-identification-checklist-osh-risks-associated-small-scale-solar-energy, Erişim Tarihi. March 1, 2023.

European Agency for Safety and Health at Work [EU-OSHA]. (2013b). Green jobs and occupational safety and health: foresight on new and emerging risks associated with new technologies by 2020; April, 2013. https://osha.europa.eu/en/publications/green-jobs-and-occupational-safety-and-health-foresight-new-and-emerging-risks/view. Erişim Tarihi. March 1, 2023.

European Agency for Safety and Health at Work. (2022). Workers' safety and health in green jobs. https://osha-europa-eu.translate.goog/tr/emerging-risks/green-jobs?_x_tr_sl=en&_x_tr_tl=tr&_x_tr_hl=tr&_x_tr_pto=sc, Erişim Tarihi. March 1, 2023.

Fernandez, C. A., Moore, K., McClure, L. A., et al. (2017). Occupational psychosocial hazards among the emerging US green collar workforce. *Journal of Occupational and Environmental Medicine, 59*(1), 1–5. https://doi.org/10.1097/JOM.0000000000000903

Gabriele, G., Leon-Perez, J., & Arenas, A. (2015). Are bullying behaviors tolerated in some cultures? Evidence for a curvilinear relationship between workplace bullying and job satisfaction among Italian workers. *Journal of Business Ethics, 131*(1), 227–237.

Gray, R. (2017). The biggest energy challenges facing humanity BBC. https://www.bbc.com/future/article/20170313-the-biggest-energy-challenges-facing-humanity. Erişim Tarihi. February 22, 2023.

Harding, F. C., James, A. T., & Robertson, H. E. (2018). The engineering challenges of CO2 storage. *Journal of Power and Energy, 232*, 17–26. https://doi.org/10.1177/0957650918756542

Hoboubi, N., Choobineh, A., Ghanavati, F. K., et al. (2017). The impact of job stress and job satisfaction on workforce productivity in an Iranian petrochemical industry. *Safety and Health at Work, 8*(1), 67–71.

Hole, J. A., & Pande, M. (2009). Worker productivity, occupational health, safety and environmental issues in thermal power plant. In *2009 IEEE international conference on industrial engineering and engineering management* (pp. 1082–1086).

Howard, A., Antczak, R., & Albertsen, K. (2022). Third European survey of enterprises on new and emerging risks (ESENER 2019): Overview report how European workplaces manage safety and health. EU-OSHA's Network of Focal Points.

International Energy Agency. (2020). Power systems in transition challenges and opportunities ahead for electricity security. https://iea.blob.core.windows.net/assets/cd69028a-da78-4b47-b1bf-7520cdb20d70/Power_systems_in_transition.pdf, Erişim Tarihi. February 22, 2023.

International Energy Agency. (2022). World energy outlook 2022. https://iea.blob.core.windows.net/assets/830fe099-5530-48f2-a7c1-11f35d510983/WorldEnergyOutlook2022.pdf, Erişim Tarihi. February 22, 2023.

ILO. (2019). Working on a warmer planet: the effect of heat stress on productivity and decent work; July, 2019. Available from: https://www.ilo.org/global/publications/books/WCMS_711919/lang%2D%2Den/index.htm#:~:text=The%20phenomenon%20of%20heat%20stress,major%20consequences%20of%20global%20warming.&text=This%20report%20shows%20the%20impact,all%20countries%20in%20the%20world. Erişim Tarihi. March 1, 2023.

Jolaolu, I. O.. (2020). Work-Lif ork-life balance of rotational we balance of rotational workers in the Nigerian Ener ers in the Nigerian energy sector. Doctoral Dissertation, Walden University, (pp. 1–130).

Kahraman, Z., & Yürüten Özdemir, K. (2022). Nükleer Enerjinin Riskleri ve Nükleer Santrallerde İş Sağlığı ve Güvenliği, Karaelmas Journal of Occupational Health and Safety, c. 6, sayı. 1, ss. 53-65, Nis. 2022, https://doi.org/10.33720/kisgd.1090500.

Karanikas, N., Steele, S., Bruschi, K., Robertson, C., Kass, J., Popovich, A., & MacFadyen, C. (2021). Occupational health hazards and risks in the wind industry. *Energy Reports, 7*, 3750–3759. https://doi.org/10.1016/j.egyr.2021.06.066

Khan, M. H. (2020). Stress management in the field workers of ZIN field Balochistan. *Electronic Research Journal of Social Sciences and Humanities, 2*(1), 31–36.

Koruca, H. İ., Aydın, L., & Demir, H. B. (2015). Bir Mobilya Fabrikasında İşgörenlerin Fiziksel ve Psikolojik İş Yükü Seviyelerinin Belirlenmesi ve Uygun İstasyonlarda Çalıştırılması. *Süleyman Demirel Üniversitesi Mühendislik Bilimleri ve Tasarım Dergisi, 3*(3), 523–531.

Kumar, G. V., & Janakiran, B. (2017). Theories of work-life balance—a conceptual review. *International Research Journal of Management and Commerce, 4*(9), 184–192.

Kuru, E. (2017). Gürültü şiddetinin yaşam kalitesine etkileri. Bildiri özeti. İAÜ – SBF Odyoloji Bölümü 1. Öğrenci Sempozyumu. (Er. Tar. 21.06. 2018) http://odyolojisempozyumu.aydin.edu.tr/.

Le, T. N., Straatman, L. V., Lea, J., & Westerberg, B. (2017). Current insights in noise-induced hearing loss: A literature review of the underlying mechanism, pathophysiology, asymmetry, and management options. *Journal of Otolaryngology-Head & Neck Surgery, 46*(1), 41.

Leka, S., Jain, A., & Ve Lerouge, L. (2017). Work-related psychosocial risks: Key definitions and an overview of the policy context in Europe. In L. Lerouge (Ed.), *Psychosocial risks in labour and social security law içinde* (pp. 1–12). Springer.

Li, F., Zhao, D., Qing, T., & Kuang, Y. (2021). Identification of occupational stress factors of commissioning workers in nuclear power plants based on a bottom-up survey design and factor analysis. *Journal of Nuclear Science and Technology, 58*(6), 714–724. https://doi.org/10.1080/00223131.2020.1858988

Lucas, H., Pinnington, S., & Cabeza, L. F. (2018). Education and training gaps in the renewable energy sector. *Solar Energy, 173*(July), 449–455. https://doi.org/10.1016/j.solener.2018.07.061

Mosadeghrad, M. A. (2014). Occupational stress and its consequences: Implications for health policy and management. *Leadership in Health Services, 27*(3), 224–239. https://doi.org/10.1108/LHS-07-2013-0032

Narban, J. S., Singh Narban, B. P., & Singh, J. (2016). A conceptual study on occupational stress (job stress/work stress) and its impacts. *IJARIIE, 2*(1), 47–56.

Organization, WH. (2019). *Global action plan on physical activity 2018–2030: More active peoplefor a healthier world*. World Health Organization.

Özcan, E. C., Varlı, E., & Eren, T. (2017). Hidroelektrik Santrallerde Vardiya Çizelgeleme Problemleri İçin Hedef Programlama Yaklaşımı. *Bilişim Teknolojileri Dergisi, 10*(4), 363–370. https://doi.org/10.17671/gazibtd.347609

Pradhan, R. K., Jena, L. K., & Kumari, I. G. (2016). Effect of work-life balance on organizational citizenship behavior. Role of organizational commitment. *Global Business Review, 17*, 15–29. https://doi.org/10.1177/0972150916631071

Prostect. (2019). Health and safety in the energy sector. https://prospect-org-uk.translate.goog/about/health-and-safety-in-the-energy-sector/?_x_tr_sl=en&_x_tr_tl=tr&_x_tr_hl=tr&_x_tr_pto=sc, Erişim Tarihi. March 1, 2023.

Rahmadani, R. (2016). Pengaruh Bullying di Tempat Kerja terhadap Mental Health dan Kepuasan Kerja pada divisi after sales PT XYZ. Thesis Fakultas Ekonomi dan Bisnis. Universitas Indonesia.

Samaniego-Rascón, D., Gameiro da Silva, M. C., Ferreira, A. D., & Cabanillas-Lopez, R. E. (2019). Solar energy industry workers under climate change: a risk assessment of the level of heat stress experienced by a worker based on measured data. *Safety Science, 118*, 33–47. https://doi.org/10.1016/j.ssci.2019.04.042

Seitz, J. (2018). How do differing degrees of working-time autonomy and overtime affect workers Well-being? A multilevel approach using data from the German SocioEconomic Panel (SOEP). *German Journal of Human Resources Management, 32*, 177–194. https://doi.org/10.1177/2397002218780630

Shigemura, J., Tanigawa, T., & Saito, N. (2012). Psychological distress in workers at the Fukushima nuclear power plants. *JAMA The Journal Of The American Medical Association, 308*(7), 667–669. https://doi.org/10.1001/jama.2012.9699

Sen, A., Mohankar, A. S., Khamaj, A., & Karmakar, S. (2021). Emerging OSH issues in installation and maintenance of floating solar photovoltaic projects and their link with sustainable development goals. *Risk Management and Healthcare Policy, 14*, 1939–1957. https://doi.org/10.2147/RMHP.S304732

Shan, D., & Neis, B. (2020). Employment-related mobility, regulatory weakness and potential fatigue-related safety concerns in short-sea seafaring on Canada's Great Lakes and St. Lawrence Seaway: Canadian seafarers' experiences. *Safety Science, 121*, 165–176.

Shanfa Yu, G. G. (2008). Psychosocial work environment and well-being: a cross-sectional study at a thermal power plant in China. *Occupational Health, 50*(2), 155–162. https://doi.org/10.1539/joh.l7098

Sindhu, S. P., Vijay, N., & Sunil, L. (2016). Recognition and prioritization of challenges in growth of solar energy using analytical hierarchy process: Indian outlook. *Energy, 100*, 332–348. https://doi.org/10.1016/j.energy.2016.01.091

SolarPower Europe. (2019). Operation & maintenance best practice guidelines; December, 2019. https://www.solarpowereurope.org/om-best-practice-guidelines-version-4-0/, Erişim Tarihi. March 10, 2023.

Stockwell, B. (2020). Considerations when implementing shift work in nuclear operations. United States.

Strietska-Ilina, O., Hofmann, C., Haro-Duran, M., & Jeon, S. (2011). Skills for green jobs: A global view: Synthesis report based on 21 country studies. Geneva, Switzerland.

Taoussi, A. A., Yassine, A. A., Malloum, M. S. M., et al. (2022). Effects of noise exposure among industrial workers in power plants of the National Electricity Company in N'Djamena, Chad. *Egyptian Journal of Otolaryngology, 38*, 63. https://doi.org/10.1186/s43163-022-00253-7

Tepe, S., Eti, S., & Kahraman, Z. (2022). İklim Değişikliğinin Psikososyal Riskler Açısından İş Kazalarına Etkisinin VAR Yöntemi ile Değerlendirilmesi: İklim Değişikliği & İş Kazaları İlişkisi. *Trk Dergisi, 3*(2), 1–24. https://doi.org/10.5281/zenodo.7497279

Torquati, L., Mielke, G. I., Brown, W. J., Burton, N. W., & Kolbe-Alexander, T. L. (2019). Shift work and poor mental health: A meta-analysis of longitudinal studies. *American Journal of Public Health, 109*(11), 13–20. https://doi.org/10.2105/AJPH.2019.305278

Tünay Doğan, N. N.. (2019). Enerji Sektöründe Çalışma Bozucu Davranışlarının Tipolojisi: Kaynakları, Çeşitleri ve Önlenmesi, Yüksek Lisans Tezi, Pamukkale Üniversitesi Sosyal Bilimler Enstitüsü, Denizli.

Van der Klauw, M., Hengel, K. O., Roozeboom, M. B., Koppes, L. L., & Venema, A. (2016). Occupational accidents in The Netherlands: Incidence, mental harm, and their relationship with psychosocial factors at work. *International Journal of Injury Control and Safety Promotion, 23*(1), 79–84.

Visser, M., Mills, M., Heyse, L., Wittek, R., & Bollettino, V. (2016). Work-life balance among humanitarian aid workers. *Nonprofit and Voluntary Sector Quarterly, 46*, 1191–1213. https://doi.org/10.1177/0899764076834890

Webster, J., Cabeças, J. M. M., Kuhl, K., Liddle, M., Ellwood, P., Smith, P., Gervais, R., Bradbrook, S. D., & Väänänen, V. (2013). Occupational safety and health in the wind energy sector. European Risk Observatory-Publications Office of the European Union. https://osha.europa.eu/en/publications/reports/occupational-safety-and-health- in-the-wind-energy-sector.

Wahyuningtias, I. A., Nasri, S. M., Erwandi, D., Lestari, F., & Kadir, A. (2023). Comparison of workplace bullying, distress level and satisfaction with life: Cross-sectional study on power plant energy in Indonesia. *PREPOTIF: Journal Kesehatan Masyarakat, 5*(1), 387–399. https://doi.org/10.31004/prepotif.v5i1.1644

Yıldız, Ö., & Ulucak, R. (2022). Yenilenebilir enerji kaynaklarının yayılımında sosyoekonomik faktörlerin etkisi. *Erciyes Üniversitesi İktisadi ve İdari Bilimler Fakültesi Dergisi, 61*, 245–280. https://doi.org/10.18070/erciyesiibd.974563

Young-Ran, Y., & Jeong-Won, H. (2016). Effect of nurses' organizational culture, workplace bullying and. *International Journal of Bio-Science and Bio-Technology, 8*(1), 372–380.

Zamanian, Z., Gharepoor, S., & Dehghani, M. (2010). Effects of electromagnetic fields on mental health of the staff employed in gas power plants, Shiraz, 2009. *Pakistan Journal of Biological Sciences, 13*(19), 956–960. https://doi.org/10.3923/pjbs.2010.956.960

Zincirkıran, M., Çelik, G. M., Ceylan, A. K., & Emhan, A. (2015). İşgörenlerin Örgütsel Bağlılık, İşten Ayrılma Niyeti, İş Stresi ve İş Tatmininin Örgütsel Performans Üzerindeki Etkisi: Enerji Sektöründe Bir Araştırma. *Finans Politik ve Ekonomik Yorumlar, 600*, 59–71.

Zhang, L. (2004). Research on human reliability analysis technique in probabilistic safety assessment. PhD. Hunan University, in Chinese.

Chapter 7
Analysis of Modern Methodological Approaches to Assessing the Energy Efficiency of Enterprises and Industrial Complexes

Solomon Eghosa Uhunamure and Tsangyao Chang

7.1 Introduction

Types of energy intensity depending on the object under consideration: production energy intensity of manufacturing products (products) (the object of research is an enterprise), energy intensity of national income, energy intensity of gross domestic product (the object of research is a federal subject or federation), and total energy intensity of manufacturing products (the object of research is a technological process).

The textbook "Energy Analysis. Methodology and basic information support" highlights the following indicators of energy use efficiency:

- Efficiency of the power plant
- Energy efficiency factor
- Energy efficiency factor by individual types and parameters of energy carriers
- Specific (actual) energy consumption

According to the authors of the manual, all these indicators are local in nature and are aimed at determining the amount of energy consumption for manufacturing products based on a given technological scheme, at best on the scale of a particular enterprise.

In relation to products, equipment, materials, fuel and energy resources, and their technological processes, indicators of their energy efficiency are used to

S. E. Uhunamure (✉)
University of Cape Town, Cape Town, South Africa
e-mail: solomon.uhunamure@uct.ac.za

T. Chang
Feng Chia University, Taichung City, Taiwan
e-mail: tychang@fcu.edu.tw

H. Dinçer et al. (eds.), *Decision Making in Interdisciplinary Renewable Energy Projects*, Contributions to Management Science,
https://doi.org/10.1007/978-3-031-51532-3_7

characterize energy saving: fuel consumption efficiency (for products used for their direct functional purpose), energy efficiency of transmission (storage) for products and processes, and energy intensity of production (for processes) (Eti et al., 2022). Standard indicators of energy efficiency of products are established with instructions for the requirements for the permissible change in the normalized values of indicators for the period of normal operation of this product. Energy efficiency indicators of products are classified by groups of homogeneous products, by the type of energy resources (energy carriers) used, and by methods for determining indicators.

7.2 Literature Review

Most often, energy efficiency is interpreted as the ratio of the useful effect (saving energy resources) to energy consumption, considering that the greater the useful effect per unit of energy consumption, the higher the energy efficiency (Dinçer et al., 2023a, 2023b, 2023c, 2023d, 2023e, 2023f). At the same time, energy efficiency is evaluated comprehensively, since it includes not only measures to increase the efficiency of production processes but also organizational and technical efforts related to optimizing the interaction of structural links within the production chain and improving the efficiency of enterprise management (Nie et al. 2020; Denisova et al., 2019; Nyangarika et al., 2019a, 2019b; Huang et al., 2021a, 2021b; Mikhaylov, 2018, 2022a; Mikhaylov et al., 2019; Conteh et al., 2021; Sediqi et al., 2022, Khan et al., 2021; Bhuiyan et al., 2022a; Liu et al., 2021a, 2021b; Daniali et al., 2021; Moiseev et al., 2023a, 2023b; An et al., 2022).

Directly related to energy efficiency is energy saving, which reduces the amount of energy consumed but at the same time preserves the production program and the quality of products (Martínez et al., 2023). Although it is also possible that the introduction of a new, more energy-intensive technology, which provides higher product quality, gives an increase in energy efficiency due to increased revenue and additional profit (Mikhaylov et al., 2022a; Mikhaylov, 2021a; Varyash et al., 2020; Zhao et al., 2021; An & Mikhaylov, 2020, 2021; Alwaelya et al., 2021; Yumashev & Mikhaylov, 2020; Yumashev et al., 2020; Mutalimov et al., 2021).

The decomposition analysis considers the ratio of energy intensity and actual energy consumption, which is used as an energy efficiency characteristic to reduce energy intensity, while taking into account technical energy efficiency and energy conservation. In this regard, the primary task is to provide an economic justification of the energy demand for the creation, renewal, or replacement of a particular energy supply system, in which the central place is given to saving energy resources. When developing measures to use reserves for saving fuel and energy resources, it is necessary to specify the comparison base, according to which the savings will be determined. The authors propose to consider the amount of energy consumption required to achieve the corresponding economic result in the future while maintaining the basic energy intensity, as a base with which the expected savings are correlated (Mikhaylov et al., 2023b; Wang et al. 2023; He et al., 2023; Irshad et al., 2023;

Iqbal et al., 2023; Aksoy et al., 2023; Dayong et al., 2020; Nyangarika et al., 2018; Danish et al., 2020, 2021; An et al., 2021; Uyeh et al., 2021; Tamashiro et al., 2021, 2023; Shaikh et al., 2021).

Several related indicators of energy efficiency of fuel and energy resources should be the resultant indicators (Carayannis et al., 2023). As an advantage, we note that this approach comprehensively reflects energy efficiency, but, unfortunately, it does not fully reflect the interaction of the internal environment of the enterprise with its external environment. In the context of this study, the so-called concept of key performance indicators (KPIs) is of undoubted interest, within the framework of which many foreign companies that monitor energy efficiency indicators successfully manage them (Eti et al., 2023). This approach allows you to optimize processes, reduce production costs, comply with regulatory requirements, and thus develop sustainably (An et al., 2019a, 2019b, 2020a, 2020b; 2020c; Mikhaylov, 2020a, 2020b, 2020c; Mikhaylov & Tarakanov, 2020; Moiseev et al., 2020, 2021; Gura et al., 2022; Dooyum et al., 2020; Mikhaylov et al., 2020; Mikhaylov, 2020a, 2020b, 2020c).

7.3 Results

The input data for the forecasting process will be as follows:

1. Global economic development prospects, market conditions, pricing of key material resources, and advanced energy-efficient technologies
2. Results of a multilevel assessment of PC energy efficiency in the current and retrospective periods
3. Portfolio of measures and energy-efficient PC development projects identified based on the results of the second-stage implementation (block 2)

If we talk about possible scenarios of economic development that affect the PC's activities, they do not depend on this activity in any way and are external to the PC. Indicators that characterize the state of the external environment, in particular, include the following:

1. Dynamics of changes in GDP, inflows of investment in fixed assets of the domestic economy, as well as the economies of countries-markets for PC products
2. Promising energy-efficient technologies for core economic activities
3. Volume of effective demand for PC products and market division indicators
4. Forecast prices for the main types of fuel and energy used at PC enterprises, as well as for alternative energy sources
5. Indicators of the development of commodity markets by volume and price
6. Price values for the main types of PC products (both national and global averages)

The demand for sustainability reporting based on key performance indicators is constantly growing. At the same time, the amount of data required is significantly smaller compared to the amount of data from information systems in the field of energy. Therefore, energy efficiency management involves collecting, analyzing, and combining data to generate key energy efficiency indicators. In addition, it is fairly assumed that continuous monitoring and analysis of the energy efficiency of an enterprise or production complex (PC) requires its parallel financial assessment.

The sustainable development of the enterprise (PC) is inextricably linked with the innovative transformation of business processes. Therefore, priority should be given to key indicators that reflect the improvement process, while using best practices that can help you find new ways to improve energy efficiency and thereby improve the overall efficiency of your business.

It is proposed to assess the energy efficiency of an enterprise at three levels, considering equipment, technologies, and the enterprise as a whole. Increasing energy efficiency when using more modern equipment is ensured by increasing its efficiency and reducing losses by improving the organization of production, etc. The same result can be achieved by changing the operating principles of the technological installation (Niu et al., 2023). At the enterprise level, energy efficiency can be improved by changing the structure of production, varying the length of the technological chain covered. From this we can conclude that by improving technologies, optimizing the set of energy resources consumed and ways of organizing production, an enterprise (PC) has a positive impact on the integral indicators of its activities, which include revenue, profit, and prime cost (Aksoy et al., 2023).

Since 1997, the International Energy Agency (IEA) has been developing groups of energy indicators to study the dynamics of energy consumption and analyze factors affecting energy consumption and carbon dioxide emissions. The IEA was the first to propose to study energy efficiency indicators depending on their aggregation in the form of a pyramid of energy efficiency indicators: from the most detailed at the base of the pyramid to aggregated at its top.

The following specific features of projects based on energy efficiency indicators can be identified:

- Accounting of energy consumption over time
- Accounting for cross-country differences
- Preference of physical indicators over cost ones
- Analysis of developments in the field of energy efficiency

Use benchmarking instead of analyzing the best available technologies. The IEA is based on industry and structural indicators (activities within the industry). The more disaggregated an indicator is, the more understandable the factors that affect it become. The IEA uses decomposition analysis to show how factors in each of the indicators affected the final energy consumption.

Energy efficiency indicators perform a variety of functions, from monitoring energy efficiency to analyzing and evaluating new technologies. However, the usefulness and effectiveness of using energy efficiency indicators depends on a number of conditions, in particular, the availability and quality of information (Ai et al., 2023).

In addition, the Asia Pacific Research Center also applies the energy efficiency pyramid to disaggregate energy efficiency assessment levels. When the level of aggregation decreases, the influence of changing structural effects and other factors also decreases (Bhuiyan et al., 2021; Dong et al., 2021, Mikhaylov, 2021b; Baboshkin et al., 2022a, 2022b; Barykin et al., 2022a; Liu et al., 2022a, 2022b; Bhuiyan et al., 2022b; Danish et al., 2022a, 2022b; Saqib et al., 2021; Mukhametov et al., 2021, Candila et al., 2021; Mikhaylov & Grilli, 2022; Li et al., 2022).

Moving down the pyramid improves understanding of the many factors that influence more aggregated energy efficiency measurements and ultimately affect other variables, such as national energy consumption. The amount of information required (at the bottom of the pyramid) increases significantly, and obtaining information becomes much more time-consuming (Wan et al., 2023).

To describe energy efficiency, the Asia Pacific Research Center uses three types of indicators: economic (cost) indicators, technical and economic (physical) indicators, and distribution indicators (Yüksel & Dinçer, 2023). Cost indicators are characterized by the ratio of energy consumption to indicators of economic activity in monetary units. GDP and value-added information for all regions is calculated using purchasing power parity to reflect differences in overall price levels. Using purchasing power parity (PPP) instead of the exchange rate increases the GRP of regions with a low standard of living, thereby reducing their energy intensity (Khan et al., 2022; Dinçer et al., 2022a, 2022b; Badr et al., 2022; Barykin et al., 2022b; Mehta et al., 2022; Kalinina et al. 2022; Shaikh et al., 2022; Mikhaylov et al., 2022b, 2023a; Nyangarika et al., 2022; Mikhaylov, 2022b).

Tax incentives for investments in green technologies involve the use of measures and means to ensure the regulatory impact of tax mechanisms on the development and implementation of technologies in the field of environmental management, energy conservation, and environmental protection. Important tax and financial motivation of corporate management and investors (Dinçer et al., 2023a, 2023b; Ivleva et al., 2023; Aissou et al., 2023; Mikhaylov, 2023).

The choice and transformation of tax instruments depends on the objectives of fiscal policy, interrelated with the socioeconomic and environmental policy of the state and the characteristics of countries and territories. The set of green tax incentives aimed at reducing the tax burden on certain types of entrepreneurship is small and differs in the countries of the world.

7.4 Conclusion

According to the Asia Pacific Research Center's methodology, energy intensity at purchasing power parity is more relevant, as it reflects the real level of economic activity (Kou et al., 2023a, 2023b). Using purchasing power parity narrows the energy intensity gap between regions. Thus, the comparison result is significantly improved (Han et al., 2023). The key advantage of using energy efficiency cost indicators is that they are comparable within production processes, since they have comparable cost denominators, regardless of different production. As energy efficiency indicators are constructed at more aggregated levels, economic indicators have become very common (Moiseev et al., 2023a, 2023b).

Although cost indicators are attractive, they have a number of limitations:

- Cost indicators do not consider structural changes in economic systems or within the economy.
- Cost indicators include a number of non-energy impacts, such as price effects.
- Not all sectors of the economy are sufficiently represented in economic terms, such as the housing and private transport sectors.
- Differences in aggregation and measurement methods between economic systems may also limit the analysis (Yüksel et al., 2023). Industry-level and lower-level energy efficiency analysis can reduce these problems, but the problem is exacerbated by multi-industry aggregate analysis (Xu et al., 2023).

References

Ai, R., Zheng, Y., Yüksel, S., & Dinçer, H. (2023). Investigating the components of fintech ecosystem for distributed energy investments with an integrated quantum spherical decision support system. *Financial Innovation, 9*(1), 27.

Aissou, A., Ahmed, H. S., Hussein, M. E., Hassan, M. R., Mikhaylov, A., & Mahmoud, E. (2023). Optimizing reliability of flow networks using RWGA subject to total capacity constraint. *Applied Mathematics and Information Sciences, 7, 4*, 559–565. https://doi.org/10.18576/amis/170404

Aksoy, T., Yüksel, S., Dınçer, H., Hacioglu, U., Mikhaylov, A., & Muyeen, S. M. (2023). Assessing the electricity production capacities of emerging markets for the sustainable investment. *IEEE Access, 11*, 24574–24590.

Alwaelya, S. A., Yousif, N. B. A., & Mikhaylov, A. (2021). Emotional development in preschoolers and socialization. *Early Child Development and Care, 191*, 16.

An, J., & Mikhaylov, A. (2021). Analysis of energy projects financial efficiency and renewable energy generation in Russia. *Finance: Theory and Practice, 25*(5), 79–91.

An, J., & Mikhaylov, A. (2020). Russian energy projects in South Africa. *Journal of Energy in Southern Africa, 31*(3), 58–64.

An, J., Mikhaylov, A., & Jung, S.-U. (2020a). The strategy of South Korea in the global oil market. *Energies, 13*(10), 2491.

An, J., Mikhaylov, A., & Kim, K. (2020b). Machine learning approach in heterogeneous group of algorithms for transport safety-critical system. *Applied Sciences, 10*(8), 2670.

An, J., Mikhaylov, A., & Richter, U. H. (2020c). Trade war effects: Evidence from sectors of energy and resources in Africa. *Heliyon, 6*, e05693. https://doi.org/10.1016/j.heliyon.2020.e05693

An, J., Mikhaylov, A., & Jung, S.-U. (2021). A linear programming approach for Robust network revenue management in the airline industry. *Journal of Air Transport Management, 91*, 101979.

An, J., Mikhaylov, A., & Moiseev, N. (2019a). Oil price predictors: Machine learning approach. *International Journal of Energy Economics and Policy, 9*(5), 1–6.

An, J., Mikhaylov, A., & Sokolinskaya, N. (2019b). Machine learning in economic planning: Ensembles of algorithms. *Journal of Physics: Conference Series, 1353*, 012126.

An, J., Mikhaylov, A., Dinçer, H., & Yüksel, S. (2022). Economic modelling of electricity generation: Long short-term memory and Q-rung orthopair fuzzy sets. *Heliyon, 8*, e10759.

Badr, A. M. N. A. A., Mohamed, T. I., Osman, N. A. Q., & Mikhaylov, A. (2022). A review of social media website users' interaction paths with governmental accounts during the COVID-19 pandemic. *Informatics, 9*, 50. https://doi.org/10.3390/informatics9030050

Barykin, S., Mikheev, A., Kiseleva, E., Putikhin, Y., Alekseeva, N., & Mikhaylov, A. (2022a). An empirical analysis of Russian regions' debt sustainability. *Economies, 10*(4), 296.

Barykin, S. E., Kapustina, I. V., Sergeev, S. M., Daniali, S. M., Kopteva, L. A., Semenova, G. N., Pryadko, I. P., Mikhaylov, A., Baboshkin, P., Datsyuk, P., & Senjyu, T. (2022b). Financial logistics models based on systematic approach improving management solutions. *F1000Research, 11*, 570. https://doi.org/10.12688/f1000research.111252.1

Bhuiyan, M. A., Dinçer, H., Yüksel, S., Mikhaylov, A., Danish, M. S. S., Pinter, G., Uyeh, D. D., & Stepanova, D. (2022a). Economic indicators and bioenergy supply in developed economies: QROF-DEMATEL and random forest models. *Energy Reports, 8*(2022), 561–570.

Bhuiyan, M. A., Zhang, Q., Khare, V., Mikhaylov, A., Pinter, G., & Huang, X. (2022b). Renewable energy consumption and economic growth nexus—a systematic literature review. *Frontiers in Environmental Science., 10*, 878394. https://doi.org/10.3389/fenvs.2022.878394

Bhuiyan, M. A., An, J., Mikhaylov, A., Moiseev, N., & Danish, M. S. S. (2021). Renewable energy deployment and COVID-19 measures for sustainable development. *Sustainability, 13*, 4418.

Candila, V., Maximov, D., Mikhaylov, A., Moiseev, N., Senjyu, T., & Tryndina, N. (2021). On the relationship between oil and exchange rates of oil-exporting and oil-importing countries: From the great recession period to the Covid-19 era. *Energies, 14*, 8065.

Carayannis, E., Kostis, P., Dinçer, H., & Yüksel, S. (2023). Quality function deployment-oriented strategic outlook to sustainable energy policies based on quintuple innovation helix. *Journal of the Knowledge Economy*, 1–19.

Conteh, F., Takahashi, H., Hemeida, A. M., Krishnan, N., Mikhaylov, A., & Senjyu, T. (2021). Hybrid grid-connected renewable power generation for sustainable electricity supply in Sierra Leone: Case study Lungi Town. *Sustainability, 13*(18).

Daniali, S. M., Barykin, S. E., Kapustina, I. V., Khortabi, F. M., Sergeev, S. M., Kalinina, O. V., Mikhaylov, A., Veynberg, R., Zasova, L., & Senjyu, T. (2021). Predicting volatility index according to technical index and economic indicators on the basis of deep learning algorithm. *Sustainability, 13*, 14011.

Danish, M. S. S., Bhattacharya, A., Stepanova, D., Mikhaylov, A., Grilli, M. L., Khosravy, M., & Senjyu, T. (2020). A systematic review of metal oxide applications for energy and environmental sustainability. *Metals, 10*(12), 1604.

Danish, M. S. S., Senjyu, T., Sabory, N. R., Khosravy, M., Grilli, M. L., Mikhaylov, A., & Majidi, H. (2021). A forefront framework for sustainable aquaponics modeling and design. *Sustainability, 13*, 9313.

Danish, M. S., Senjyu, T. S., Sabory, N. R., & Mikhaylov, A. (2022a). Lessons from COVID-19 conferring environmental re-engineering opportunity. In M. Danish & T. Senjyu (Eds.), *Eco-friendly and agile energy strategies and policy development* (pp. 99–118). IGI Global. https://doi.org/10.4018/978-1-7998-9502-2.ch005

Danish, M. S., Senjyu, T. S., Sabory, N. R., & Mikhaylov, A. (2022b). Interplay consequences of COVID-19 on global environmental sustainability. In M. Danish & T. Senjyu (Eds.), *Eco-friendly and agile energy strategies and policy development*. IGI Global. https://doi.org/10.4018/978-1-7998-9502-2.ch003

Dayong, N., Mikhaylov, A., Bratanovsky, S., Shaikh, Z. A., & Stepanova, D. (2020). Mathematical modeling of the technological processes of catering products production. *Journal of Food Process Engineering, 43*(2), 1–9.

Denisova, V., Mikhaylov, A., & Lopatin, E. (2019). Blockchain infrastructure and growth of global power consumption. *International Journal of Energy Economics and Policy, 9*(4), 22–29.

Dinçer, H., Yüksel, S., Mikhaylov, A., Pinter, G., & Shaikh, Z. A. (2022a). Analysis of renewable-friendly smart grid technologies for the distributed energy investment projects using a hybrid picture fuzzy rough decision-making approach. *Energy Reports, 8*, 11466–11477. https://doi.org/10.1016/j.egyr.2022.08.275

Dinçer, H., Yüksel, S., Mikhaylov, A., Barykin, S. E., Aksoy, T., & Hacioğlu, U. (2022b). Analysis of environmental priorities for green project investments using an integrated

q-rung orthopair fuzzy modelling. *IEEE Access, 10*, 50996–51007. https://doi.org/10.1109/ACCESS.2022.3174058

Dinçer, H., Yüksel, S., Hacioglu, U., Yilmaz, M. K., & Delen, D. (2023a). Development of a sustainable corporate social responsibility index for performance evaluation of the energy industry: A hybrid decision-making methodology. *Resources Policy, 85*, 103940.

Dinçer, H., Eti, S., Gökalp, Y., Yüksel, S., & Çelebi, B. (2023b). The role of small-scale wind turbines in increasing the competitiveness of hospitals. In *Handbook of research on quality and competitiveness in the healthcare services sector* (pp. 336–352). IGI Global.

Dinçer, H., Yüksel, S., Mikhaylov, A., Muyeen, S. M., Chang, T., Barykin, S., & Kalinina, O. (2023c). CO2 emissions integrated fuzzy model: A case of seven emerging economies. *Energy Reports, 9*, 5741–5751.

Dinçer, H., Yüksel, S., Aksoy, T., Hacıoğlu, Ü., Mikhaylov, A., & Pinter, G. (2023d). Analysis of solar module alternatives for efficiency-based energy investments with hybrid 2-tuple IVIF modeling. *Energy Reports, 10*, 61–71.

Dinçer, H., Eti, S., Yüksel, S., Özdemir, S., Yílmaz, A. E., & Ergün, E. (2023e). Integrating data mining and fuzzy decision-making techniques for analyzing the key minimizing factors of carbon emissions. *Journal of Intelligent & Fuzzy Systems, 45*, 1–17.

Dinçer, H., Eti, S., Aksoy, T., Yüksel, S., Hacioglu, U., Mikhaylov, A., & Muyeen, S. M. (2023f). Analysis of environmental impact for material production investments using a novel soft computing methodology. *IEEE Access, 11*, 37987–38001. https://doi.org/10.1109/ACCESS.2023.3266524

Dong, B., Ikonnikova, I., Rogulin, R., Sakulyeva, T., & Mikhaylov, A. (2021). Environmental economic approach to optimization of transport communication in megacities. *Journal of Environmental Science and Health, Part A, 56*(6), 660–666.

Dooyum, U. D., Mikhaylov, A., & Varyash, I. (2020). Energy security concept in Russia and South Korea. *International Journal of Energy Economics and Policy, 10*(4), 102–107.

Eti, S., Dinçer, H., Gökalp, Y., Yüksel, S., & Kararoğlu, D. (2023). Identifying key issues to handle the inflation problem in the healthcare industry caused by energy prices: An evaluation with decision-making models. In *Managing inflation and supply chain disruptions in the global economy* (pp. 162–178).

Eti, S., Dinçer, H., Yüksel, S., & Gökalp, Y. (2022). Analysis of the suitability of the solar panels for hospitals: A new fuzzy decision-making model proposal with the T-spherical TOP-DEMATEL method. *Journal of Intelligent & Fuzzy Systems, Preprint*, 1–13.

Gura, D., Mikhaylov, A., Glushkov, S., Zaikov, M., & Shaikh, Z. A. (2022). Model for estimating power dissipation along the interconnect length in single on-chip topology. *Evolutionary Intelligence, 15*, 2369–2373. https://doi.org/10.1007/s12065-020-00407-7

Han, Z., Gong, L., Chen, H., & Yüksel, S. (2023). The role of mineral and oil resources in achieving a clean environment in emerging economies: An asymmetric analysis. *Resources Policy, 86*, 104099.

He, H., Huang, Y., Nakadomari, A., Masrur, H., Krishnan, N., Hemeida, A. M., Mikhaylov, A., Omine, E., & Senjyu, T. (2023). Potential and economic viability of green hydrogen production from seawater electrolysis using renewable energy in remote Japanese islands. *Renewable Energy, 202*, 1436–1447. https://doi.org/10.1016/j.renene.2022.12.046

Huang, Y., Masrur, H., Shigenobu, R., Hemeida, A. M., Mikhaylov, A., & Senjyu, T. (2021a). A comparative design of a campus microgrid considering a multi-scenario and multi-objective approach. *Energies, 14*, 2853.

Huang, Y., Yona, A., Takahashi, H., Hemeida, A. M., Mandal, P., Mikhaylov, A., Senjyu, T., & Lotfy, M. E. (2021b). Energy management system optimization of drug store operate electric vehicles charging station. *Sustainability, 14*, 1217836.

Iqbal, K. M. J., Khan, M. I., Mikhaylov, A., Shah, A. A., Yadykin, V., Filho, W. L., Tariq, M. A. U. R., & Ullah, W. (2023). Modeling principles, criteria and indicators to assess water sector governance for climate compatibility and sustainability. *Frontiers in Environmental Science, 11*, 989930.

Irshad, A. S., Ludin, G. A., Masrur, H., Ahmadi, M., Yona, A., Mikhaylov, A., Krishnan, N., & Senjyu, T. (2023). Optimization of grid-photovoltaic and battery hybrid system with most technically efficient PV technology after the performance analysis. *Renewable Energy, 207*, 714–730.

Ivleva, S., Hajimajeed, M., Kartashova, O., & Mikhaylov, A. (2023). A comparative study of regional variations in common visual impairments among school-age children. *Electronic Journal of General Medicine, 20*(5), em504. https://doi.org/10.29333/ejgm/13264

Kalinina, O. V., Barykin, S. E., Sergeev, S. M., Semenova, G. N., Fatkullina, A., Mikhaylov, A., & De La Poza, E. (2022). Smart city perspectives in post-pandemic governance: Externalities reduction policy. *F1000Research, 11*, 1032. https://doi.org/10.12688/f1000research.123195.1

Khan, A. A., Laghari, A. A., Gadekallu, T. R., Shaikh, Z. A., Javed, A. R., Rashid, M., Estrela, V. V., & Mikhaylov, A. (2022). A drone-based data management and optimization using metaheuristic algorithms and blockchain smart contracts in a secure fog environment. *Computers and Electrical Engineering, 102*, 108234. https://doi.org/10.1016/j.compeleceng.2022.108234

Khan, A. A., Shaikh, Z. A., Baitenova, L., Mutaliyeva, L., Moiseev, N., Mikhaylov, A., Laghari, A. A., Idris, S. A., & Alshazly, H. (2021). QoS-ledger: Smart contracts and metaheuristic for secure quality-of-service and cost-efficient scheduling of medical-data processing. *Electronics, 10*, 3083.

Kou, G., Pamucar, D., Dinçer, H., & Yüksel, S. (2023a). From risks to rewards: A comprehensive guide to sustainable investment decisions in renewable energy using a hybrid facial expression-based fuzzy decision-making approach. *Applied Soft Computing, 142*, 110365.

Kou, G., Yuksel, S., & Dincer, H. (2023b). A facial expression and expert recommendation fuzzy decision-making approach for sustainable business investments within the metaverse world. *Applied Soft Computing*, 110849.

Li, J., Dinçer, H., Yüksel, S., Mikhaylov, A., & Barykin, S. E. (2022). Bipolar q-ROF hybrid decision making model with golden cut for analyzing the levelized cost of renewable energy alternatives. *IEEE Access, 10*, 35900. https://doi.org/10.1109/ACCESS.2022.3168315

Liu, L., Kato, T., Mandal, P., Mikhaylov, A., Hemeida, A. M., & Senjyu, T. (2021a). Multi-area power system load frequency control by using model predictive control by introducing renewable energy sources. In *2021 International conference on science & contemporary technologies (ICSCT)* (pp. 1–6). https://doi.org/10.1109/ICSCT53883.2021.9642521

Liu, L., Kato, T., Mandal, P., Mikhaylov, A., Hemeida, A. M., Takahashi, H., & Senjyu, T. (2021b). Two cases studies of model predictive control approach for hybrid renewable energy systems. *AIMS Energy, 9*(6), 1241–1259.

Liu, Z.-J., Panfilova, E., Mikhaylov, A., & Kurilova, A. (2022a). COVID-19 crisis impact on the stability between parties in crowdfunding and crowdsourcing. *Wireless Personal Communications, 122*, 915–930.

Liu, Z., Panfilova, E., Mikhaylov, A., & Kurilova, A. (2022b). Assessing stability in the relationship between parties in crowdfunding and crowdsourcing projects during the COVID-19 crisis. *Journal of Global Information Management (JGIM), 30*(4), 1–18. https://doi.org/10.4018/JGIM.297905

Martínez, L., Dinçer, H., & Yüksel, S. (2023). A hybrid decision making approach for new service development process of renewable energy investment. *Applied Soft Computing, 133*, 109897.

Mehta, R., Verghese, J., Mahajan, S., Barykin, S., Bozhuk, S., Kozlova, N., Kapustina, I., Mikhaylov, A., Naumova, E., & Dedyukhina, N. (2022). Consumers' behavior in conversational commerce marketing based on messenger chatbots. *F1000Research, 11*(647). https://doi.org/10.12688/f1000research.122037.1

Mikhaylov, A., Dinçer, H., & Yüksel, S. (2023a). Analysis of financial development and open innovation oriented fintech potential for emerging economies using an integrated decision-making approach of MF-X-DMA and golden cut bipolar q-ROFSs. *Financial Innovation, 9*(1), 10. https://doi.org/10.1186/s40854-022-00234-w

Mikhaylov, A., Dinçer, H., Yüksel, S., Pinter, G., & Shaikh, Z. A. (2023b). Bitcoin mempool growth and trading volumes: Integrated approach based on QROF multi-SWARA and aggregation operators. *Journal of Innovation & Knowledge, 8*(3), 100378.
Mikhaylov, A. (2018). The yield of the carry trade strategy. *Finance: Theory and Practice, 22*(3), 52–63. https://doi.org/10.26794/2587-5671-2018-22-3-52-63
Mikhaylov, A. (2020a). Cryptocurrency market analysis from the open innovation perspective. *Journal of Open Innovation: Technology, Market, and Complexity, 6*(4), 197.
Mikhaylov, A. (2020b). Geothermal energy development in Iceland. *International Journal of Energy Economics and Policy, 10*(4), 31–35.
Mikhaylov, A. (2020c). Cryptocurrency market development: Hurst method. *Finance: Theory and Practice, 24*(3), 81–91.
Mikhaylov, A. (2021a). Development of Friedrich von Hayek's theory of private money and economic implications for digital currencies. *Terra Economicus, 19*(1), 53–62.
Mikhaylov, A. (2021b). Lichens as indicators of atmospheric pollution in urban ecosystems. *Israel Journal of Ecology & Evolution, 67*(1–2), 60–68. https://doi.org/10.1163/22244662-bja10016
Mikhaylov, A. (2022a). Efficiency of renewable energy plants in Russia. *Anais da Academia Brasileira de Ciências, 94*(4), e20191226. https://doi.org/10.1590/0001-3765202220191226
Mikhaylov, A. (2022b). Sustainable development and renewable energy: A new view to a global problem. *Energies, 15*, 1397. https://doi.org/10.3390/en15041397
Mikhaylov, A. (2023). Understanding the risks associated with wallets, depository services, trading, lending, and borrowing in the crypto space. *Journal of Infrastructure, Policy and Development, 6*, 2.
Mikhaylov, A., & Grilli, M. L. (2022). Machine learning methods and sustainable development: Metal oxides and multilayer metal-oxides. *Metals, 12*(4).
Mikhaylov, A., & Tarakanov, S. (2020). Development of Levenberg-Marquardt theoretical approach for electric network. *Journal of Physics: Conference Series, 1515*, 052006.
Mikhaylov, A., Bhatti, I. M., Dinçer, H., & Yüksel, S. (2022a). Integrated decision recommendation system using iteration-enhanced collaborative filtering, golden cut bipolar for analyzing the risk-based oil market spillovers. *Computational Economics*. https://doi.org/10.1007/s10614-022-10341-8
Mikhaylov, A., Yumashev, A., & Kolpak, E. (2022b). Quality of life, anxiety and depressive disorders in patients with extrasystolic arrhythmia. *Archives of Medical Science, 18*(2). https://doi.org/10.5114/aoms.2020.101359
Mikhaylov, A., Burakov, D., & Didenko, V. (2019). Relationship between oil price and macroeconomic indicators in Russia. *Finance: Theory and Practice, 23*(2), 105–116. https://doi.org/10.26794/2587-5671-2019-23-2-105-116
Mikhaylov, A., Moiseev, N., Aleshin, K., & Burkhardt, T. (2020). Global climate change and greenhouse effect. *Entrepreneurship and Sustainability Issues, 7*(4), 2897–2913. https://doi.org/10.9770/jesi.2020.7.4(21)
Moiseev, N., Mikhaylov, A., Dinçer, H., & Yüksel, S. (2023a). Effects of market capitalization shocks for open innovation models in e-commerce industry: Golden cut QROF MCDM analysis. *Financial Innovation, 9*(1), 32.
Moiseev, N., Mikhaylov, A., Dinçer, H., & Yüksel, S. (2023b). Market capitalization shock effects on open innovation models in e-commerce: Golden cut q-rung orthopair fuzzy multicriteria decision-making analysis. *Financial Innovation, 9*(1), 55.
Moiseev, N., Mikhaylov, A., Varyash, I., & Saqib, A. (2020). Investigating the relation of GDP per capita and corruption index. *Entrepreneurship and Sustainability Issues, 8*(1), 780–794.
Moiseev, N., Sorokin, A., Zvezdina, N., Mikhaylov, A., Khomyakova, L., & Danish, M. S. S. (2021). Credit risk theoretical model on the base of DCC-GARCH in time-varying parameters framework. *Mathematics, 9*, 2423. https://doi.org/10.3390/math9192423
Mukhametov, A., Bekhorashvili, A., Avdeenko, A., & Mikhaylov, A. (2021). The impact of growing legume plants under conditions of Biologization and soil cultivation on Chernozem fertility and productivity of rotation crops. *Legume Research, 44*, 12.

Mutalimov, V., Kovaleva, I., Mikhaylov, A., & Stepanova, D. (2021). Assessing regional growth of small business in Russia. *Entrepreneurial Business and Economics Review, 9*(3), 119–133. https://doi.org/10.15678/EBER.2021.090308

Nie, D., Panfilova, E., Samusenkov, V., & Mikhaylov, A. (2020). E-learning financing models in Russia for sustainable development. *Sustainability, 12*(11), 4412.

Niu, X., Yüksel, S., & Dinçer, H. (2023). Emission strategy selection for the circular economy-based production investments with the enhanced decision support system. *Energy, 274*, 127446.

Nyangarika, A., Mikhaylov, A., Muyeen, S. M., Yadykin, V., Mottaeva, A., Pryadko, I., Barykin, S., Fomenko, N., Rykov, G., & Shvandar, K. (2022). Energy stability and decarbonization in developing countries: Random Forest approach for forecasting of crude oil trade flows and macro indicators. *Frontiers in Environmental Science, 10*, 1031343. https://doi.org/10.3389/fenvs.2022.1031343

Nyangarika, A., Mikhaylov, A., & Richter, U. (2019a). Influence oil price towards economic indicators in Russia. *International Journal of Energy Economics and Policy, 9*(1), 123–129.

Nyangarika, A., Mikhaylov, A., & Richter, U. (2019b). Oil price factors: Forecasting on the base of modified auto-regressive integrated moving average model. *International Journal of Energy Economics and Policy, 9*(1), 149–159.

Nyangarika, A., Mikhaylov, A., & Tang, B.-J. (2018). Correlation of oil prices and gross domestic product in oil producing countries. *International Journal of Energy Economics and Policy, 8*(5), 42–48.

Saqib, A., Chan, T.-H., Mikhaylov, A., & Lean, H. H. (2021). Are the responses of sectoral energy imports asymmetric to exchange rate volatilities in Pakistan? Evidence from recent foreign exchange regime. *Frontiers in Energy Research, 9*, 614463. https://doi.org/10.3389/fenrg.2021.614463

Sediqi, M. M., Nakadomari, A., Mikhaylov, A., Krishnan, N., Lotfy, M. E., Yona, A., & Senjyu, T. (2022). Impact of time-of-use demand response program on optimal operation of Afghanistan real power system. *Energies, 15*, 296.

Shaikh, Z. A., Kraikin, A., Mikhaylov, A., & Pinter, G. (2022). Forecasting stock prices of companies producing solar panels using machine learning methods. *Complexity, 2022*, 9186265. https://doi.org/10.1155/2022/9186265

Shaikh, Z. A., Moiseev, N., Mikhaylov, A., & Yüksel, S. (2021). Facile synthesis of copper oxide-cobalt oxide/nitrogen-doped carbon (Cu2O-Co3O4/CN) composite for efficient water splitting. *Applied Sciences, 11*, 9974. https://doi.org/10.3390/app11219974

Tamashiro, K., Omine, E., Krishnan, N., Mikhaylov, A., Hemeida, A. M., & Senjyu, T. (2023). Optimal components capacity based multi-objective optimization and optimal scheduling based MPC-optimization algorithm in smart apartment buildings. *Energy & Buildings, 278*, 112616. https://doi.org/10.1016/j.enbuild.2022.112616

Tamashiro, K., Alharbi, T., Mikhaylov, A., Hemeida, A. M., Krishnan, N., Lotfy, M. E., & Senjyu, T. (2021). Investigation of home energy management with advanced direct load control and optimal scheduling of controllable loads. *Energies, 14*, 7314. https://doi.org/10.3390/en14217314

Uyeh, D. D., Asem-Hiablie, S., Park, T., Kim, K.-M., Mikhaylov, A., Woo, S., & Ha, Y. (2021). Could Japonica rice be an alternative variety for increased global food security and climate change mitigation? *Food, 10*, 1869. https://doi.org/10.3390/foods10081869

Varyash, I., Mikhaylov, A., Moiseev, N., & Aleshin, K. (2020). Triple bottom line and corporate social responsibility performance indicators for Russian companies. *Entrepreneurship and Sustainability Issues, 8*(1), 313–331.

Wan, Q., Miao, X., Wang, C., Dinçer, H., & Yüksel, S. (2023). A hybrid decision support system with golden cut and bipolar q-ROFSs for evaluating the risk-based strategic priorities of fintech lending for clean energy projects. *Financial Innovation, 9*(1), 1–25.

Wang, Q., Guo, J., Li, R., Mikhaylov, A., & Moiseev, N. (2023). Does technical assistance alleviate energy poverty in sub-Saharan African countries? A new perspective on spatial spillover effects of technical assistance. *Energy Strategy Reviews, 45*, 101047. https://doi.org/10.1016/j.esr.2022.101047

Xu, X., Yüksel, S., & Dinçer, H. (2023). An integrated decision-making approach with golden cut and bipolar q-ROFSs to renewable energy storage investments. *International Journal of Fuzzy Systems, 25*(1), 168–181.

Yüksel, S., & Dinçer, H. (2023). Sustainability analysis of digital transformation and circular industrialization with quantum spherical fuzzy modeling and golden cuts. *Applied Soft Computing, 138*, 110192.

Yüksel, S., Dinçer, H., & Uluer, G. S. (2023). Socio-economic indicators of nuclear energy investments: A study on countries. In *Inclusive developments through socio-economic indicators: New theoretical and empirical insights* (pp. 33–44). Emerald Publishing Limited.

Yumashev, A., & Mikhaylov, A. (2020). Development of polymer film coatings with high adhesion to steel alloys and high wear resistance. *Polymer Composites, 41*(7), 2875–2880.

Yumashev, A., Ślusarczyk, B., Kondrashev, S., & Mikhaylov, A. (2020). Global indicators of sustainable development: Evaluation of the influence of the human development index on consumption and quality of energy. *Energies, 13*, 2768.

Zhao, Q., Cherkasov, S., Avdeenko, A., Kondratenko, L., & Mikhaylov, A. (2021). Integral estimate of the added value of agricultural crops in the synergetic agriculture on the example of Vicia faba (Vicia faba L.). *Polish Journal of Environmental Studies, 30*(5), 1–9. https://doi.org/10.15244/pjoes/131949

Chapter 8
An Analysis of Alternative Energy Sources and Applications in Maritime Transportation with a Strategic Management Approach

Filiz Mızrak and Gonca Reyhan Akkartal

8.1 Introduction

The maritime industry plays a crucial role in the global economy by facilitating international trade and transportation of goods. However, the industry's impact on the environment cannot be overlooked. Maritime transportation is responsible for a significant share of global greenhouse gas emissions, air pollution, and marine pollution. Therefore, the adoption of green strategies and operational strategies has become increasingly necessary to promote sustainability and reduce the negative impacts of maritime transportation on the environment. In this context, this literature review aims to explore the use of green strategies and operational strategies in the maritime industry and their potential to promote sustainable development (Jović et al., 2020).

The literature suggests that alternative energy sources, such as wind, solar, and hydrogen fuel cells, can play a crucial role in reducing greenhouse gas emissions and promoting energy efficiency in the maritime sector (Burel et al., 2013). However, the adoption of these technologies requires significant investments in infrastructure, technology, and training. In addition to alternative energy sources, the adoption of green and operational strategies can also help organizations improve their environmental performance and reduce costs. Green strategies, such as Environmental Management Systems and Life Cycle Assessment, can help organizations identify areas for improvement in their environmental performance, while operational strategies, such as Sustainable Supply Chain Management, can help organizations reduce their carbon footprint and enhance their reputation. Overall, the adoption of green and operational strategies can lead to new business opportunities, reduce

F. Mızrak (✉) · G. R. Akkartal
Logistics Management, Beykoz University, İstanbul, Turkey
e-mail: filizmizrak@beykoz.edu.tr; gonca.akkartal@medipol.edu.tr

H. Dinçer et al. (eds.), *Decision Making in Interdisciplinary Renewable Energy Projects*, Contributions to Management Science,
https://doi.org/10.1007/978-3-031-51532-3_8

environmental impacts, and contribute to achieving global environmental goals (Priadi, 2022).

Alternative energy systems are increasingly being adopted in ship technology to reduce the environmental impacts of maritime transportation. Wind, solar, and hydrogen fuel cells are among the most promising alternative energy sources in the maritime industry. Wind turbines can be installed on ships to harness wind energy and reduce fuel consumption, while solar panels can be used to generate electricity onboard. Hydrogen fuel cells can also be used to power ships, as they produce zero emissions and have high-energy efficiency. The adoption of these technologies requires significant investments, but they can also provide significant cost savings in the long run by reducing fuel consumption and emissions. Therefore, investment decisions in the maritime industry should consider the potential benefits of alternative energy systems in terms of reducing environmental impacts, improving energy efficiency, and reducing costs. Furthermore, the adoption of alternative energy systems can also enhance the reputation of organizations and attract customers who prioritize environmental responsibility, thereby contributing to the social and economic well-being of the maritime industry (Kołakowski et al., 2022).

In this context, minimizing the negative effects of ships on the environment emerges as an important field. The aim of the study is to shed light into the use of alternative energy systems in ship technology and contribute to investment decisions of maritime industry. In the second part of the study, after the introduction, content of operational strategies and green strategies have been explained and the severity of the emission problem has been stressed with a detailed literature review. Furthermore, solar energy and wind energy have been suggested as alternative resources in the maritime industry with specific examples. Finally, onshore supply system has been explained benefiting from latest studies in the literature.

8.2 Conceptual Framework

This section of the study provides an overview of the most common operational studies and green strategies adopted by logistics companies, with a specific focus on the maritime industry. It highlights the severity of the carbon emission problem within the maritime sector, emphasizing the need for sustainable practices.

Operational strategies are essential for organizations to optimize their day-to-day operations, achieve efficiency, and maintain a competitive edge (Iris & Lam, 2019). Several common operational strategies are widely employed across industries. One such strategy is lean management, which focuses on reducing waste and maximizing value-added activities. By streamlining processes, eliminating non-value-added tasks, and promoting continuous improvement, lean management aims to enhance productivity and resource utilization (Chahal & Narwal, 2017). Another common operational strategy is just-in-time (JIT) inventory management, which emphasizes minimizing inventory levels and maintaining a smooth flow of materials. JIT ensures that materials and resources arrive precisely when needed, reducing carrying costs and inventory-related inefficiencies. Additionally, total quality management (TQM) is a strategy that focuses on continuous quality improvement throughout all aspects of an organization. TQM involves establishing robust quality control processes, fostering a culture of accountability, and continuously seeking customer feedback to drive quality enhancements (Oakland & Marosszeky, 2006). These operational strategies, among others, are widely adopted due to their proven effectiveness in optimizing operations, enhancing productivity, and delivering value to customers. Overall, the literature suggests that operational strategies can have significant benefits for organizations across various industries. However, the choice of strategy should be based on the specific needs and goals of the organization, as well as the industry and competitive landscape.

Green strategies have become increasingly important in the maritime industry as a means to mitigate environmental impact and promote sustainability (Eti et al., 2022). Several common green strategies are widely employed by maritime companies (Martínez et al., 2023). One prevalent strategy is the adoption of alternative fuels and energy sources, such as liquefied natural gas (LNG), hydrogen, or biofuels, to reduce greenhouse gas emissions. Another common green strategy is the implementation of energy-efficient technologies and practices on board vessels, including optimized hull designs, advanced propulsion systems, and energy management systems (Dinçer et al., 2023a, 2023b, 2023c, 2023d, 2023e, 2023f). Additionally, the maritime industry often focuses on improving waste management and implementing recycling initiatives to minimize pollution and marine debris (Carayannis et al., 2023). Embracing digitalization and data-driven technologies is another common green strategy, enabling enhanced operational efficiency, optimized routes, and reduced fuel consumption (Eti et al., 2023). Furthermore, the industry emphasizes compliance with international regulations, such as the International Maritime Organization's (IMO) sulfur cap and greenhouse gas emissions reduction strategies, to promote sustainable practices globally. By adopting these green strategies, the maritime industry is working toward a more sustainable future, reducing its ecological footprint, and contributing to global efforts in combatting climate change (Zheng et al., 2022).

In the next part of the study, the severity of the emission problem will be illustrated referring to the maritime industry.

8.3 Maritime Industry

The maritime industry plays a crucial role in global trade and transportation, but it also contributes significantly to carbon emissions and climate change. As one of the largest emitters of greenhouse gases, the industry faces pressing challenges in reducing its environmental footprint. Carbon emissions from ships are primarily generated by the burning of fossil fuels for propulsion, resulting in the release of carbon dioxide and other harmful pollutants into the atmosphere. These emissions have adverse effects on air quality and marine ecosystems and contribute to global warming. The International Maritime Organization (IMO) has recognized the need to address this issue and has implemented regulations such as the IMO 2020 sulfur cap and the upcoming IMO greenhouse gas emissions reduction strategy. The maritime industry is now focusing on adopting cleaner technologies, utilizing alternative fuels, improving vessel efficiency, and implementing emission reduction measures to mitigate its carbon footprint. Collaboration among industry stakeholders, governments, and international organizations is crucial to effectively tackle the carbon emission problem in the maritime sector and move toward a more sustainable future (Xing et al., 2020).

Compared to other types of transportation in the world, maritime transportation is the most preferred type of transportation because it is more economical and can carry a larger amount of cargo at once. About 93% of the shipping process on ships is provided by diesel machines. About 67% of these machines are four-stroke and 26% are two-stroke diesel machines. In some studies, the diesel engine utilization rate of the entire maritime transport fleet in the world is stated as 99% (Jahangiri et al., 2018). The main reasons why diesel engines are preferred in ship propulsion systems can be listed as low-maintenance repair and spare parts costs and cheaper fuel used. The end products of combustion of fossil fuels used in main engines and generators constitute exhaust emissions. Therefore, almost all the exhaust emissions from ships originate from diesel engines used on ships (Xing et al., 2020).

Fossil fuels used in ships can be listed as heavy fuels, low-sulfur heavy fuel, marine distilled fuels, and recently ultralow sulfur marine gas oil (MGO) in accordance with international rules. Generally, heavy fuel is used in ship diesel engines with low-speed high-fuel consumption capacity, and marine distilled fuel is generally used in medium- and high-speed ship diesel engines. Apart from these, the use of natural gas, which has fewer polluting properties, on ships has recently started to become widespread. To reduce sulfur oxide (SOX) emissions, ultralow sulfur-containing marine gas oil (MGO) has recently been used in ship diesel engines as required by international rules (Zhao et al., 2021).

With the expansion of the world trade volume, the number of ships used in maritime transportation has also increased. Naturally, there has been a significant increase in fossil fuel consumption and, accordingly, in exhaust emissions. Exhaust emissions from a ship's main engine most commonly include NOX, SOX, CO2, PM, HC, and CO gases. Therefore, it is crucial to look for alternative energy sources or improve application to take precautions against further pollution (Tichavska et al., 2019).

Solar energy is the radiant energy released from the conversion of hydrogen gas in the sun to helium and the fusion reactions that occur during this transformation. Energy in various wavelengths is emitted from the sun, which is a thermonuclear reactor, and only two billionth of the energy emitted from the entire surface of the sun comes to the earth. Radiation energy comes from the sun to the earth in 8 min (Yang et al., 2020). Not all the solar radiation reaches the Earth's surface, 35% of it is reflected by the Earth's atmosphere, 45% reaches the earth's surface by passing through the atmosphere, while 20% of the remaining radiation is kept in the atmosphere. With the increase in energy prices and global climate change, the importance given to the use of renewable energy sources is increasing rapidly, and the most important of the known renewable energy sources has been solar energy (Yuan et al., 2018).

It is necessary to examine whether solar energy systems are applicable on ships, there are some existing factors that may affect this applicability, and the factors in question can be listed as follows (Nguyen & Dong, 2020);

- Availability of adequate deck area exposed to the sun
- Availability of a suitable grid-connected PV solar system
- Techno-economic selection of existing solar panels
- Scientific preparation of the system layout

The sun deck area plays an important role in the amount of solar energy absorbed by the solar panels, and the deck area varies depending on the functions consisting of various factors such as ship type, ship characteristics, structure, and arrangement of onboard machinery. The more solar panels placed on the deck, the higher the amount of sunlight that these panels can collect, so that the collected energy can fill the batteries faster and provide the necessary energy to the system completely (Zhu et al., 2017). On the other hand, the amount of energy obtained is important in addition to the large deck area, the cloudiness of the air, and the low-angle and low-intensity light level in the winter months. Charging time ranges from 4 to 16 h on a battery, depending on surface area and light conditions. Solar energy can be calculated based on geographic location, solar panel area, and panel efficiency (Wang et al., 2019).

The economic importance of the onboard solar energy system is to reduce fossil fuel consumption and make it a power source for electrical devices. Thus, the initial investment, maintenance and repair costs of fossil fuel-powered auxiliary machines will be avoided. The application of this system to the ship is aimed at both cost reduction and minimization of emissions (Manickavasagam et al., 2019). By storing the energy obtained from solar panels in batteries, the energy required for lighting, air conditioning, and some light electric vehicles can be met with the energy coming from the solar panel while cruising or while the ship is in port, without the need for generators to operate. In this way, a more environmentally friendly ship design could be made by saving fuel (Krčum et al., 2018).

With the increase in the world population and the progress of industrialization accompanying technological developments, the need for energy is increasing day by day. However, it is known that there are concerns about traditional-based energy sources, which are mainly used to meet this need, and it harms the environment. At this point, the contribution of renewable energy sources to electricity production is an important issue as it can be a solution. In addition, in recent years, it has been seen from industrial investments and applications as well as literature studies that there is an interest in the use of solar and wind energy-based power generation systems (Msigwa et al., 2022).

The investments made in the installed power-oriented wind energy in the last 10 years prove that there will be a serious interest in the future energy policy and that wind energy will play an even more effective role. Electricity generation from wind power not only provides economic benefits but also benefits the protection of the ecological environment. On the other hand, it is one of the effective factors in reducing or eliminating the countries' dependence on raw materials and energy from foreign countries. It is seen that the effective and efficient use of this type of energy to meet the growing energy demand is a very important issue and has been investigated in detail in the literature studies. It is also predicted that wind energy will play an even more active role in the future (Sadorsky, 2021).

The wind has a large size and scattered energy and is in constant motion on the Earth. Wind energy is an economical, clean, and renewable energy source that can contribute significantly to energy production (Yüksel et al., 2023). Today, the energy cost per kWh produced by wind energy is close to the cost of energy produced by coal (Xu et al., 2023). With a new technology to be developed in the field of wind energy, it is predicted that the amount of production will increase much more with the reduction of costs (Rehmatulla et al., 2017).

Wind turbines, which are used to convert the aerodynamic energy of the wind into mechanical energy, are the key devices in the energy conversion phase (Han et al., 2023). The mechanical energy obtained by wind turbines is converted into electrical energy using an electrical generator (Kou et al., 2023a, 2023b). The electrical energy produced is transferred to the power unit in order to be converted into a form suitable for the consumer's demand (Moiseev et al., 2023). Wind energy is also called wind energy conversion system in the commercial and scientific world due to these conversion stages (Lundquist et al., 2019).

For centuries, world trade was conducted in the oceans only by ships powered by wind energy. But over time, fossil fuel-powered vehicles were invented and wind and sun, known as renewable energy, were ignored (Yüksel & Dinçer, 2023). However, today it is clear that more use of fossil energy sources, which harm nature, will bring the end of life, so it is an alternative to fossil fuel vehicles that make our modern standard of living possible. What we are looking for is that we should turn to renewable energies. For this purpose, it is obvious that it is necessary to turn to renewable energy solutions such as wind and solar energy to power boats, freighters, and other ships (Gutierrez-Romero et al., 2019).

It should not be forgotten that this system is possible with globalization and economic growth. Looking at the history of ship designs, it is possible to design ships that can work with wind energy (Wan et al., 2023). For many years, scientists have worked on wind turbines to design various turbines, and in our age, these turbines can be turned into systems that power ships with a superior design that will further develop them (Sorensen et al., 2017).

Wind energy, one of the renewable energy types, is preferred in both conventional sails and propulsion systems in modern designs. These designs include the fletner rotors, the kite system, the drag sails, the soft sails, the wing sails, and the wind turbines. In propulsion systems installed through these devices, wind energy often does not serve as the main propulsion element of a typical ship (Ai et al., 2023). However, in bad weather conditions with high wind intensity, they can enable the ship to reach the desired speed by saving fuel without giving full sailing. A ship driven entirely by wind energy, on the other hand, can have a maximum carrying capacity of 3,000 to 10,000 tons in size, to be used in certain coastal voyages (Tillig & Ringsberg, 2020).

Toward the end of the twentieth century, the high increases in oil prices made wind energy a focus of great interest for commercial ships (Kumar et al., 2019a, 2019b). Some ships were built either new or converted. These include the 1600 DWT tanker Shin Aitoku Maru and the 26,000 dwt bulk carrier Usuki Pioneer. Japanese engineers designed and incorporated rigid sails into ships to reduce the fuel consumption. Japan Machinery Development Association has proven that rigid sails reduce fuel consumption by designing Japanese ships such as Shin Aitoku Maru and Usuki Pioneer equipped with rigid sails. Thanks to these sails, it has been observed that the fuel consumption has decreased by about 10%–30%. Various rigid sails have been applied to a few small ships, but they have not been widely used on large ships due to the many engineering and operational difficulties (Prasanth et al., 2021).

Alternative energy applications are becoming increasingly important in the shipping industry, as companies look to reduce their carbon footprint and improve their sustainability practices. One example of alternative energy application is the use of wind power, which can be harnessed through the installation of wind turbines on ships or through the use of sails (Niu et al., 2023). Another example is the use of solar energy, which can be used to power onboard systems or as a supplementary source of energy. Additionally, the use of hydrogen fuel cells is becoming more popular, providing a clean energy source for onboard power and propulsion systems. Furthermore, the use of biofuels, such as biodiesel or bioethanol, can be a more sustainable alternative to traditional fossil fuels. These alternative energy sources can be integrated with operational strategies, such as optimizing cargo loading or reducing ship speed, to further improve efficiency and reduce emissions. By adopting these alternative energy applications, the shipping industry can contribute to global efforts to reduce greenhouse gas emissions and promote sustainability (Pan et al., 2021).

One of the most significant benefits of alternative energy applications on ships is the reduction in greenhouse gas emissions. Shipping is responsible for a significant amount of global emissions, and the adoption of these alternative energy sources can help to mitigate this impact. For example, wind power can reduce fuel consumption and emissions by up to 30%, while solar energy can reduce emissions from onboard power and lighting systems. Additionally, the use of hydrogen fuel cells can provide a clean, efficient power source with no emissions other than water vapor (Yang et al., 2020). These alternative energy sources also have the potential to save costs in the long run, as they can reduce the reliance on expensive and volatile fossil fuels. Furthermore, as governments continue to introduce stricter emissions regulations, the adoption of alternative energy applications on ships can help companies to meet these standards and avoid penalties. Overall, the adoption of alternative energy applications is a critical step toward creating a more sustainable and efficient shipping industry.

There are also some challenges associated with the adoption of alternative energy applications on ships (Aksoy et al., 2023). For example, the upfront cost of installation and maintenance can be high, and the available technology may not yet be able to meet the energy demands of larger vessels. Additionally, some alternative energy sources, such as biofuels, can compete with food production, potentially creating ethical concerns. However, as technology continues to develop and the demand for sustainable shipping practices increases, these challenges are being addressed and overcome. In fact, many leading companies in the shipping industry are already investing in alternative energy applications, recognizing the potential benefits in terms of both environmental impact and long-term cost savings. As the industry continues to shift toward sustainable practices, the adoption of alternative energy applications will play a critical role in achieving this goal (Nguyen et al., 2021).

Onshore energy supply systems are becoming increasingly popular in the shipping industry as companies look for ways to reduce their environmental impact and improve their energy efficiency. An onshore energy supply system is a shore-based electrical infrastructure that allows ships to connect to the electrical grid while they are docked, eliminating the need for onboard generators. This can significantly reduce fuel consumption and emissions as well as reduce the noise and air pollution associated with running generators while in port. Onshore energy supply systems can also reduce the costs associated with operating onboard generators, as they provide a more cost-effective and reliable source of energy. Furthermore, as many ports and harbors around the world are implementing stricter emissions regulations, the use of onshore energy supply systems can help companies to meet these standards and avoid penalties. Overall, onshore energy supply systems are an important step toward creating a more sustainable and efficient shipping industry (Beriro et al., 2022).

In addition to reducing emissions and operating costs, onshore energy supply systems can also provide other benefits for the shipping industry. For example, they can improve the quality of life for those living near ports and harbors by reducing noise and air pollution. Additionally, they can reduce the risk of spills or other environmental incidents that may occur when ships are refueling or transferring oil or other hazardous materials. Furthermore, the use of onshore energy supply systems can help to promote sustainable development by encouraging the use of renewable energy sources, such as wind and solar power. As the demand for sustainable shipping practices continues to grow, the use of onshore energy supply systems is likely to become even more widespread. Many ports and harbors around the world are already investing in these systems, recognizing the potential benefits they offer for both the environment and the economy (Kumar et al., 2019a, 2019b).

However, there are also some challenges associated with the adoption of onshore energy supply systems in the shipping industry. For example, the installation and maintenance costs can be high, and some ports may not have the necessary infrastructure in place to support these systems. Additionally, the availability of renewable energy sources, such as wind or solar power, may be limited in some areas, which could make it more difficult to implement onshore energy supply systems that rely on these sources. However, as technology continues to develop and costs come down, these challenges are likely to become less significant. Overall, onshore energy supply systems are an important tool in the effort to create a more sustainable and efficient shipping industry, and their adoption is likely to continue to grow in the coming years (Gutierrez-Romero et al., 2019).

The process of implementing an onshore energy supply system typically involves several key steps. The first step is to assess the energy needs of the ships that will be using the system, as well as the availability of renewable energy sources in the area. Once these factors have been evaluated, the next step is to design and install the necessary infrastructure, which may include electrical cables, transformers, and switchgear. The installation process may involve significant construction work and may require permits or approvals from local authorities. Once the infrastructure is in place, ships can connect to the onshore energy supply system by plugging into the electrical outlets provided at the dock (Wang et al., 2021). The system may be operated by a third-party provider or by the port authority itself. Finally, ongoing maintenance and monitoring are necessary to ensure that the system remains in good working order and that any issues are addressed promptly. Overall, the process of implementing an onshore energy supply system can be complex and may require significant upfront investment, but it offers a promising solution for reducing emissions and promoting sustainability in the shipping industry (Kizielewicz & Skrzeszewska, 2021).

8.4 Conclusion

Analysis of alternative energy sources and their applications in maritime transportation with a strategic management approach highlights the immense potential for a more sustainable future in the industry. The study has examined various alternative energy sources, including wind, solar, and hydrogen and their viability in propelling vessels and supporting auxiliary systems. By adopting a strategic management perspective, maritime companies can effectively evaluate the feasibility, benefits, and challenges associated with alternative energy solutions. The findings emphasize the need for proactive planning, stakeholder collaboration, and continuous innovation to overcome infrastructure limitations, technological barriers, and cost considerations. Implementing alternative energy sources in maritime transportation holds the promise of significant environmental benefits, such as reduced greenhouse gas emissions and enhanced energy efficiency. This analysis underscores the importance of strategic decision-making, industry collaboration, and regulatory support in driving the adoption and implementation of alternative energy sources in maritime transportation. By embracing sustainable practices and alternative energy solutions, the maritime industry can play a crucial role in mitigating climate change and contributing to a greener and more sustainable future.

Moreover, the strategic management approach emphasizes the long-term advantages of incorporating alternative energy sources in maritime transportation. By reducing dependence on fossil fuels, the industry can mitigate the environmental impact of its operations and contribute to global efforts in combating climate change. The analysis highlights that alternative energy sources not only offer potential environmental benefits but can also lead to enhanced operational efficiency and cost savings over time. By strategically integrating these energy sources into vessel designs, propulsion systems, and auxiliary services, maritime companies can optimize their operations while minimizing their carbon footprint.

One effective application solution in the context of alternative energy sources and maritime transportation is the development of onshore energy supply systems. Onshore energy supply systems involve establishing renewable energy infrastructure near ports and terminals to provide clean and reliable power for ships during their stay in port. These systems can include solar power installations, wind turbines, or hydrogen production facilities, among others. By utilizing onshore energy supply systems, ships can reduce their reliance on fossil fuels and tap into clean and sustainable energy sources while docked. This not only helps to significantly lower emissions and minimize the environmental impact of maritime operations but also offers economic benefits in terms of cost savings and improved energy efficiency. Onshore energy supply systems serve as an important application solution that supports the transition to a greener and more sustainable maritime industry.

In conclusion, the analysis of alternative energy sources and applications in maritime transportation with a strategic management approach reveals a clear path toward a more sustainable future for the industry. By embracing alternative energy solutions, maritime companies can not only reduce their carbon emissions but also enhance their operational efficiency and cost-effectiveness. However, successful implementation requires strategic planning, stakeholder collaboration, and ongoing innovation. With concerted efforts from all industry stakeholders, the maritime sector can lead the way in adopting sustainable practices and becoming a driving force for positive environmental change in the global transportation landscape.

References

Ai, R., Zheng, Y., Yüksel, S., & Dinçer, H. (2023). Investigating the components of fintech ecosystem for distributed energy investments with an integrated quantum spherical decision support system. *Financial Innovation, 9*(1), 27.

Aksoy, T., Yüksel, S., Dinçer, H., Hacioğlu, Ü., Mikhaylov, A., & Muyeen, S. M. (2023). Assessing the electricity production capacities of emerging markets for the sustainable investments. *IEEE Access, 11*, 24574–24590.

Beriro, D., Nathanail, J., Salazar, J., Kingdon, A., Marchant, A., Richardson, S., et al. (2022). A decision support system to assess the feasibility of onshore renewable energy infrastructure. *Renewable and Sustainable Energy Reviews, 168*, 112771.

Burel, F., Taccani, R., & Zuliani, N. (2013). Improving sustainability of maritime transport through utilization of liquefied natural gas (LNG) for propulsion. *Energy, 57*, 412–420.

Carayannis, E., Kostis, P., Dinçer, H., & Yüksel, S. (2023). Quality function deployment-oriented strategic outlook to sustainable energy policies based on quintuple innovation helix. *Journal of the Knowledge Economy*, 1–19.

Chahal, V., & Narwal, M. (2017). An empirical review of lean manufacturing and their strategies. *Management Science Letters, 7*(7), 321–336.

Dinçer, H., Yüksel, S., Hacioglu, U., Yilmaz, M. K., & Delen, D. (2023a). Development of a sustainable corporate social responsibility index for performance evaluation of the energy industry: A hybrid decision-making methodology. *Resources Policy, 85*, 103940.

Dinçer, H., Eti, S., Gökalp, Y., Yüksel, S., & Çelebi, B. (2023b). The role of small-scale wind turbines in increasing the competitiveness of hospitals. In *Handbook of research on quality and competitiveness in the healthcare services sector* (pp. 336–352). IGI Global.

Dinçer, H., Yüksel, S., Mikhaylov, A., Muyeen, S. M., Chang, T., Barykin, S., & Kalinina, O. (2023c). CO2 emissions integrated fuzzy model: A case of seven emerging economies. *Energy Reports, 9*, 5741–5751.

Dinçer, H., Yüksel, S., Aksoy, T., Hacıoğlu, Ü., Mikhaylov, A., & Pinter, G. (2023d). Analysis of solar module alternatives for efficiency-based energy investments with hybrid 2-tuple IVIF modeling. *Energy Reports, 10*, 61–71.

Dinçer, H., Eti, S., Yüksel, S., Özdemir, S., Yílmaz, A. E., & Ergün, E. (2023e). Integrating data mining and fuzzy decision-making techniques for analyzing the key minimizing factors of carbon emissions. *Journal of Intelligent & Fuzzy Systems, 45*, 1–17.

Dinçer, H., Eti, S., Aksoy, T., Yüksel, S., Hacioglu, U., Mikhaylov, A., & Muyeen, S. M. (2023f). Analysis of environmental impact for material production investments using a novel soft computing methodology. *IEEE Access, 11*, 37987–38001. https://doi.org/10.1109/ACCESS.2023.3266524

Eti, S., Dinçer, H., Gökalp, Y., Yüksel, S., & Kararoğlu, D. (2023). Identifying key issues to handle the inflation problem in the healthcare industry caused by energy prices: An evaluation with decision-making models. In *Managing inflation and supply chain disruptions in the global economy* (pp. 162–178).

Eti, S., Dinçer, H., Yüksel, S., & Gökalp, Y. (2022). Analysis of the suitability of the solar panels for hospitals: A new fuzzy decision-making model proposal with the T-spherical TOP-DEMATEL method. *Journal of Intelligent & Fuzzy Systems, Preprint*, 1–13.

Gutierrez-Romero, J. E., Esteve-Pérez, J., & Zamora, B. (2019). Implementing onshore power supply from renewable energy sources for requirements of ships at berth. *Applied Energy, 255*, 113883.

Han, Z., Gong, L., Chen, H., & Yüksel, S. (2023). The role of mineral and oil resources in achieving a clean environment in emerging economies: An asymmetric analysis. *Resources Policy, 86*, 104099.

Iris, Ç., & Lam, J. S. L. (2019). A review of energy efficiency in ports: Operational strategies, technologies and energy management systems. *Renewable and Sustainable Energy Reviews, 112*, 170–182.

Jahangiri, S., Nikolova, N., & Tenekedjiev, K. (2018). An improved emission inventory method for estimating engine exhaust emissions from ships. *Sustainable Environment Research, 28*(6), 374–381.

Jović, M., Tijan, E., Žgaljić, D., & Aksentijević, S. (2020). Improving maritime transport sustainability using blockchain-based information exchange. *Sustainability, 12*(21), 8866.

Kizielewicz, J., & Skrzeszewska, K. (2021). Identifying actions to prepare electricity infrastructure in seaports for future power supplying cruise ships with energy from land. *Energies, 14*(23), 8173.

Kołakowski, P., Gil, M., Wróbel, K., & Ho, Y. S. (2022). State of play in technology and legal framework of alternative marine fuels and renewable energy systems: A bibliometric analysis. *Maritime Policy & Management, 49*(2), 236–260.

Kou, G., Pamucar, D., Dinçer, H., & Yüksel, S. (2023a). From risks to rewards: A comprehensive guide to sustainable investment decisions in renewable energy using a hybrid facial expression-based fuzzy decision-making approach. *Applied Soft Computing, 142*, 110365.

Kou, G., Yuksel, S., & Dincer, H. (2023b). A facial expression and expert recommendation fuzzy decision-making approach for sustainable business investments within the metaverse world. *Applied Soft Computing*, 110849.

Krčum, M., Gudelj, A., & Tomas, V. (2018). Optimal design of ship's hybrid power system for efficient energy. *Transactions on Maritime Science, 7*(01), 23–32.
Kumar, J., Kumpulainen, L., & Kauhaniemi, K. (2019a). Technical design aspects of harbour area grid for shore to ship power: State of the art and future solutions. *International Journal of Electrical Power & Energy Systems, 104*, 840–852.
Kumar, J., Memon, A. A., Kumpulainen, L., Kauhaniemi, K., & Palizban, O. (2019b). Design and analysis of new harbour grid models to facilitate multiple scenarios of battery charging and onshore supply for modern vessels. *Energies, 12*(12), 2354.
Lundquist, J. K., DuVivier, K. K., Kaffine, D., & Tomaszewski, J. M. (2019). Costs and consequences of wind turbine wake effects arising from uncoordinated wind energy development. *Nature Energy, 4*(1), 26–34.
Manickavasagam, K., Thotakanama, N. K., & Puttaraj, V. (2019). Intelligent energy management system for renewable energy driven ship. *IET Electrical Systems in Transportation, 9*(1), 24–34.
Martínez, L., Dinçer, H., & Yüksel, S. (2023). A hybrid decision making approach for new service development process of renewable energy investment. *Applied Soft Computing, 133*, 109897.
Moiseev, N., Mikhaylov, A., Dinçer, H., & Yüksel, S. (2023). Market capitalization shock effects on open innovation models in e-commerce: Golden cut q-rung orthopair fuzzy multicriteria decision-making analysis. *Financial Innovation, 9*(1), 55.
Msigwa, G., Ighalo, J. O., & Yap, P. S. (2022). Considerations on environmental, economic, and energy impacts of wind energy generation: Projections towards sustainability initiatives. *Science of the Total Environment*, 157755.
Nguyen, H. P., Hoang, A. T., Nizetic, S., Nguyen, X. P., Le, A. T., Luong, C. N., et al. (2021). The electric propulsion system as a green solution for management strategy of CO2 emission in ocean shipping: A comprehensive review. *International Transactions on Electrical Energy Systems, 31*(11), e12580.
Nguyen, X. P., & Dong, V. H. (2020). A study on traction control system for solar panel on vessels. In AIP conference proceedings (Vol. 2235, 1, 020016). AIP Publishing LLC.
Niu, X., Yüksel, S., & Dinçer, H. (2023). Emission strategy selection for the circular economy-based production investments with the enhanced decision support system. *Energy, 274*, 127446.
Oakland, J. S., & Marosszeky, M. (2006). *Total quality in the construction supply chain*. Routledge.
Pan, P., Sun, Y., Yuan, C., Yan, X., & Tang, X. (2021). Research progress on ship power systems integrated with new energy sources: A review. *Renewable and Sustainable Energy Reviews, 144*, 111048.
Prasanth, K., Prakash, M. S., & Sivaprasad, K. (2021). Optimization of wind sail using computational fluid dynamics simulation. *International Journal of Vehicle Structures & Systems, 13*(4), 477–481.
Priadi, A. A. (2022). Optimalization of smart technologies in improving sustainable maritime transportation. In IOP conference series: Earth and environmental science (Vol. 972, 1, 012084). IOP Publishing.
Rehmatulla, N., Parker, S., Smith, T., & Stulgis, V. (2017). Wind technologies: Opportunities and barriers to a low carbon shipping industry. *Marine Policy, 75*, 217–226.
Sadorsky, P. (2021). Wind energy for sustainable development: Driving factors and future outlook. *Journal of Cleaner Production, 289*, 125779.
Sorensen, A. J., Skjetne, R., Bo, T., Miyazaki, M. R., Johansen, T. A., Utne, I. B., & Pedersen, E. (2017). Toward safer, smarter, and greener ships: Using hybrid marine power plants. *IEEE Electrification Magazine, 5*(3), 68–73.
Tichavska, M., Tovar, B., Gritsenko, D., Johansson, L., & Jalkanen, J. P. (2019). Air emissions from ships in port: Does regulation make a difference? *Transport Policy, 75*, 128–140.
Tillig, F., & Ringsberg, J. W. (2020). Design, operation and analysis of wind-assisted cargo ships. *Ocean Engineering, 211*, 107603.
Wan, Q., Miao, X., Wang, C., Dinçer, H., & Yüksel, S. (2023). A hybrid decision support system with golden cut and bipolar q-ROFSs for evaluating the risk-based strategic priorities of fintech lending for clean energy projects. *Financial Innovation, 9*(1), 1–25.

Wang, H., Oguz, E., Jeong, B., & Zhou, P. (2019). Life cycle and economic assessment of a solar panel array applied to a short route ferry. *Journal of Cleaner Production, 219*, 471–484.

Wang, L., Liang, C., Shi, J., Molavi, A., Lim, G., & Zhang, Y. (2021). A bilevel hybrid economic approach for optimal deployment of onshore power supply in maritime ports. *Applied Energy, 292*, 116892.

Xing, H., Spence, S., & Chen, H. (2020). A comprehensive review on countermeasures for CO2 emissions from ships. *Renewable and Sustainable Energy Reviews, 134*, 110222.

Xu, X., Yüksel, S., & Dinçer, H. (2023). An integrated decision-making approach with golden cut and bipolar q-ROFSs to renewable energy storage investments. *International Journal of Fuzzy Systems, 25*(1), 168–181.

Yang, R., Yuan, Y., Ying, R., Shen, B., & Long, T. (2020). A novel energy management strategy for a ship's hybrid solar energy generation system using a particle swarm optimization algorithm. *Energies, 13*(6), 1380.

Yuan, Y., Wang, J., Yan, X., Li, Q., & Long, T. (2018). A design and experimental investigation of a large-scale solar energy/diesel generator powered hybrid ship. *Energy, 165*, 965–978.

Yüksel, S., & Dinçer, H. (2023). Sustainability analysis of digital transformation and circular industrialization with quantum spherical fuzzy modeling and golden cuts. *Applied Soft Computing, 138*, 110192.

Yüksel, S., Dinçer, H., & Uluer, G. S. (2023). Socio-economic indicators of nuclear energy investments: A study on countries. In *Inclusive developments through socio-economic indicators: New theoretical and empirical insights* (pp. 33–44). Emerald Publishing Limited.

Zhao, J., Wei, Q., Wang, S., & Ren, X. (2021). Progress of ship exhaust gas control technology. *Science of the Total Environment, 799*, 149437.

Zheng, W., Li, B., & Song, D. (2022). The optimal green strategies for competitive ocean carriers under potential regulation. *European Journal of Operational Research, 303*(2), 840–856.

Zhu, Y., Zhou, S., Feng, Y., Hu, Z., & Yuan, L. (2017). Influences of solar energy on the energy efficiency design index for new building ships. *International Journal of Hydrogen Energy, 42*(30), 19389–19394.

Chapter 9
The Cost of Output in the Energy Industry in Russia During Integration Processes in Eurasia

Alexey Mikhaylov

9.1 Introduction

The joint project of ADEME and the International Energy Agency, the Technical Service for Energy Efficiency Strategy and Indicators, mainly focuses on analyzing global energy efficiency trends and the interaction of energy efficiency strategies of different countries. This project has received contributions from over 70 WEC member countries with technical support from ENTERDATA (Ali et al., 2023). The Technical Service for Energy Efficiency Strategy and Indicators provides similar indicators for assessing energy efficiency such as energy intensity, technical and economic indicators, and energy saving indicators (Ai et al., 2023). The two main types of energy intensity distinguished by this project are primary energy intensity and final energy intensity (Aksoy et al., 2023). Primary energy intensity determines the amount of energy required to produce one unit of GDP, while final energy intensity is a better indicator for assessing energy efficiency at the final level (Wan et al., 2023). It is related to the energy consumed per unit of GDP by end-users for energy needs (Bhuiyan et al., 2021; Dong et al., 2021; Mikhaylov, 2021b, 2021c; Barykin et al., 2022b; Liu et al., 2022a, 2022b; Bhuiyan et al., 2022b; Danish et al., 2022a, 2022b; Saqib et al., 2021; Mukhametov et al., 2021; Candila et al., 2021; Mikhaylov & Grilli, 2022; Li et al., 2022).

The importance of energy efficiency has grown significantly over the years, resulting in a desire to compare energy efficiency indicators not only within countries but also across them. While such comparisons are desirable, they can be challenging to construct due to a variety of factors that impact energy efficiency

A. Mikhaylov (✉)
Institute of China and Modern Asia of the Russian Academy of Sciences, Moscow, Russia

H. Dinçer et al. (eds.), *Decision Making in Interdisciplinary Renewable Energy Projects*, Contributions to Management Science,
https://doi.org/10.1007/978-3-031-51532-3_9

measurements (Niu et al., 2023). These factors include differences in economic structures, resource bases, technology prevalence, measurement methods, and geographical and climatic factors. The ODYSSEE database was developed by the French Agency for Environment and Energy Management (ADEME) in 1992 as part of a project within the United Nations Foundation. The database was created to monitor national achievements in energy efficiency. The ODYSSEE databases are still regularly updated, and the scope has expanded to include more than 30 EU member states (Dinçer et al., 2023a, 2023b; Ivleva et al., 2023; Aissou et al., 2023; Mikhaylov, 2023).

In Russia, the Center for Efficient Energy Use (CENEF) is actively developing and implementing energy efficiency assessment methods. CENEF's work has practical applications, such as evaluating the State Program "Energy Saving and Energy Efficiency Improvement for the period up to 2020" and developing a report for the World Bank. CENEF identifies five interconnected groups of energy efficiency indicators (Nie et al. 2020; Mikhaylov et al., 2023b; Mikhaylov, 2021a; Varyash et al., 2020; Zhao et al., 2021; An & Mikhaylov, 2020, 2021; Alwaelya et al., 2021; Yumashev & Mikhaylov, 2020; Yumashev et al., 2020; Mutalimov et al., 2021).

The given content discusses various indicators that reflect the efficiency of energy use in different aspects of the economy (Yüksel & Dinçer, 2023). These indicators include energy intensity of GDP, electricity intensity of GDP, heat capacity of GDP, gas intensity of GDP, and more. Additionally, there are indicators that reflect energy efficiency in different sectors and in the production of products or services. The ENERGYBAL-GEM model developed by CENEF helps in determining national energy efficiency targets and target values of indicators for individual sectors (Moiseev et al., 2023a, 2023b). CENEF methods use various indicators of energy efficiency in the industry such as reducing energy intensity of industrial products, specific energy consumption, and relative indicators of specific energy consumption for the most energy-intensive products (Khan et al., 2022; Dinçer et al., 2022a, 2022b; Badr et al., 2022; Barykin et al., 2022a; Mehta et al., 2022; Kalinina et al. 2022; Shaikh et al., 2021; Mikhaylov et al., 2022a, 2022b; Nyangarika et al., 2022).

9.2 Literature Review

The specialized literature related to the assessment and analysis of energy efficiency of production processes can be categorized into two approaches, each having its own unique characteristics, advantages, and disadvantages. The first approach suggests assessing energy efficiency using economic, technical, and economic indicators along with the degree of implementation of energy-efficient technologies (Kou et al., 2023a, 2023b). One of the methods of the World Energy Council is an example of this approach. The second approach classifies energy efficiency indicators according to the type of activity (Han et al., 2023). Several methods, such as those of the Asia Pacific Research Center, the International Energy Agency Indicators project, the French Environmental Agency ADEME project, the Technical Service for Strategy and Energy Efficiency of the World Energy Consulate, and one of the methods of the Lawrence Berkeley National Laboratory, follow this approach (Chang et al., 2023; Denisova et al., 2019; Eti et al., 2023b; Nyangarika et al., 2019a, 2019b; Huang et al., 2021a, 2021b; Mikhaylov, 2018, 2022b; Mikhaylov et al., 2019; Conteh et al., 2021; Sediqi et al., 2022; Khan et al., 2021; Bhuiyan et al., 2022a; Liu et al., 2021a, 2021b; Daniali et al., 2021; Moiseev et al., 2023a, 2023b).

These indicators are evaluated at three different levels: the production complex as a whole, the manufacturing of products, and the technological process used to create a specific type of product. The approach used involves calculating indicators based on energy and power consumption for primary and secondary energy resources. When using primary energy resources, the method does not include a separate indicator for heat and electricity losses. In contrast, when using secondary (final) energy resources, an indicator for heat and electricity losses is included. This approach allows for an assessment of energy efficiency as a whole as well as an analysis of each production process from fuel preparation to final product processing. Conducting a separate analysis of each process is preferable since it allows for the evaluation of each process's contribution to the final result and identifies areas that need improvement (Mikhaylov et al., 2023a; Wang et al. 2023; He et al., 2023; Irshad et al., 2023; Iqbal et al., 2023; Aksoy et al., 2023).

To determine cost indicators accurately, it is essential to control price variations. The cost of production is always denoted in a specific currency, which is subject to fluctuations in exchange rates and production costs (Xu et al., 2023). These fluctuations affect the unit of measurement and thereby impact energy efficiency calculations. A change in the valuation of value units due to inflation or exchange rate deviations increases (Yüksel et al., 2023). Inflation raises the cost estimate related to the output unit if all other factors remain constant. For instance, when making international comparisons and using the exchange rate to convert the unit value to a total equivalent, changes in the exchange rate can also increase or decrease the nominal value of the output (Carayannis et al., 2023). In both situations, the ability to compare energy efficiency indicators can be simplified only by introducing regulation to eliminate the impact of internal price fluctuations (An et al., 2024; An and Mikhaylov, 2024; Jargalsaikhan et al., 2024; Eti et al., 2024; Yuksel et al., 2024; Rahadian et al., 2024; Mikhaylov, 2023; Datsyuk et al., 2023; Khan et al., 2023; Dayong et al., 2020; Mikhaylov et al., 2018; Nyangarika et al., 2018; Danish et al., 2020, 2021; An et al., 2021; Uyeh et al., 2021; Tamashiro et al., 2021, 2023; Shaikh et al., 2021; An et al., 2019a, 2019b; 2020a, 2020b, 2020c, 2022; Mikhaylov & Tarakanov, 2020; Moiseev et al., 2020; Moiseev et al., 2021; Gura et al., 2022; Dooyum et al., 2020; Mikhaylov et al., 2020; Mikhaylov, 2020a, 2020b, 2020c).

9.3 The Cost of Output in the Energy Industry

The energy industry is a complex system that involves the production, distribution, and consumption of energy. It includes different sources of energy such as fossil fuels, nuclear energy, and renewable energy. The energy industry is crucial for the functioning of our daily lives, including powering our homes, businesses, and transportation systems. However, the energy industry also has a significant impact on the environment and the economy. One of the challenges facing the energy industry is the need for cost-effective output (Dinçer et al., 2023a, 2023b, 2023c, 2023d, 2023e, 2023f). Cost-effective output refers to the ability to produce energy at a low cost while maintaining high efficiency and reliability. Achieving cost-effective output is essential for energy companies to remain competitive and profitable. It also benefits consumers by keeping energy costs low and stable.

Electricity cost comparison involves comparing the cost of energy from different electricity suppliers. This can be done using energy comparison websites or by contacting different electricity suppliers directly. When comparing the cost of energy, it is important to consider the unit price of electricity. This is usually measured in kilowatt-hours (kWh). The unit price of electricity can vary depending on the supplier and the type of energy source. For example, the unit price of electricity from renewable sources may be higher than the unit price of electricity from fossil fuels (Martínez et al., 2023).

Understanding the cost of energy production is essential for comparing the cost of energy from different suppliers. The cost of energy production can vary depending on the source of energy. For example, the cost of producing electricity from fossil fuels may be different from the cost of producing electricity from renewable sources (Eti et al., 2023a). The cost per kWh by source can help consumers understand the cost of producing electricity from different sources. This information can be found on the energy supplier's website or by contacting the supplier directly. It is important to note that the cost of energy production is just one factor to consider when choosing an energy supplier. Other factors such as service quality, contract terms, and customer service should also be considered.

When choosing an energy supplier, there are several factors to consider. These include the cost of energy, the quality of service, and the terms and conditions of the energy supply contract. The cost of energy is an important factor to consider. Consumers should compare the cost of energy from different suppliers to find the one that offers the best value for money. However, it is also important to consider the quality of service and the terms and conditions of the energy supply contract.

The quality of service can vary depending on the supplier. Consumers should choose a supplier that offers reliable service and good customer service. The terms and conditions of the energy supply contract should also be considered. This includes the length of the contract, payment options, and any additional charges or fees. Navigating the energy industry can be challenging. However, there are several tips that consumers can use to reduce costs and find the best energy supplier. One tip is to compare different energy suppliers regularly (Eti et al., 2022). This can help consumers find the best value for money and encourage competition in the energy industry.

Another tip is to reduce energy usage. This can be done by using energy-efficient appliances, turning off lights and appliances when not in use, and reducing heating and cooling usage. Consumers can also consider renewable energy sources such as solar panels or wind turbines. While these options may require an initial investment, they can result in long-term cost savings. There are several energy comparison tools and resources available to consumers. These include comparison websites, government energy comparison websites, and energy-saving calculators. Comparison websites allow consumers to compare different energy suppliers and find the one that offers the best value for money. Government energy comparison websites provide impartial information on different energy suppliers and their tariffs. Energy-saving calculators can help consumers estimate their energy usage and find ways to reduce energy consumption.

In addition to energy comparison tools and resources, there are also energy comparison services available. These services can help consumers find the best energy supplier and manage their energy usage. Energy comparison services can provide information on different energy suppliers, help consumers switch to a new supplier, and provide advice on reducing energy consumption. These services can be particularly useful for consumers who are new to the energy industry or who do not have the time or expertise to compare different energy suppliers themselves.

9.4 Conclusion

Physical indicators have certain limitations that need to be considered. One such restriction is related to the aggregation of information by physical indicators. When the denominator is defined as a physical unit, it becomes difficult to compare physical indicators with other units of measurement unless a common denominator is adjusted. To analyze the time series of industry energy consumption, the Laspeyres indicative analysis can be applied. This helps to calculate the relative influence of each of the conditions over time. The indexes begin to build with the choice of the base year, and then the ratio is taken, which is identical to the one discussed above. Each parameter in the numerator changes over time, while the remaining parameters in the numerator and denominator remain at the level of the base year. The result is an index that measures the relative impact of changing factors on total energy consumption.

References

Ai, R., Zheng, Y., Yüksel, S., & Dinçer, H. (2023). Investigating the components of fintech ecosystem for distributed energy investments with an integrated quantum spherical decision support system. *Financial Innovation, 9*(1), 27.

Aissou, A., Ahmed, H. S., Hussein, M. E., Hassan, M. R., Mikhaylov, A., & Mahmoud, E. (2023). Optimizing reliability of flow networks using RWGA subject to Total capacity constraint. *Applied Mathematics and Information Sciences, 7*(4), 559–565. https://doi.org/10.18576/amis/170404

Aksoy, T., Yüksel, S., Dınçer, H., Hacioglu, U., Mikhaylov, A., & Muyeen, S. M. (2023). Assessing the electricity production capacities of emerging markets for the sustainable investment. *IEEE Access, 11*, 24574–24590.

Ali, K., Jianguo, D., & Kirikkaleli, D. (2023). How do energy resources and financial development cause environmental sustainability? *Energy Reports, 9*, 4036–4048.

Alwaelya, S. A., Yousif, N. B. A., & Mikhaylov, A. (2021). Emotional development in preschoolers and socialization. *Early Child Development and Care, 191*, 16.

An, J., & Mikhaylov, A. (2020). Russian energy projects in South Africa. *Journal of Energy in Southern Africa, 31*(3), 58–64.

An, J., & Mikhaylov, A. (2021). Analysis of energy projects financial efficiency and renewable energy generation in Russia. *Finance: Theory and Practice, 25*(5), 79–91.

An, J., & Mikhaylov, A. (2024). Technology-Based Forecasting Approach for recognizing tradeoff between time-to-market reduction and devising a scheduling process in open innovation management. *Journal of Open Innovation: Technology, Market, and Complexity, 10*(1), 100207. https://doi.org/10.1016/j.joitmc.2024.100207

An, J., Mikhaylov, A., Dinçer, H., & Yüksel, S. (2022). Economic modelling of electricity generation: Long short-term memory and Q-rung orthopair fuzzy sets. *Heliyon, 8*, e10759.

An, J., Mikhaylov, A., & Sokolinskaya, N. (2019a). Machine learning in economic planning: Ensembles of algorithms. *Journal of Physics: Conference Series, 1353*, 012126.

An, J., Mikhaylov, A., & Moiseev, N. (2019b). Oil price predictors: Machine learning approach. *International Journal of Energy Economics and Policy, 9*(5), 1–6.

An, J., Mikhaylov, A., & Richter, U. H. (2020a). Trade war effects: Evidence from sectors of energy and resources in Africa. *Heliyon, 6*, e05693. https://doi.org/10.1016/j.heliyon.2020.e05693

An, J., Mikhaylov, A., & Jung, S.-U. (2020b). The strategy of South Korea in the global oil market. *Energies, 13*(10), 2491.

An, J., Mikhaylov, A., & Kim, K. (2020c). Machine learning approach in heterogeneous group of algorithms for transport safety-critical system. *Applied Sciences, 10*(8), 2670.

An, J., Mikhaylov, A., & Jung, S.-U. (2021). A linear programming approach for Robust network revenue management in the airline industry. *Journal of Air Transport Management, 91*, 101979.

An, J., Mikhaylov, A., & Chang, T. (2024). Relationship between the popularity of a platform and the price of NFT assets. *Finance Research Letters, 61*(3), 105057. https://doi.org/10.1016/j.frl.2024.105057

Badr, A. M. N. A. A., Mohamed, T. I., Osman, N. A. Q., & Mikhaylov, A. (2022). A review of social media website users' interaction paths with governmental accounts during the COVID-19 pandemic. *Informatics, 9*, 50. https://doi.org/10.3390/informatics9030050

Barykin, S. E., Kapustina, I. V., Sergeev, S. M., Daniali, S. M., Kopteva, L. A., Semenova, G. N., Pryadko, I. P., Mikhaylov, A., Baboshkin, P., Datsyuk, P., & Senjyu, T. (2022a). Financial logistics models based on systematic approach improving management solutions. *F1000Research, 11*, 570. https://doi.org/10.12688/f1000research.111252.1

Barykin, S., Mikheev, A., Kiseleva, E., Putikhin, Y., Alekseeva, N., & Mikhaylov, A. (2022b). An empirical analysis of Russian regions' debt sustainability. *Economies, 10*(4), 296.

Bhuiyan, M. A., An, J., Mikhaylov, A., Moiseev, N., & Danish, M. S. S. (2021). Renewable energy deployment and COVID-19 measures for sustainable development. *Sustainability, 13*, 4418.

Bhuiyan, M. A., Dinçer, H., Yüksel, S., Mikhaylov, A., Danish, M. S. S., Pinter, G., Uyeh, D. D., & Stepanova, D. (2022a). Economic indicators and bioenergy supply in developed economies: QROF-DEMATEL and random forest models. *Energy Reports, 8*(2022), 561–570.

Bhuiyan, M. A., Zhang, Q., Khare, V., Mikhaylov, A., Pinter, G., & Huang, X. (2022b). Renewable energy consumption and economic growth nexus—a systematic literature review. *Frontiers in Environmental Science, 10*, 878394. https://doi.org/10.3389/fenvs.2022.878394

Candila, V., Maximov, D., Mikhaylov, A., Moiseev, N., Senjyu, T., & Tryndina, N. (2021). On the relationship between oil and exchange rates of oil-exporting and oil-importing countries: From the great recession period to the Covid-19 era. *Energies, 14*, 8065.

Carayannis, E., Kostis, P., Dinçer, H., & Yüksel, S. (2023). Quality function deployment-oriented strategic outlook to sustainable energy policies based on quintuple innovation helix. *Journal of the Knowledge Economy*, 1–19.

Chang, H.-W., Chang, T., Xiang, F., Mikhaylov, A., & Grigorescu, A. (2023a). Revisiting R&D intensity and CO2 emissions link in the USA using time varying granger causality test: 1870~2020. *Heliyon, 9*(10), e20319. https://doi.org/10.1016/j.heliyon.2023.e20319

Conteh, F., Takahashi, H., Hemeida, A. M., Krishnan, N., Mikhaylov, A., & Senjyu, T. (2021). Hybrid grid-connected renewable power generation for sustainable electricity supply in Sierra Leone: Case study Lungi Town. *Sustainability, 13*(18).

Daniali, S. M., Barykin, S. E., Kapustina, I. V., Khortabi, F. M., Sergeev, S. M., Kalinina, O. V., Mikhaylov, A., Veynberg, R., Zasova, L., & Senjyu, T. (2021). Predicting volatility index according to technical index and economic indicators on the basis of deep learning algorithm. *Sustainability, 13*, 14011.

Danish, M. S. S., Bhattacharya, A., Stepanova, D., Mikhaylov, A., Grilli, M. L., Khosravy, M., & Senjyu, T. (2020). A systematic review of metal oxide applications for energy and environmental sustainability. *Metals, 10*(12), 1604.

Danish, M. S. S., Estrella-Pajulas, L. L., Alemaida, I. M. A., Grilli, M. L., Mikhaylov, A., & Senjyu, T. (2022). Green synthesis of silver oxide nanoparticles for photocatalytic environmental remediation and biomedical applications. *Metals, 12*(4).

Danish, M. S. S., Senjyu, T., Sabory, N. R., Khosravy, M., Grilli, M. L., Mikhaylov, A., & Majidi, H. (2021). A forefront framework for sustainable aquaponics modeling and design. *Sustainability, 13*, 9313.

Danish, M. S., Senjyu, T. S., Sabory, N. R., & Mikhaylov, A. (2022a). Lessons from COVID-19 conferring environmental re-engineering opportunity. In M. Danish & T. Senjyu (Eds.), *Eco-friendly and agile energy strategies and policy development* (pp. 99–118). IGI Global. https://doi.org/10.4018/978-1-7998-9502-2.ch005

Danish, M. S., Senjyu, T. S., Sabory, N. R., & Mikhaylov, A. (2022b). Interplay consequences of COVID-19 on global environmental sustainability. In M. Danish & T. Senjyu (Eds.), *Eco-friendly and agile energy strategies and policy development*. IGI Global. https://doi.org/10.4018/978-1-7998-9502-2.ch003

Datsyuk, P., Mikhaylov, A., & Pinter, G. (2023). Group correlation analysis of socio-economic indicators and the volume of collected E-waste in European countries. *Environmental Research Communications, 5*(11), 115004. https://doi.org/10.1088/2515-7620/ad0850

Dayong, N., Mikhaylov, A., Bratanovsky, S., Shaikh, Z. A., & Stepanova, D. (2020). Mathematical modeling of the technological processes of catering products production. *Journal of Food Process Engineering, 43*(2), 1–9.

Denisova, V., Mikhaylov, A., & Lopatin, E. (2019). Blockchain infrastructure and growth of global power consumption. *International Journal of Energy Economics and Policy, 9*(4), 22–29.

Dinçer, H., Yüksel, S., Mikhaylov, A., Pinter, G., & Shaikh, Z. A. (2022a). Analysis of renewable-friendly smart grid technologies for the distributed energy investment projects using a hybrid picture fuzzy rough decision-making approach. *Energy Reports, 8*, 11466–11477. https://doi.org/10.1016/j.egyr.2022.08.275

Dinçer, H., Yüksel, S., Mikhaylov, A., Barykin, S. E., Aksoy, T., & Hacioğlu, U. (2022b). Analysis of environmental priorities for green project investments using an integrated q-rung orthopair fuzzy modelling. *IEEE Access, 10*, 50996–51007. https://doi.org/10.1109/ACCESS.2022.3174058

Dinçer, H., Yüksel, S., Hacioglu, U., Yilmaz, M. K., & Delen, D. (2023a). Development of a sustainable corporate social responsibility index for performance evaluation of the energy industry: A hybrid decision-making methodology. *Resources Policy, 85*, 103940.

Dinçer, H., Eti, S., Gökalp, Y., Yüksel, S., & Çelebi, B. (2023b). The role of small-scale wind turbines in increasing the competitiveness of hospitals. In *Handbook of research on quality and competitiveness in the healthcare services sector* (pp. 336–352). IGI Global.

Dinçer, H., Yüksel, S., Mikhaylov, A., Muyeen, S. M., Chang, T., Barykin, S., & Kalinina, O. (2023c). CO2 emissions integrated fuzzy model: A case of seven emerging economies. *Energy Reports, 9*, 5741–5751.

Dinçer, H., Yüksel, S., Aksoy, T., Hacıoğlu, Ü., Mikhaylov, A., & Pinter, G. (2023d). Analysis of solar module alternatives for efficiency-based energy investments with hybrid 2-tuple IVIF modeling. *Energy Reports, 10*, 61–71.

Dinçer, H., Eti, S., Yüksel, S., Özdemir, S., Yflmaz, A. E., & Ergün, E. (2023e). Integrating data mining and fuzzy decision-making techniques for analyzing the key minimizing factors of carbon emissions. *Journal of Intelligent & Fuzzy Systems, 45*, 1–17.

Dinçer, H., Eti, S., Aksoy, T., Yüksel, S., Hacioglu, U., Mikhaylov, A., & Muyeen, S. M. (2023f). Analysis of environmental impact for material production investments using a novel soft computing methodology. *IEEE Access, 11*, 37987–38001. https://doi.org/10.1109/ACCESS.2023.3266524

Dong, B., Ikonnikova, I., Rogulin, R., Sakulyeva, T., & Mikhaylov, A. (2021). Environmental economic approach to optimization of transport communication in megacities. *Journal of Environmental Science and Health, Part A, 56*(6), 660–666.

Dooyum, U. D., Mikhaylov, A., & Varyash, I. (2020). Energy security concept in Russia and South Korea. *International Journal of Energy Economics and Policy, 10*(4), 102–107.

Eti, S., Dinçer, H., Gökalp, Y., Yüksel, S., & Kararoğlu, D. (2023a). Identifying key issues to handle the inflation problem in the healthcare industry caused by energy prices: An evaluation with decision-making models. In *Managing inflation and supply chain disruptions in the global economy* (pp. 162–178).

Eti, S., Dincer, H., Yuksel, S., Uslu, Y. D., Gokalp, Y., Kalkavan, H., Mikhaylov, A., & Pinter, G. (2023b). Determination of priority criteria in green building transformation: An analysis on the service industry. *Research in Globalization, 7*(12), 100164. https://doi.org/10.1016/j.resglo.2023.100164

Eti, S., Yüksel, S., Dinçer, H., Kalkavan, H., Hacioglu, U., Mikhaylov, A., Danish, M. S. S., & Pinter, G. (2024). Assessment of technical and financial challenges for renewable energy project alternatives. *Cleaner Engineering and Technology, 18*, 100719. https://doi.org/10.1016/j.clet.2023.100719

Eti, S., Dinçer, H., Yüksel, S., & Gökalp, Y. (2022). Analysis of the suitability of the solar panels for hospitals: A new fuzzy decision-making model proposal with the T-spherical TOP-DEMATEL method. *Journal of Intelligent & Fuzzy Systems, Preprint*, 1–13.

Gura, D., Mikhaylov, A., Glushkov, S., Zaikov, M., & Shaikh, Z. A. (2022). Model for estimating power dissipation along the interconnect length in single on-chip topology. *Evolutionary Intelligence, 15*, 2369–2373. https://doi.org/10.1007/s12065-020-00407-7

Han, Z., Gong, L., Chen, H., & Yüksel, S. (2023). The role of mineral and oil resources in achieving a clean environment in emerging economies: An asymmetric analysis. *Resources Policy, 86*, 104099.

He, H., Huang, Y., Nakadomari, A., Masrur, H., Krishnan, N., Hemeida, A. M., Mikhaylov, A., Omine, E., & Senjyu, T. (2023). Potential and economic viability of green hydrogen production from seawater electrolysis using renewable energy in remote Japanese islands. *Renewable Energy, 202*, 1436–1447. https://doi.org/10.1016/j.renene.2022.12.046

Huang, Y., Masrur, H., Shigenobu, R., Hemeida, A. M., Mikhaylov, A., & Senjyu, T. (2021a). A comparative design of a campus microgrid considering a multi-scenario and multi-objective approach. *Energies, 14*, 2853.

Huang, Y., Yona, A., Takahashi, H., Hemeida, A. M., Mandal, P., Mikhaylov, A., Senjyu, T., & Lotfy, M. E. (2021b). Energy management system optimization of drug store operate electric vehicles charging station. *Sustainability, 14*, 1217836.

Iqbal, K. M. J., Khan, M. I., Mikhaylov, A., Shah, A. A., Yadykin, V., Filho, W. L., Tariq, M. A. U. R., & Ullah, W. (2023). Modeling principles, criteria and indicators to assess water sector governance for climate compatibility and sustainability. *Frontiers in Environmental Science, 11*, 989930.

Irshad, A. S., Ludin, G. A., Masrur, H., Ahmadi, M., Yona, A., Mikhaylov, A., Krishnan, N., & Senjyu, T. (2023). Optimization of grid-photovoltaic and battery hybrid system with most technically efficient PV technology after the performance analysis. *Renewable Energy, 207*, 714–730.

Ivleva, S., Hajimajeed, M., Kartashova, O., & Mikhaylov, A. (2023). A comparative study of regional variations in common visual impairments among school-age children. *Electronic Journal of General Medicine, 20*(5), em504. https://doi.org/10.29333/ejgm/13264

Jargalsaikhan, N., Ueda, S., Masahiro, F., Matayoshi, H., Mikhaylov, A., Byambaa, S., & Senjyu, T. (2024). Exploring influence of air density deviation on power production of wind energy conversion system: Study on correction method. *Renewable Energy, 2120*, 119636. https://doi.org/10.1016/j.renene.2023.119636

Kalinina O.V., Barykin S.E., Sergeev S.M., Semenova G.N, Fatkullina A, Mikhaylov A, De La Poza E. (2022). Smart city perspectives in post-pandemic governance: Externalities reduction policy. F1000Research, 11, 1032. https://doi.org/10.12688/f1000research.123195.1

Khan, A. A., Laghari, A. A., Gadekallu, T. R., Shaikh, Z. A., Javed, A. R., Rashid, M., Estrela, V. V., & Mikhaylov, A. (2022). A drone-based data management and optimization using metaheuristic algorithms and blockchain smart contracts in a secure fog environment. *Computers and Electrical Engineering, 102*, 108234. https://doi.org/10.1016/j.compeleceng.2022.108234

Khan, A. A., Shaikh, Z. A., Baitenova, L., Mutaliyeva, L., Moiseev, N., Mikhaylov, A., Laghari, A. A., Idris, S. A., & Alshazly, H. (2021). QoS-ledger: Smart contracts and metaheuristic for secure quality-of-service and cost-efficient scheduling of medical-data processing. *Electronics, 10*, 3083.

Khan, R. U., Saqib, A., Abbasi, M. A., Mikhaylov, A., & Pinter, G. (2023). Green leadership, environmental knowledge sharing, and sustainable performance in manufacturing industry: Application from upper echelon theory. *Sustainable Energy Technologies and Assessments, 60*, 103540. https://doi.org/10.1016/j.seta.2023.103540

Kou, G., Pamucar, D., Dinçer, H., & Yüksel, S. (2023a). From risks to rewards: A comprehensive guide to sustainable investment decisions in renewable energy using a hybrid facial expression-based fuzzy decision-making approach. *Applied Soft Computing, 142*, 110365.

Kou, G., Yuksel, S., & Dincer, H. (2023b). A facial expression and expert recommendation fuzzy decision-making approach for sustainable business investments within the metaverse world. *Applied Soft Computing*, 110849.

Li, J., Dinçer, H., Yüksel, S., Mikhaylov, A., & Barykin, S. E. (2022). Bipolar q-ROF hybrid decision making model with golden cut for analyzing the levelized cost of renewable energy alternatives. *IEEE Access, 10*, 35900. https://doi.org/10.1109/ACCESS.2022.3168315

Liu, L., Kato, T., Mandal, P., Mikhaylov, A., Hemeida, A. M., & Senjyu, T. (2021a). Multi-area power system load frequency control by using model predictive control by introducing renewable energy sources. In *2021 International conference on science & contemporary technologies (ICSCT)* (pp. 1–6). https://doi.org/10.1109/ICSCT53883.2021.9642521

Liu, L., Kato, T., Mandal, P., Mikhaylov, A., Hemeida, A. M., Takahashi, H., & Senjyu, T. (2021b). Two cases studies of model predictive control approach for hybrid renewable energy systems. *AIMS Energy, 9*(6), 1241–1259.

Liu, Z.-J., Panfilova, E., Mikhaylov, A., & Kurilova, A. (2022a). COVID-19 crisis impact on the stability between parties in crowdfunding and crowdsourcing. *Wireless Personal Communications, 122*, 915–930.
Liu, Z., Panfilova, E., Mikhaylov, A., & Kurilova, A. (2022b). Assessing stability in the relationship between parties in crowdfunding and crowdsourcing projects during the COVID-19 crisis. *Journal of Global Information Management (JGIM), 30*(4), 1–18. https://doi.org/10.4018/JGIM.297905
Martínez, L., Dinçer, H., & Yüksel, S. (2023). A hybrid decision making approach for new service development process of renewable energy investment. *Applied Soft Computing, 133*, 109897.
Mehta, R., Verghese, J., Mahajan, S., Barykin, S., Bozhuk, S., Kozlova, N., Kapustina, I., Mikhaylov, A., Naumova, E., & Dedyukhina, N. (2022). Consumers' behavior in conversational commerce marketing based on messenger chatbots. *F1000Research, 11*(647). https://doi.org/10.12688/f1000research.122037.1
Mikhaylov, A. (2018). The yield of the carry trade strategy. *Finance: Theory and Practice, 22*(3), 52–63. https://doi.org/10.26794/2587-5671-2018-22-3-52-63
Mikhaylov, A. (2020a). Cryptocurrency market analysis from the open innovation perspective. *Journal of Open Innovation: Technology, Market, and Complexity, 6*(4), 197.
Mikhaylov, A. (2020b). Geothermal energy development in Iceland. *International Journal of Energy Economics and Policy, 10*(4), 31–35.
Mikhaylov, A. (2020c). Cryptocurrency market development: Hurst method. *Finance: Theory and Practice, 24*(3), 81–91.
Mikhaylov, A. (2021a). Development of Friedrich von Hayek's theory of private money and economic implications for digital currencies. *Terra Economicus, 19*(1), 53–62.
Mikhaylov, A. (2021b). Lichens as indicators of atmospheric pollution in urban ecosystems. *Israel Journal of Ecology & Evolution, 67*(1–2), 60–68. https://doi.org/10.1163/22244662-bja10016
Mikhaylov, A. (2022a). Efficiency of renewable energy plants in Russia. *Anais da Academia Brasileira de Ciências, 94*(4), e20191226. https://doi.org/10.1590/0001-3765202220191226
Mikhaylov, A. (2022b). Sustainable development and renewable energy: A new view to a global problem. *Energies, 15*, 1397. https://doi.org/10.3390/en15041397
Mikhaylov, A. (2023). Understanding the risks associated with wallets, depository services, trading, lending, and borrowing in the crypto space. *Journal of Infrastructure, Policy and Development, 7*(3), 2223. https://doi.org/10.24294/jipd.v7i3.2223
Mikhaylov, A., Bhatti, I. M., Dinçer, H., & Yüksel, S. (2022a). Integrated decision recommendation system using iteration-enhanced collaborative filtering, golden cut bipolar for analyzing the risk-based oil market spillovers. *Computational Economics*. https://doi.org/10.1007/s10614-022-10341-8
Mikhaylov, A., Yumashev, A., & Kolpak, E. (2022b). Quality of life, anxiety and depressive disorders in patients with extrasystolic arrhythmia. *Archives of Medical Science, 18*(2). https://doi.org/10.5114/aoms.2020.101359
Mikhaylov, A., Burakov, D., & Didenko, V. (2019). Relationship between oil price and macroeconomic indicators in Russia. *Finance: Theory and Practice, 23*(2), 105–116. https://doi.org/10.26794/2587-5671-2019-23-2-105-116
Mikhaylov, A., Dinçer, H., Yüksel, S., Pinter, G., & Shaikh, Z. A. (2023a). Bitcoin mempool growth and trading volumes: Integrated approach based on QROF multi-SWARA and aggregation operators. *Journal of Innovation & Knowledge, 8*(3), 100378.
Mikhaylov, A., Dinçer, H., & Yüksel, S. (2023b). Analysis of financial development and open innovation oriented fintech potential for emerging economies using an integrated decision-making approach of MF-X-DMA and golden cut bipolar q-ROFSs. *Financial Innovation, 9*(1), 10. https://doi.org/10.1186/s40854-022-00234-w
Mikhaylov, A., & Grilli, M. L. (2022). Machine learning methods and sustainable development: Metal oxides and multilayer metal-oxides. *Metals, 12*(4).

Mikhaylov, A., Moiseev, N., Aleshin, K., & Burkhardt, T. (2020). Global climate change and greenhouse effect. *Entrepreneurship and Sustainability Issues, 7*(4), 2897–2913. https://doi.org/10.9770/jesi.2020.7.4(21)

Mikhaylov, A., Sokolinskaya, N., & Nyangarika, A. (2018). Optimal carry trade strategy based on currencies of energy and developed economies. *Journal of Reviews on Global Economics, 7*, 582–592.

Mikhaylov, A., & Tarakanov, S. (2020). Development of Levenberg-Marquardt theoretical approach for electric network. *Journal of Physics: Conference Series, 1515*, 052006.

Moiseev, N., Mikhaylov, A., Dinçer, H., & Yüksel, S. (2023a). Effects of market capitalization shocks for open innovation models in e-commerce industry: Golden cut QROF MCDM analysis. *Financial Innovation, 9*(1), 32.

Moiseev, N., Mikhaylov, A., Dinçer, H., & Yüksel, S. (2023b). Market capitalization shock effects on open innovation models in e-commerce: Golden cut q-rung orthopair fuzzy multicriteria decision-making analysis. *Financial Innovation, 9*(1), 55.

Moiseev, N., Mikhaylov, A., Varyash, I., & Saqib, A. (2020). Investigating the relation of GDP per capita and corruption index. *Entrepreneurship and Sustainability Issues, 8*(1), 780–794.

Moiseev, N., Sorokin, A., Zvezdina, N., Mikhaylov, A., Khomyakova, L., & Danish, M. S. S. (2021). Credit risk theoretical model on the base of DCC-GARCH in time-varying parameters framework. *Mathematics, 9*, 2423. https://doi.org/10.3390/math9192423

Mukhametov, A., Bekhorashvili, A., Avdeenko, A., & Mikhaylov, A. (2021). The impact of growing legume plants under conditions of biologization and soil cultivation on Chernozem fertility and productivity of rotation crops. *Legume Research, 44*, 12.

Mutalimov, V., Kovaleva, I., Mikhaylov, A., & Stepanova, D. (2021). Assessing regional growth of small business in Russia. *Entrepreneurial Business and Economics Review, 9*(3), 119–133. https://doi.org/10.15678/EBER.2021.090308

Nie, D., Panfilova, E., Samusenkov, V., & Mikhaylov, A. (2020). E-learning financing models in Russia for sustainable development. *Sustainability, 12*(11), 4412.

Niu, X., Yüksel, S., & Dinçer, H. (2023). Emission strategy selection for the circular economy-based production investments with the enhanced decision support system. *Energy, 274*, 127446.

Nyangarika, A., Mikhaylov, A., Muyeen, S. M., Yadykin, V., Mottaeva, A., Pryadko, I., Barykin, S., Fomenko, N., Rykov, G., & Shvandar, K. (2022). Energy stability and decarbonization in developing countries: Random Forest approach for forecasting of crude oil trade flows and macro indicators. *Frontiers in Environmental Science, 10*, 1031343. https://doi.org/10.3389/fenvs.2022.1031343

Nyangarika, A., Mikhaylov, A., & Richter, U. (2019a). Influence oil price towards economic indicators in Russia. *International Journal of Energy Economics and Policy, 9*(1), 123–129.

Nyangarika, A., Mikhaylov, A., & Richter, U. (2019b). Oil price factors: Forecasting on the base of modified auto-regressive integrated moving average model. *International Journal of Energy Economics and Policy, 9*(1), 149–159.

Nyangarika, A., Mikhaylov, A., & Tang, B.-J. (2018). Correlation of oil prices and gross domestic product in oil producing countries. *International Journal of Energy Economics and Policy, 8*(5), 42–48.

Rahadian, D., Firli, A., Dinçer, H., Yüksel, S., Mikhaylov, A., & Ecer, F. (2024). A hybrid neuro fuzzy decision-making approach to the participants of derivatives market for fintech investors in emerging economies. *Financial Innovation, 10*(1), 37. https://doi.org/10.1186/s40854-023-00563-6

Saqib, A., Chan, T.-H., Mikhaylov, A., & Lean, H. H. (2021). Are the responses of sectoral energy imports asymmetric to exchange rate volatilities in Pakistan? Evidence from recent foreign exchange regime. *Frontiers in Energy Research, 9*, 614463. https://doi.org/10.3389/fenrg.2021.614463

Sediqi, M. M., Nakadomari, A., Mikhaylov, A., Krishnan, N., Lotfy, M. E., Yona, A., & Senjyu, T. (2022). Impact of time-of-use demand response program on optimal operation of Afghanistan real power system. *Energies, 15*, 296.

Shaikh, Z. A., Kraikin, A., Mikhaylov, A., & Pinter, G. (2022). Forecasting stock prices of companies producing solar panels using machine learning methods. *Complexity, 2022*, 9186265. https://doi.org/10.1155/2022/9186265

Shaikh, Z. A., Moiseev, N., Mikhaylov, A., & Yüksel, S. (2021). Facile synthesis of copper oxide-cobalt oxide/nitrogen-doped carbon (Cu2O-Co3O4/CN) composite for efficient water splitting. *Applied Sciences., 11*, 9974. https://doi.org/10.3390/app11219974

Tamashiro, K., Alharbi, T., Mikhaylov, A., Hemeida, A. M., Krishnan, N., Lotfy, M. E., & Senjyu, T. (2021). Investigation of home energy management with advanced direct load control and optimal scheduling of controllable loads. *Energies, 14*, 7314. https://doi.org/10.3390/en14217314

Tamashiro, K., Omine, E., Krishnan, N., Mikhaylov, A., Hemeida, A. M., & Senjyu, T. (2023). Optimal components capacity based multi-objective optimization and optimal scheduling based MPC-optimization algorithm in smart apartment buildings. *Energy & Buildings, 278*, 112616. https://doi.org/10.1016/j.enbuild.2022.112616

Uyeh, D. D., Asem-Hiablie, S., Park, T., Kim, K.-M., Mikhaylov, A., Woo, S., & Ha, Y. (2021). Could Japonica rice be an alternative variety for increased global food security and climate change mitigation? *Food, 10*, 1869. https://doi.org/10.3390/foods10081869

Varyash, I., Mikhaylov, A., Moiseev, N., & Aleshin, K. (2020). Triple bottom line and corporate social responsibility performance indicators for Russian companies. *Entrepreneurship and Sustainability Issues, 8*(1), 313–331.

Wan, Q., Miao, X., Wang, C., Dinçer, H., & Yüksel, S. (2023). A hybrid decision support system with golden cut and bipolar q-ROFSs for evaluating the risk-based strategic priorities of fintech lending for clean energy projects. *Financial Innovation, 9*(1), 1–25.

Wang, Q., Guo, J., Li, R., Mikhaylov, A., & Moiseev, N. (2023). Does technical assistance alleviate energy poverty in sub-Saharan African countries? A new perspective on spatial spillover effects of technical assistance. *Energy Strategy Reviews, 45*, 101047. https://doi.org/10.1016/j.esr.2022.101047

Xu, X., Yüksel, S., & Dinçer, H. (2023). An integrated decision-making approach with golden cut and bipolar q-ROFSs to renewable energy storage investments. *International Journal of Fuzzy Systems, 25*(1), 168–181.

Yüksel, S., & Dinçer, H. (2023). Sustainability analysis of digital transformation and circular industrialization with quantum spherical fuzzy modeling and golden cuts. *Applied Soft Computing, 138*, 110192.

Yüksel, S., Dinçer, H., & Uluer, G. S. (2023). Socio-economic indicators of nuclear energy investments: A study on countries. In *Inclusive developments through socio-economic indicators: New theoretical and empirical insights* (pp. 33–44). Emerald Publishing Limited.

Yüksel, S., Eti, S., Dinçer, H., Gokalp, Y., Yavuz, D., Mikhaylov, A., & Pinter, G. (2024). Prioritizing the indicators of energy performance management: A novel fuzzy decision-making approach for G7 service industries. *Environmental Research Communications, 6*(1), 015003. https://doi.org/10.1088/2515-7620/ad1c07

Yumashev, A., & Mikhaylov, A. (2020). Development of polymer film coatings with high adhesion to steel alloys and high wear resistance. *Polymer Composites, 41*(7), 2875–2880.

Yumashev, A., Ślusarczyk, B., Kondrashev, S., & Mikhaylov, A. (2020). Global indicators of sustainable development: Evaluation of the influence of the human development index on consumption and quality of energy. *Energies, 13*, 2768.

Zhao, Q., Cherkasov, S., Avdeenko, A., Kondratenko, L., & Mikhaylov, A. (2021). Integral estimate of the added value of agricultural crops in the synergetic agriculture on the example of Vicia faba (Vicia faba L.). *Polish Journal of Environmental Studies, 30*(5), 1–9. https://doi.org/10.15244/pjoes/131949

Chapter 10
Modeling the Long-Term Relationships between Renewable Energy and the Human Capital Index on Economic Growth in the German Economy

Şeyma Bozkaya and Feride Duran

10.1 Introduction

The increase in the amount of production that occurs with the increase in productivity in production factors is expressed as economic growth The development measures of the countries vary depending on the size of their production. Therefore, economic growth is an important indicator in determining the level of development of societies. Stable growth of economies is important for closing the development gap of countries. Therefore, determining the factors affecting economic growth well and developing policies accordingly will lead to the closure of the development gap between countries. While capital accumulation, labor force, and natural resources are considered sufficient to explain economic growth in the development stages of economies, these factors are insufficient to explain growth in today's information and technology age. As a matter of fact, as the relations between the countries' economies develop, factors such as technology level, human capital, and energy come to the fore in addition to physical capital in the explanation of economic growth (Turna & Ceylan, 2022: 226).

The aim of all countries, large and small, in the world is to achieve sustainable economic growth and development. In the carry out of this aim, energy has become very important as well as production factors such as labor and capital, because energy is both a basic element in all stages of the production process and a great indicator of welfare. In other words, the energy factor has a very important share as it is both the input and the output of the production process (Apaydın et al., 2019:119). In addition, situations such as energy crises, population growth,

Ş. Bozkaya (✉)
Independent Researcher, Kayseri, Turkey

F. Duran
Independent Researcher, Konya, Turkey

H. Dinçer et al. (eds.), *Decision Making in Interdisciplinary Renewable Energy Projects*, Contributions to Management Science,
https://doi.org/10.1007/978-3-031-51532-3_10

industrialization, and technological improvement in the world in recent years increase the need for energy day by day (Türkmen et al., 2018: 30).

With the industrial revolution, nonrenewable energy has become the most used energy source, and this situation continues today. These types of energy constitute more than 87% of the primary energy used in the world (Lyman & Economist, 2016). This situation increases environmental pollution, and the world's desire for a more livable environment is gradually disappearing. In order to prevent this, the importance of environmentally friendly energies is increasing in the world. For this reason, environmental technologies have become very important in the world recently. Because in order to achieve the goal of sustainable growth, it is extremely important to replace clean and polluting energies (Li & Lin, 2016, Shahbaz et al., 2022).

In terms of both reducing environmental pollution and energy efficiency, the increase in the use of renewable energy sources has an important effect on economic growth. In recent years, it is known that countries that want to grow have increased these environmentally friendly energy investments. One of these countries is Germany. The Energiewende[1] transition process of the German economy has accelerated since 1968 ((Renn & Marshall, 2016: 231). Since then, the share of renewable energy investments in the German economy has gradually increased with the encouraging policies followed. Especially since 1998, the rate of renewable energies in German energy has been constantly increasing as Energiewende has made significant progress toward the set targets. As of 2018, almost one third of Germany's energy needs have been obtained from renewable energy sources, and R&D and technology capacities have been increased. It is foreseen that the increase in direct added value obtained from renewable energy will continue. It is estimated that the added value, which was 18 billion Euros in 2012, will increase to approximately 25 billion Euros in 2030 (depending on the installation rates). As of 2018, 334,000 people were employed in the renewable energy sector in Germany, and the increase in its contribution to employment is expected to continue in the coming years (Erkök, 2022:305–306). The rate of renewable energy in the German economy is growing day by day, and these policies encourage world economies to invest in this field.

Improvements in investments in renewable energy sources and increases in usage capacities in the German economy increase the importance of this country in the energy literature. In this study, it is aimed to investigate the effects of renewable energy consumption, technological innovation and human capital on the German economy. The study consists of three parts following the introduction part. In the second part of the study, examples from the literature explaining the relationship between renewable energy consumption and growth are given. The third part of the study includes the findings obtained as a result of the model, methodology, and

[1] It is the transition policies of Germany to environmentally friendly, reliable and energy efficient energy sources.

econometric analyzes used. Finally, the study was concluded by giving policy recommendations and discussions based on the findings obtained as a result of the analysis.

10.2 Literature Review

It is known that studies examining the relationship between energy consumption and economic growth have increased in the literature with the 1970s. When the studies conducted since this date are examined, it is seen that the studies differ in terms of country group, method, and period range. In this study, it is important in terms of examining the German economy in terms of both the variables used and the fact that the period is new and up-to-date. With the increasing interest in renewable resources in the world in recent years and the size of the German economy's increasing use of renewable energy, it will be important to determine the effect of the use of renewable resources on the growth in the German economy. On the other hand, it is also significant to be able to determine the effects of technological innovation, physical capital, and human capital index elements on growth in addition to renewable energy in the study. For these purposes, in this part of the study, examples from the current literature were tried to be examined on the basis of time series analysis.

Yildirim et al. (2012) determined only one causal relationship from energy consumption derived from biomass waste to GDP in the USA using datasets from 1949 to 2010, 1960 to 2010, and 1970 to 2010. They also did not find any causal relationship between GDP and renewable energy. Lin and Moubarak (2014) found that there is a bidirectional causality relationship between renewable energy consumption and economic growth in the long run in China between 1977 and 2011 and that renewable energy increases economic growth. Shahbaz et al. (2015) found that renewable energy consumption increased economic growth in Pakistan between 1972 and 2011 and there was a bidirectional causality between them. Rafindadi and Ozturk (2017) showed that renewable energy consumption increased economic growth in Germany between 1971 and 2013 and there was a bidirectional causality relationship between the variables.

Alper (2018) determined that the increase in the use of renewable energy in Turkey during the 1990–2017 period increased economic growth and there is a unidirectional causality relationship from economic growth to renewable energy use. Apaydın et al. (2019) concluded that between 1965 and 2017, a 1% increase in renewable energy consumption in Turkey increased economic growth by approximately 0.4%, while a 1% decrease reduced growth by 0.7. Naımoglu (2021) showed that in the 1990–2019 period, renewable energy consumption in the German economy increased economic growth both in the short and long term, while the increase in energy losses decreased the economic growth. Gyimah et al. (2022) determined that the increase in renewable energy consumption between 1990 and 2015 in Ghana had a positive effect on economic growth.

10.3 Data, Model Description, Methodology, and Econometric Application

This study examines the effects of primary energy consumption, renewable energy consumption, technological innovation, physical capital, and human capital index on growth in the German economy. The analysis period of the study covers the years 1990–2021. The analysis period was determined in this way due to the limitation of the human capital index data. Germany (Ernst & Young European Attractiveness Survey (2013))[1], which ranks fourth in the world production ranking and has the largest production volume in Europe, is also an important exporter. With its qualified workforce, it also has a high capacity to produce high-value products. Therefore, since Germany's economy has a significant importance in world production, it was determined as the sample country. In this country, which has a large production volume, the gross domestic product (GDP (constant 2015 US$)) is used as the growth indicator. In addition, Germany is the world's fifth largest energy consumer, and a large part of it is imported. At the same time, Germany supports the use of renewable energy. The variable of renewable energy consumption (exajoules (input-equivalent) (includes renewable power (apart from hydro, which is reported seperately) and biofuels)), which is accepted as an environmentally friendly production factor in production processes, has been added to the model. In addition to renewable energy consumption, primary energy consumption is also included in the model. This country, which is an engineering country, ranks third after the USA and Japan in the triple patent ranking and is one of the leading countries in technology (web.archive.org 2023). Therefore, technological innovation (measured by patent applications (including residents and nonresidents)) variable has been used, since technology is a country where technology has a great importance in production processes. In addition, since it is a country with a very high human capital quality, the human capital index (life expectancy, quality of education, level of standard of living) variable was preferred. Finally, the physical capital (constant 2015 US$) variable, which has a significant place in the production processes, is included in the model. Economic growth, physical capital, and technological innovation variables were obtained from the World Bank. The database of renewable energy consumption is British Petroleum's World Energy 2021 (n.d.), while the database of the human capital center is UNDP (1990). Natural logarithmic transformations of all variables were used. The model to be used in the study is as follows:

$$\mathrm{lnEG}_t = \beta_0 + \beta_1 \mathrm{lnREC}_t + \beta_2 \mathrm{lnHDI}_t + \beta_3 \mathrm{lnFC}_t + \beta_4 \mathrm{lnTIN}_t + \beta_5 \mathrm{lnEC}_t + u_t \quad (10.1)$$

One of the first and most important stages of time series analysis is the unit root test stage, in which it is determined whether the series subject to empirical analysis contain a unit root. In this context, the augmented Dickey-Fuller (ADF) test was applied to the series in order to examine whether the series contain unit roots or not. The ADF unit root test is an improved version of the Dickey-Fuller (DF) test created by Dickey and Fuller in 1979. In their study conducted by Dickey and Fuller in

1981, the augmented Dickey-Fuller (ADF) test was developed by adding lagged values to the dependent variables of the regression equations of the DF unit root test. The regression equations of the ADF test are formulated in three models (Dickey & Fuller, 1981; 1057–1072):

$$\Delta Y_t = pY_{t-1} + \sum_{i=1}^{n} \theta_i \Delta Y_{t-i} + e_t \left(\text{Fixed and Trending}\right) \tag{10.2}$$

$$\Delta Y_t = \alpha\, pY_{t-1} + \sum_{i=1}^{n} \theta_i \Delta Y_{t-1} + e_t \left(\text{Fixed}\right) \tag{10.3}$$

$$\Delta Y_t = \alpha + \beta T + pY_{t-1} + \sum_{i=1}^{n} \theta_i \Delta Y_{t-i} + e_t \left(\text{Fixed and Trending}\right) \tag{10.4}$$

In time series analysis, the relations between the variables should be handled within the framework of economic theory and supported by applied tests. The causality results obtained by causality tests may be inconsistent if the linear combinations of the series, which are not stationary but are equally integrated, are stationary. Therefore, the existence of a long-term relationship between two or more nonstationary series is examined by cointegration tests (Barışık & Demircioğlu, 2006: 76–78). In this context, Pesaran et al. (2001)'s the limit test (auto-regressive distributive lag (ARDL)) was used. The aim of the ARDL bounds test is to determine the optimum lag length by considering the different lags of each variable in the model and to investigate the existence of a cointegration relationship between this lag length and the variables (Delice & Emre, 2011:18). The fact that the variables to be used in the ARDL model are integrated at the level I(0) or stationary at the first difference does not pose any problem for the application of the bounds test. This method uses the error correction model. Therefore, it produces statistically more consistent results compared to classical cointegration tests. The advantage of the error correction model is that it contains information about both short- and long-term relationships between variables (Akel & Gazel, 2014: 30–31). In the boundary test, which examines the long-term relationship between the variables, the way to make estimations for the short- and long-term coefficients is opened when the test statistic reached exceeds the upper critical limit. The unconstrained error correction model (UECM) for the ARDL bounds testing approach is as follows:

$$\begin{aligned} \Delta \text{lnEG} = {} & \alpha_0 + \sum_{i=1}^{m} \beta_{1i} \Delta \text{lnEG}_{t-i} + \sum_{i=0}^{n} \beta_{2i} \Delta \text{lnEC}_{t-i} + \sum_{i=0}^{m} \beta_{3i} \Delta \text{lnFC}_{t-i} \\ & + \sum_{i=0}^{n} \beta_{4i} \Delta \text{lnHDI}_{t-i} + \sum_{i=0}^{n} \beta_{5i} \Delta \text{lnERC}_{t-i} + \sum_{i=0}^{m} \beta_{6i} \Delta \text{lnTIN}_{t-i} s \\ & + \delta_1 \text{lnEG}_{t-1} + \delta_2 \text{lnEC}_{t-1} + \delta_3 \text{lnFC}_{t-1} + \delta_4 \text{lnHDI}_{t-1} \\ & + \delta_5 \text{lnREC}_{t-1} + \delta_4 \text{lnTIN}_{t-1} + \varepsilon_i \end{aligned} \tag{10.5}$$

α represents the constant term, while Δ expresses the difference operator, and ε_i expresses the error term. After estimating the regression equation in this equation, a long-term relationship is examined with the Wald test (F statistic). The hypotheses for this test are as follows:

$$H_0 : \delta_1 = \delta_2 = \delta_3 = \delta_4 = 0 (\text{There is no cointegration})$$

$$H_0 : \delta_1 \neq \delta_2 \neq \delta_3 \neq \delta_4 \neq 0 (\text{There is cointegration})$$

For this study, the ARDL (*m*, *n*, *m*, *n*, *n*, *m*) model in Eq. 10.5 was created in order to estimate the long-term coefficients based on the equation numbered 1.

$$\begin{aligned} \text{lnEG}_t = \alpha_0 + \sum_{i=1}^{m} \alpha_{1i} \text{lnEG}_{t-i} + \sum_{i=0}^{n} \alpha_{2i} \text{lnEC}_{t-i} + \sum_{i=0}^{m} \alpha_{3i} \text{lnFC}_{t-i} + \sum_{i=0}^{n} \alpha_{3-4i} \text{lnHDI}_{t-i} \\ + \sum_{i=0}^{n} \alpha_{5i} \text{lnREC}_{t-i} + \sum_{i=0}^{m} \alpha_{6i} \text{lnREC}_{t-i} + \varepsilon_i \end{aligned} \tag{10.6}$$

Table 10.1 shows statistics for unit root, appropriate lag length for the ARDL model, ARDL cutoff test, and diagnostic test. The first part of the table includes the ADF unit root test results. Fixed and fixed+trend options of ADF unit root test were used. According to the results of the unit root test, it was observed that the variables became stationary at the mixed level. The mixed level of integration of the variables led to the application of the ARDL model. After deciding to apply the ARDL model, it is necessary to determine the appropriate lag length for this model. In the second part of Table 10.1, the results of the appropriate lag length test are given. In this step, the variables are tested with different delay combinations, and the first line with the most asterisks (*) shows the appropriate delay length according to the information criteria (AIC, SIC or HQ). As a result of the test, the appropriate lag length was determined as 1 according to the AIC value.

After determining the appropriate lag length, a test is performed to understand the existence of cointegration between the variables. This test is done regardless of whether the variables in question are I(0), I(1) or mutually cointegrated (De Vita & Abbott, 2002: 294). In the third part of Table 10.1, ARDL limit test results show whether there is a cointegration relationship between the variables. Since the F-statistic value of 6.99 shown in the table is higher than the critical values at the 1% significance level, it was concluded that there is cointegration between the variables at the 1% significance level.

In the last part of Table 10.1, test statistics examining whether there is a problem of autocorrelation, normality, and variable variance for the ARDL model established are given. The Breusch-Godfrey test is used to detect the presence of autocorrelation between error terms in a regression. In statistics, the Breusch-Godfrey test is used to evaluate the validity of some modeling assumptions found in applying regression-like models to observed data series (Goldfrey, 1978). When evaluated according to probability values, it was observed that the model had a normal distribution in the Jarquae-Bera test. At the same time, there is no autocorrelation between the variables at the 5% significance level and there is no problem of varying variance.

Table 10.1 Required test statistics for ARDL model

ADF unit root results					
Variables	**Level**		**First difference**		
	Constant (*t*-stat) [*p*-val.]	Constant+Trend (*t*-stat)[*p*-val.]	Constant (*t*-stat) [*p*-val.]	Constant+Trend (*t*-stat) [*p*-val.]	
lnEG	(−1.)[0.589]	(−4.251)[0.011**]	(−5.827) [0.000***]	(−5.760)[0.000***]	
lnFC	(−1.811) [0.368]	(−3.166)[0.109]	(−6.581) [0.000***]	(−6.649)[0.000***]	
lnEC	(−0.141) [0.935]	(−3.109)[0.121	(−7.766) [0.000***]	(−7.800)[0.000***]	
lnHDI	(−6.396) [0.000***]	(−0.324)[0.986]	(−0.938) [0.760]	(−5.930)[0.000***]	
lnREC	(−3.327) [0.022**]	(−1.373)[0.999]	(−0.997) [0.740]	(−4.425)[0.007***]	
lnTIN	(−3.005) [0.045**]	(−2.068)[0.541]	(−3.251) [0.026**]	(−3.810)[0.029**]	
Delay length suitable for ARDL model					
Lag	**LogL**	**AIC**	**SIC**	**HQ**	**LR**
0	374.4603	−24.63069	−24.39715	−24.55598	NA
1	535.7734	−33.72823	−32.31703*	−33.26997*	258.1011*
2	562.9384	−33.86256*	−31.29370	−33.04076	34.40895
ARDL limit test results					
K	5%			1%	
4	I(0)–I(1) 2.14–3.34			I(0)–I(1) 2.82–4.21	
Calculated F-stat	6.99***				
Diagnostic tests					
Statistics		**Obs*R-squared**		**Prob chi-square**	
Normality (Jargue-Bera)		0.350		0.839	
Autocorrelation (Breusch-Godfrey Serial Correlation LM Test)		0.047		0.828	
Varying variance (Breusch-Pagan-Godfrey)		9.514		0.3008	

Note: **, *** denotes significance at 5% and 1% significance levels, respectively. k represents the number of arguments

Table 10.2 shows the ARDL long-term coefficient results.

Table 10.2 shows the results of the selected ARDL (1, 0, 1, 0, 0, 1) model and the ARDL long-term coefficient results. In the table, the probability value of F expresses the significance of the model. In addition, R^2 states that the independent variables explain the dependent variable by 99%. According to ARDL long-term estimation results, a positive relationship was found between the dependent variable and all variables except technological innovation (lnTIN). In this context, a 1% increase in primary energy consumption (lnEC) increases economic growth (lnEG) by 0.4%. A 1% increase in physical capital (lnFC) increases economic growth by 0.6%. An

Table 10.2 ARDL (1, 0, 1, 0, 0, 1) test results

Variables	Cofficient	*t*-statistic	*p*-value
lnEG (−1)	0.634	5.241	0.000***
lnFC	0.326	10.149	0.000***
lnFC (−1)	−0.097	−1.659	0.116
lnHDI	0.691	2.057	0.051***
lnEC	0.182	2.502	0.019**
lnREC	0.011	3.930	0.000***
lnTIN	0.073	1.302	0.206
lnTIN(−1)	−0.147	−2.616	0.019**
Constant	3.660	2.047	0.015**
adjR²: 0.99			
F-Statistics: 725.76			
Prob (F-statistic): 0.000*			
ARDL long-run coefficient results			
Variables	**Cofficient**	***t*-stats**	***p*-value**
lnEC	0.499	1.896	0.070*
lnFC	0.617	6.013	0.000***
lnHDI	1.891	4.248	0.000***
lnREC	0.032	2.225	0.036**
lnTIN	−0.201	−2.416	0.024**

Note: ***, **, * indicate 1%, 5%, and 10% significance, respectively. The lag length is automatically selected as 1 based on the AIC information criterion. The selected model is ARDL (1, 0, 1, 0, 0, 1)

increase of 1% in the human capital variable constitutes an increase of 1.8% in economic growth. Likewise, there is a positive relationship between renewable energy consumption (lnREC) and economic growth, which is the dependent variable. Accordingly, a 1% increase in renewable energy consumption creates an increase of 0.03% in economic growth. However, it has been observed that there is a negative relationship between the technological innovation variable and the economic growth variable. It was concluded that a 1% increase in technological innovation led to a 0.2% decrease in economic growth.

10.4 Conclusion and Discussion

This study investigates how renewable energy consumption, technological innovation, human capital index, and physical capital affect growth in Germany, one of the largest economies in the world. The analysis period covers the years 1999–2021. The ARDL method was used as a long-term coefficient estimator. The empirical results show that all independent variables (human capital index, renewable energy consumption, primary energy consumption, and physical capital) except technological innovation have a positive effect on economic growth.

In line with the information obtained from empirical applications, the positive effect of primary energy consumption on economic growth shows that energy consumption has a supportive effect on economic growth. This result is an expected finding especially for the German economy. This result supports the argument that energy policies and energy efficiency measures have the potential to promote economic growth. The result that physical capital has increased economic growth shows that increasing infrastructure investments and means of production have a positive effect on economic growth. It supports the economic theory that physical capital accumulation can increase productive capacity and promote economic growth. The positive effect of human capital shows that investments in human capital such as education, health services, and human resources development can strengthen economic growth. The return of investments in human capital is relatively delayed compared to other production factors, but it is generally accepted that the effect of this factor has a stronger and long-term effect. Increasing human capital supports economic growth by increasing labor productivity and innovation capacity. Considering that Germany is a country known for its investments in human capital, this result came out as expected. The fact that renewable energy consumption, which is accepted as environmentally friendly, has a small positive effect on growth shows that policies toward renewable energy sources should be encouraged. In this way, it is predicted that while environmental sustainability can be increased, economic growth can also be supported. The negative impact of technological innovation on growth is an interesting and unexpected finding. However, the impact of technological innovation on economic growth can be complex. This negative relationship suggests that it may be due to high investment costs at the beginning of technological innovation or some temporary difficulties in the adaptation process of new technologies. In addition, considering the effect of technological innovation on economic growth together with other factors will enable a more comprehensive evaluation of the effects of technological innovation.

When the findings are evaluated in general, they provide some clues for policy makers. It is an important point that energy policies require an economic growth policy in line with sustainability goals. In addition, it has been shown that infrastructure investments and the development of human capital can increase economic growth. On the other hand, the effect of technological innovation on economic growth in the German economy should be emphasized more, and it should be discussed in detail how and in what direction the policies in this area can be shaped.

References

Akel, V. & Gazel, S. (2014). Döviz Kurları ile BIST Sanayi Endeksi Arasındaki Eşbütünleşme İlişkisi: Bir ARDL Sınır Testi Yaklaşımı, *Erciyes Üniversitesi İktisadi ve İdari Bilimler Fakültesi Dergisi*, Sayı: 44, Temmuz, ss. 23–41.

Alper, F. Ö. (2018). Yenilenebilir enerji ve ekonomik büyüme arasındaki ilişki: 1990-2017 Türkiye örneği. *Çankırı Karatekin Üniversitesi İktisadi ve İdari Bilimler Fakültesi Dergisi, 8*(2), 223–242.

Apaydın, Ş., Güngör, A., & Taşdoğan, C. (2019). *Türkiye'de Yenilenebilir Enerji Tüketiminin Ekonomik Büyüme Üzerindeki Asimetrik Etkileri (the asymmetric effects of renewable energy consumption on economic growth in Turkey).* Journal of Mehmet Akif Ersoy University Economics and Administrative Sciences Faculty.

Barışık, S., & Demircioğlu, E. (2006). Türkiye'de Döviz Kuru Rejimi, Konvertibilete, İhracat-İthalat İlişkisi (1980-2001). *ZKÜ Sosyal Bilimler Dergisi, 2*(3), 71–84.

De Vita, G., & Ve Abbott, A. (2002). Are saving and investment cointegrated? *An ARDL Bounds Testing Approach, Economics Letters, 77*, 293–299.

Delice, G., & Emre, B. Y. (2011). Dolaysız Yabancı Sermaye Yatırımları ve Dış Ticaret Bilançosu: Türkiye Üzerine Bir Uygulama. *Uludağ Üniversitesi İktisadi ve İdari Bilimler Fakültesi Dergisi, 30*(2), 1–28.

Dickey, D. A., & Fuller, W. A. (1979). Distribution of the estimators for autoregressive time series with a unit root. *Journal of American Statistical Association, 74*(366), 427–431.

Dickey, D. A., & Fuller, W. A. (1981). Likelihood ratio statistics for autoregressive time series with a unit root. *Econometrica, 49*(4), 1057–1072.

Erkök, B. (2022). Almanya'nın Enerji Dönüşümü Politikaları: Türkiye İçin Bir Örnek. *Fiscaoeconomia, 6*(2), 294–324.

Godfrey, L. G. (1978). Testing against general autoregressive and moving average error models when the regressors include lagged dependent variables. *Econometrica: Journal of the Econometric Society*, 1293–1301.

Gyimah, J., Yao, X., Tachega, M. A., Hayford, I. S., & Opoku-Mensah, E. (2022). Renewable energy consumption and economic growth: New evidence from Ghana. *Energy, 248*, 123559.

https://web.archive.org/web/20111121032201/http://www.tatsachen-ueber deutschland.de/en/education-and-research/main-content-07/industry-strongly-engaged-in research.html. E.T. June 1, 2023.

Li, K., & Lin, B. (2016). Impact of energy technology patents in China: Evidence from a panel 514 cointegration and error correction model. *Energy Policy, 89*, 214–223.

Lin, B., & Moubarak, M. (2014). Renewable energy consumption–economic growth nexus for China. *Renewable and Sustainable Energy Reviews, 40*, 111–117.

Lyman, R., & Economist, E. (2016). *Why renewable energy cannot replace fossil fuels by 2050.* Friends of Science Society.

Naımoglu, M. (2021). Fourier Yaklaşımıyla Yenilenebilir Enerji Tüketimi Ve Enerji Kayıplarının Ekonomik Büyüme Üzerindeki Etkisi: Almanya Örneği. *Journal of Economics and Research, 2*(1), 59–68.

Pesaran, M. H., Shin, Y., & Smith Richard, J. (2001). Bound testing approaches to the analysis of long run relationships. *Journal of Applied Econometrics, 16*, 289–326.

Rafindadi, A. A., & Ozturk, I. (2017). Impacts of renewable energy consumption on the German economic growth: Evidence from combined cointegration test. *Renewable and Sustainable Energy Reviews, 75*, 1130–1141.

Renn, O., & Marshall, J. P. (2016). Coal, nuclear and renewable energy policies in Germany: From the 1950s to the "Energiewende". *Energy Policy, 99*, 224–232.

Shahbaz, M., Loganathan, N., Zeshan, M., & Zaman, K. (2015). Does renewable energy consumption add in economic growth? An application of auto-regressive distributed lag model in Pakistan. *Renewable and Sustainable Energy Reviews, 44*, 576–585.

Shahbaz, M., Song, M., Ahmad, S., & Vo, X. V. (2022). Does economic growth stimulate energy consumption? The role of human capital and R&D expenditures in China. *Energy Economics, 105*, 105662.

Türkmen, S., Özbek, S., & Karakuş, M. (2018). Türkiye'de Elektrik Tüketimi ve Ekonomik Büyüme Arasındaki İlişki: Ampirik Bir Analiz. *Kahramanmaraş Sütçü İmam Üniversitesi İktisadi ve İdari Bilimler Fakültesi Dergisi, 8*(2), 129–142.

Turna, Y., & Ceylan, R. (2022). The relationship between economic growth and physical capital, human capital and energy consumption in Turkey: Nardl approach. *Journal Of Mehmet Akif Ersoy University Economics and Administrative Sciences Faculty.*

United Nation Development Program (UNDP). (1990). *Human development report, concept and measurement of human development*. UNDP.

World Energy 2021 of British Petroleum. (n.d.)

Yildirim, E., Saraç, Ş., & Aslan, A. (2012). Energy consumption and economic growth in the USA: Evidence from renewable energy. *Renewable and Sustainable Energy Reviews, 16*(9), 6770–6774.

Chapter 11
Methodological Tools for Assessing the Energy Efficiency of the Industrial Complex

Iman Mahmoud and Emerson Guzzi Zuan Esteves

11.1 Introduction

To determine the long-term energy needs of the industrial complex, it is necessary to establish the relationship between economic growth and an increase in energy consumption. Nowadays, in the most developed countries of the world, increased attention is paid to promising development opportunities based on the "green economy" strategy (Mikhaylov et al., 2023a, 2023b; Wang et al., 2023; He et al., 2023; Irshad et al., 2023; Iqbal et al., 2023; Aksoy et al., 2023).

As a result of the analysis of leading developments in the field of "green economy", the following indicators were adopted that characterize the transition of the production complex to the concept of "green economy" (Eti et al., 2022):

- Ratio CO of CO2 emissions to primary energy consumption (g of CO2/gofcu)
- Ratio of CO2 emissions to gross revenue in comparable prices
- Ratio CO of CO2 emissions to primary electricity consumption (CO2/kWh)
- Ratio CO of CO2 emissions to final consumption of energy resources (g of CO2/gofcu) (Martínez et al., 2023)
- Carbon intensity of products: the ratio CO of CO2 emissions to output in physical terms (g of CO2/ton of nat. units)
- Share of consumption of primary energy resources produced using their own nontraditional and renewable energy sources in the total volume of consumption of their own produced primary energy resources, relative units (%)

I. Mahmoud (✉)
Faculty of Medicine for Girls, AL-Azhar University Cairo, Cairo, Egypt

E. G. Z. Esteves
The State University of Londrina, Paraná, Brazil
e-mail: emerson.esteves@uel.br

H. Dinçer et al. (eds.), *Decision Making in Interdisciplinary Renewable Energy Projects*, Contributions to Management Science,
https://doi.org/10.1007/978-3-031-51532-3_11

11.2 Literature Review

It is proposed to take the reduction of greenhouse gas emissions and pollutants as the main indicators of the green economy and sustainable development (Denisova et al., 2019; Nyangarika et al., 2019a, 2019b; Huang et al., 2021a, 2021b; Mikhaylov, 2018, 2022a, 2022b; Mikhaylov et al., 2019; Conteh et al., 2021; Sediqi et al., 2022, Khan et al., 2021; Bhuiyan et al., 2022a, 2022b; Liu et al., 2021a, 2021b; Daniali et al., 2021; Moiseev et al., 2023a, 2023b; An et al., 2022).

The Organization for Economic Cooperation and Development (OECD) focuses on key groups of indicative indicators to characterize the socioeconomic situation in the country, as well as the growth of the "green economy" (Bhuiyan et al., 2021; Dong et al., 2021, Mikhaylov, 2021b; Barykin et al., 2022a, 2022b; Liu et al., 2022a, 2022b; Danish et al., 2022a, 2022b; Saqib et al., 2021; Mukhametov et al., 2021, Candila et al., 2021; Mikhaylov & Grilli, 2022; Li et al., 2022).

As a result, leading experts propose the following indicators that measure the productivity of the environment and resources of the economy, which reflect the main aspects of the movement toward a low-carbon and resource-efficient economy: carbon and energy productivity – economic output per unit of CO2 emissions or TPES (Total primary energy supplied) (Khan et al., 2022; Dinçer et al., 2022a, 2022b; Badr et al., 2022; Barykin et al., 2022a, 2022b; Mehta et al., 2022; Kalinina et al., 2022; Shaikh et al., 2022; Mikhaylov et al., 2022a, 2022b, 2023; Nyangarika et al., 2022).

Reflects the energy and environmental efficiency of energy resources; resource productivity – economic output per unit of natural resource or used material; multifactor productivity. Increasing productivity due to efficient production and economic activities contributes to the creation of jobs and expansion of sales markets (Mikhaylov et al., 2022a, 2022b; Mikhaylov, 2021a; Varyash et al., 2020; Zhao et al., 2021; An & Mikhaylov, 2020, 2021; Alwaelya et al., 2021; Yumashev & Mikhaylov, 2020; Yumashev et al., 2020; Mutalimov et al., 2021).

It is also necessary to highlight energy efficiency indicators from the perspective of the "green economy," represented by indicators of the ODYSSEE project database ODYSSEE in the field of energy efficiency and the International Energy Agency (Nie et al., 2020; Dayong et al., 2020; Mikhaylov et al., 2018; Nyangarika et al., 2018; Danish et al., 2020, 2021; An et al., 2021; Uyeh et al., 2021; Tamashiro et al., 2021, 2023; Shaikh et al., 2021).

It tracks how many tons of CO2 are released per unit of energy delivered (Eti et al., 2023). The ESC indicator ESC shows only one side of the decarbonization problem: It is necessary to reduce the growth of energy demand and make energy resources cleaner (An et al., 2019a, 2019b; Mikhaylov, 2020a, 2020b, 2020c; Mikhaylov & Tarakanov, 2020; An et al., 2020a, 2020b, 2020c; Moiseev et al., 2020, 2021; Gura et al., 2022; Dooyum et al., 2020; Mikhaylov, 2020a, 2020b, 2020c, 2023; Dinçer et al., 2023a, 2023b; Ivleva et al., 2023; Aissou et al., 2023).

11.3 Results

After analyzing the relevant literature, the author proposes a system of indicators for assessing the energy efficiency of industrial complexes, divided into three levels of the production complex:

1. The level of the production complex as a whole (1)
2. The level of certain types of products produced in the production complex (2)
3. The level of the technological process for the production of products (3)

PC energy efficiency indicators can be measured as a whole, as well as an analysis of each production process. Such an analysis is extremely necessary, since the consideration of each process will allow us to assess its impact on the resulting energy efficiency assessment without taking into account the mutual influence of these processes.

It is proposed to consider the following sequence of production processes:

- Preparation of the material
- Production of semi-finished products
- Additional processing
- Final processing

It should be remembered that in world practice, when determining the energy intensity of products, the indicator of the volume of secondary fuel and energy resources (final energy consumption) used for production is widely used. The value of primary energy includes losses associated with the conversion of fuel to electricity, as well as losses during transmission and distribution of electricity.

The proposed methodological approach uses indicators calculated using energy consumption and power consumption for both primary energy resources and secondary (final) energy resources.

Within the framework of the developed methodology, the author offers a fairly wide set of indicators for assessing the energy efficiency of the production complex. As an example, the block structure and individual PC energy efficiency indicators for the level of the production complex as a whole are shown below, which is key in terms of evaluating the resulting indicators of its energy efficiency. At this level, energy efficiency indicators are formed in the context of three blocks:

- Energy intensity of the gross revenue of the production complex by primary energy consumption in comparable prices
- Electrical capacity of the gross revenue of the production complex by primary energy consumption
- Energy intensity of the gross revenue of the production complex by final (secondary) energy consumption in comparable prices
- Electrical capacity of the gross revenue of the production complex by final (secondary) energy consumption
- The level of heat losses in the heating networks of the production complex, %

- Share of consumption of primary energy resources produced using its own non-traditional and renewable energy sources in the total volume of consumption of its own primary energy resources produced by the production complex, relative units (%)
- Integral indicator of specific energy consumption for the output of the production complex, rel. units

Economic efficiency of energy consumption and efficiency of use of fixed assets include the following indicators:

- Share of expenses for fuel and energy resources in the gross revenue of the production complex, rel. units
- Coefficient of advance of electric energy consumption in relation to the growth of the production volume of the production complex, rel. units
- Coefficient of advance of consumption of primary energy resources in relation to the growth of the production volume of the production complex, rel. units
- Share of expenditures on measures aimed at improving energy efficiency and energy saving in the gross revenue of the production complex, rel. units (%)
- Integrated indicator of energy saving potential by product types of the production complex, identified during energy analysis or comparison with the best values of specific energy consumption of a similar product type, relative unit's environmental efficiency of fuel and energy resources consumption, which is estimated by the following indicators
- The ratio CO of CO2 emissions to the primary consumption of energy resources in the entire production complex, CO2/g cu. t
- The ratio CO of CO2 emissions to primary electricity consumption in the entire production complex, CO2/kWh
- The ratio CO of CO2 emissions to the final consumption of energy resources in the whole production complex, CO of CO2/g cu. t
- Carbon intensity of production of the production complex, rel. units (Dinçer et al., 2023a, 2023b, 2023c, 2023d, 2023e, 2023f)

The following are the stages of assessing the energy efficiency of a production complex:

1. Assessment of the energy efficiency of production in general, taking into account the interrelationships of various processes and/or systems.
2. Evaluate opportunities for optimizing individual processes and/or systems.
3. Evaluate the possibilities for optimizing individual components and components of systems (e.g., electric motors, pumps, valves).

Determining the energy efficiency of a production complex is not an easy task, since:

- A complex production complex includes more than one production unit or technological cycle.
- The production complex produces various types of products, each of which is characterized by its own energy intensity;

- It is necessary to take into account auxiliary production facilities that do not produce final products, heating, lighting, and the own needs of production complexes. When calculating the values of energy intensity indicators for manufacturing products (products), the consumption of fuel and energy resources is taken into account only for the main and auxiliary production processes (Carayannis et al., 2023). The consumption of fuel and energy resources for heating, lighting, and various own and other needs is not included in the volume of costs when calculating the values of energy intensity indicators (Yüksel et al., 2023). In this study, a system of indicators is proposed, divided into three levels of the production complex (Xu et al., 2023).

1. The level of the production complex as a whole (1).
2. The level of certain types of products produced by the production complex (2).
3. The level of the technological process for the production of products (3).

The following main blocks can be identified in the approach implementation scheme:

1. Block for multifactor multilevel assessment of PC energy efficiency
2. Block for selecting a portfolio of measures and energy-efficient PC development projects
3. Block for forming predictive strategies for energy-efficient PC development aimed at ensuring competitiveness and achieving long-term competitive advantages of PCs in the forecast period (Han et al., 2023)

The methodological approach to conducting a multifactor assessment of PC energy efficiency (block 1) is based on the use of a block-based multilevel system of indicators that characterize various aspects of PC energy efficiency in comparison with leading competitors (Kou et al., 2023a, 2023b). The results of the assessment allow us to identify shortcomings in the development of PCs in terms of energy efficiency and develop measures to overcome them. The final result of the block is a comparative assessment of the PC's energy efficiency relative to the leading competitors, as well as leading achievements in the field of energy efficiency in relation to the core economic activities in which the PC in question operates (Moiseev et al., 2023a, 2023b).

At the same time, when conducting a detailed analysis of the PC's energy efficiency, a three-level assessment is performed, in accordance with the levels indicated above. At the same time, it is often quite sufficient to evaluate only at the level of the production complex as a whole (Yüksel & Dinçer, 2023).

The result (output) of the PC energy efficiency assessment process will be strategic (target) guidelines for the development of the PC and its components from the point of view of energy efficiency, which, in turn, along with the results of the PC energy efficiency assessment, are input information for the second block is the block for selecting a portfolio of measures and energy-efficient PC development projects. Such selection is performed on the basis of solving an optimization problem under resource constraints (Wan et al., 2023).

11.4 Conclusion

The hierarchy analysis method is a systematic procedure for hierarchical representation of elements that define the essence of any problem. The essence of the method consists in decomposing the problem into increasingly simple components and further processing the sequence of judgments of the decision maker using paired comparisons. As a result, the relative degree (intensity) of interaction between elements in the hierarchy can be expressed (Ai et al., 2023). These judgments are then expressed numerically. The hierarchy analysis method includes procedures for synthesizing multiple judgments, obtaining priority criteria, and finding alternative solutions. Basic values of energy efficiency indicators can be obtained both within the complex and in comparison, with competitors: These are either the values of similar indicators of the main competitors, or targets (Niu et al., 2023). Thus, we can distinguish two approaches to determining the basic values of energy efficiency indicators:

1. Absolute values (energy efficiency indicators of the main competitors or targets).
2. Relative values: indicators for certain years are taken as basic indicators, or in accordance with the abovementioned methodology, sample statistical characteristics of the variational series of these indicators are used (Aksoy et al., 2023).

References

Ai, R., Zheng, Y., Yüksel, S., & Dinçer, H. (2023). Investigating the components of fintech ecosystem for distributed energy investments with an integrated quantum spherical decision support system. *Financial Innovation, 9*(1), 27.

Aissou, A., Ahmed, H. S., Hussein, M. E., Hassan, M. R., Mikhaylov, A., & Mahmoud, E. (2023). Optimizing reliability of flow networks using RWGA subject to total capacity constraint. *Applied Mathematics and Information Sciences, 7*(4), 559–565. https://doi.org/10.18576/amis/170404

Aksoy, T., Yüksel, S., Dınçer, H., Hacioglu, U., Mikhaylov, A., & Muyeen, S. M. (2023). Assessing the electricity production capacities of emerging markets for the sustainable investment. *IEEE Access, 11*, 24574–24590.

Alwaelya, S. A., Yousif, N. B. A., & Mikhaylov, A. (2021). Emotional development in preschoolers and socialization. *Early Child Development and Care, 191*, 16.

An, J., & Mikhaylov, A. (2020). Russian energy projects in South Africa. *Journal of Energy in Southern Africa, 31*(3), 58–64.

An, J., & Mikhaylov, A. (2021). Analysis of energy projects financial efficiency and renewable energy generation in Russia. *Finance: Theory and Practice, 25*(5), 79–91.

An, J., Mikhaylov, A., Dinçer, H., & Yüksel, S. (2022). Economic modelling of electricity generation: Long short-term memory and Q-rung orthopair fuzzy sets. *Heliyon, 8*, e10759.

An, J., Mikhaylov, A., & Jung, S.-U. (2021). A linear programming approach for Robust network revenue management in the airline industry. *Journal of Air Transport Management, 91*, 101979.

An, J., Mikhaylov, A., & Sokolinskaya, N. (2019a). Machine learning in economic planning: Ensembles of algorithms. *Journal of Physics: Conference Series, 1353*, 012126.

An, J., Mikhaylov, A., & Moiseev, N. (2019b). Oil price predictors: Machine learning approach. *International Journal of Energy Economics and Policy, 9*(5), 1–6.

An, J., Mikhaylov, A., & Richter, U. H. (2020a). Trade war effects: Evidence from sectors of energy and resources in Africa. *Heliyon, 6*, e05693. https://doi.org/10.1016/j.heliyon.2020.e05693

An, J., Mikhaylov, A., & Jung, S.-U. (2020b). The strategy of South Korea in the global oil market. *Energies, 13*(10), 2491.

An, J., Mikhaylov, A., & Kim, K. (2020c). Machine learning approach in heterogeneous group of algorithms for transport safety-critical system. *Applied Sciences, 10*(8), 2670.

Badr, A. M. N. A. A., Mohamed, T. I., Osman, N. A. Q., & Mikhaylov, A. (2022). A review of social media website users' interaction paths with governmental accounts during the COVID-19 pandemic. *Informatics, 9*, 50. https://doi.org/10.3390/informatics9030050

Barykin, S. E., Kapustina, I. V., Sergeev, S. M., Daniali, S. M., Kopteva, L. A., Semenova, G. N., Pryadko, I. P., Mikhaylov, A., Baboshkin, P., Datsyuk, P., & Senjyu, T. (2022a). Financial logistics models based on systematic approach improving management solutions. *F1000Research, 11*, 570. https://doi.org/10.12688/f1000research.111252.1

Barykin, S., Mikheev, A., Kiseleva, E., Putikhin, Y., Alekseeva, N., & Mikhaylov, A. (2022b). An empirical analysis of Russian regions' debt sustainability. *Economies, 10*(4), 296.

Bhuiyan, M. A., An, J., Mikhaylov, A., Moiseev, N., & Danish, M. S. S. (2021). Renewable energy deployment and COVID-19 measures for sustainable development. *Sustainability, 13*, 4418.

Bhuiyan, M. A., Dinçer, H., Yüksel, S., Mikhaylov, A., Danish, M. S. S., Pinter, G., Uyeh, D. D., & Stepanova, D. (2022a). Economic indicators and bioenergy supply in developed economies: QROF-DEMATEL and random forest models. *Energy Reports, 8*(2022), 561–570.

Bhuiyan, M. A., Zhang, Q., Khare, V., Mikhaylov, A., Pinter, G., & Huang, X. (2022b). Renewable energy consumption and economic growth nexus—a systematic literature review. *Frontiers in Environmental Science., 10*, 878394. https://doi.org/10.3389/fenvs.2022.878394

Candila, V., Maximov, D., Mikhaylov, A., Moiseev, N., Senjyu, T., & Tryndina, N. (2021). On the relationship between oil and exchange rates of oil-exporting and oil-importing countries: From the great recession period to the Covid-19 era. *Energies, 14*, 8065.

Carayannis, E., Kostis, P., Dinçer, H., & Yüksel, S. (2023). Quality function deployment-oriented strategic outlook to sustainable energy policies based on quintuple innovation helix. *Journal of the Knowledge Economy*, 1–19.

Conteh, F., Takahashi, H., Hemeida, A. M., Krishnan, N., Mikhaylov, A., & Senjyu, T. (2021). Hybrid grid-connected renewable power generation for sustainable electricity supply in Sierra Leone: Case study Lungi Town. *Sustainability, 13*(18).

Daniali, S. M., Barykin, S. E., Kapustina, I. V., Khortabi, F. M., Sergeev, S. M., Kalinina, O. V., Mikhaylov, A., Veynberg, R., Zasova, L., & Senjyu, T. (2021). Predicting volatility index according to technical index and economic indicators on the basis of deep learning algorithm. *Sustainability, 13*, 14011.

Danish, M. S. S., Bhattacharya, A., Stepanova, D., Mikhaylov, A., Grilli, M. L., Khosravy, M., & Senjyu, T. (2020). A systematic review of metal oxide applications for energy and environmental sustainability. *Metals, 10*(12), 1604.

Danish, M. S. S., Senjyu, T., Sabory, N. R., Khosravy, M., Grilli, M. L., Mikhaylov, A., & Majidi, H. (2021). A forefront framework for sustainable aquaponics modeling and design. *Sustainability, 13*, 9313.

Danish, M. S., Senjyu, T. S., Sabory, N. R., & Mikhaylov, A. (2022a). Lessons from COVID-19 conferring environmental re-engineering opportunity. In M. Danish & T. Senjyu (Eds.), *Eco-friendly and agile energy strategies and policy development* (pp. 99–118. IGI Global). https://doi.org/10.4018/978-1-7998-9502-2.ch005

Danish, M. S., Senjyu, T. S., Sabory, N. R., & Mikhaylov, A. (2022b). Interplay consequences of COVID-19 on global environmental sustainability. In M. Danish & T. Senjyu (Eds.), *Eco-friendly and agile energy strategies and policy development*. IGI Global. https://doi.org/10.4018/978-1-7998-9502-2.ch003

Dayong, N., Mikhaylov, A., Bratanovsky, S., Shaikh, Z. A., & Stepanova, D. (2020). Mathematical modeling of the technological processes of catering products production. *Journal of Food Process Engineering, 43*(2), 1–9.

Denisova, V., Mikhaylov, A., & Lopatin, E. (2019). Blockchain infrastructure and growth of global power consumption. *International Journal of Energy Economics and Policy, 9*(4), 22–29.

Dinçer, H., Yüksel, S., Mikhaylov, A., Pinter, G., & Shaikh, Z. A. (2022a). Analysis of renewable-friendly smart grid technologies for the distributed energy investment projects using a hybrid picture fuzzy rough decision-making approach. *Energy Reports, 8*, 11466–11477. https://doi.org/10.1016/j.egyr.2022.08.275

Dinçer, H., Yüksel, S., Mikhaylov, A., Barykin, S. E., Aksoy, T., & Hacioğlu, U. (2022b). Analysis of environmental priorities for green project investments using an integrated q-rung orthopair fuzzy modelling. *IEEE Access, 10*, 50996–51007. https://doi.org/10.1109/ACCESS.2022.3174058

Dinçer, H., Yüksel, S., Hacioglu, U., Yilmaz, M. K., & Delen, D. (2023a). Development of a sustainable corporate social responsibility index for performance evaluation of the energy industry: A hybrid decision-making methodology. *Resources Policy, 85*, 103940.

Dinçer, H., Eti, S., Gökalp, Y., Yüksel, S., & Çelebi, B. (2023b). The role of small-scale wind turbines in increasing the competitiveness of hospitals. In *Handbook of research on quality and competitiveness in the healthcare services sector* (pp. 336–352). IGI Global.

Dinçer, H., Yüksel, S., Mikhaylov, A., Muyeen, S. M., Chang, T., Barykin, S., & Kalinina, O. (2023c). CO2 emissions integrated fuzzy model: A case of seven emerging economies. *Energy Reports, 9*, 5741–5751.

Dinçer, H., Yüksel, S., Aksoy, T., Hacıoğlu, Ü., Mikhaylov, A., & Pinter, G. (2023d). Analysis of solar module alternatives for efficiency-based energy investments with hybrid 2-tuple IVIF modeling. *Energy Reports, 10*, 61–71.

Dinçer, H., Eti, S., Yüksel, S., Özdemir, S., Yflmaz, A. E., & Ergün, E. (2023e). Integrating data mining and fuzzy decision-making techniques for analyzing the key minimizing factors of carbon emissions. *Journal of Intelligent & Fuzzy Systems, 45*, 1–17.

Dinçer, H., Eti, S., Aksoy, T., Yüksel, S., Hacioglu, U., Mikhaylov, A., & Muyeen, S. M. (2023f). Analysis of environmental impact for material production investments using a novel soft computing methodology. *IEEE Access, 11*, 37987–38001. https://doi.org/10.1109/ACCESS.2023.3266524

Dong, B., Ikonnikova, I., Rogulin, R., Sakulyeva, T., & Mikhaylov, A. (2021). Environmental economic approach to optimization of transport communication in megacities. *Journal of Environmental Science and Health, Part A, 56*(6), 660–666.

Dooyum, U. D., Mikhaylov, A., & Varyash, I. (2020). Energy security concept in Russia and South Korea. *International Journal of Energy Economics and Policy, 10*(4), 102–107.

Eti, S., Dinçer, H., Gökalp, Y., Yüksel, S., & Kararoğlu, D. (2023). Identifying key issues to handle the inflation problem in the healthcare industry caused by energy prices: An evaluation with decision-making models. In *Managing inflation and supply chain disruptions in the global economy* (pp. 162–178).

Eti, S., Dinçer, H., Yüksel, S., & Gökalp, Y. (2022). Analysis of the suitability of the solar panels for hospitals: A new fuzzy decision-making model proposal with the T-spherical TOP-DEMATEL method. *Journal of Intelligent & Fuzzy Systems, Preprint*, 1–13.

Gura, D., Mikhaylov, A., Glushkov, S., Zaikov, M., & Shaikh, Z. A. (2022). Model for estimating power dissipation along the interconnect length in single on-chip topology. *Evolutionary Intelligence, 15*, 2369–2373. https://doi.org/10.1007/s12065-020-00407-7

Han, Z., Gong, L., Chen, H., & Yüksel, S. (2023). The role of mineral and oil resources in achieving a clean environment in emerging economies: An asymmetric analysis. *Resources Policy, 86*, 104099.

He, H., Huang, Y., Nakadomari, A., Masrur, H., Krishnan, N., Hemeida, A. M., Mikhaylov, A., Omine, E., & Senjyu, T. (2023). Potential and economic viability of green hydrogen production from seawater electrolysis using renewable energy in remote Japanese islands. *Renewable Energy, 202*, 1436–1447. https://doi.org/10.1016/j.renene.2022.12.046

Huang, Y., Masrur, H., Shigenobu, R., Hemeida, A. M., Mikhaylov, A., & Senjyu, T. (2021a). A comparative design of a campus microgrid considering a multi-scenario and multi-objective approach. *Energies, 14*, 2853.

Huang, Y., Yona, A., Takahashi, H., Hemeida, A. M., Mandal, P., Mikhaylov, A., Senjyu, T., & Lotfy, M. E. (2021b). Energy management system optimization of drug store operate electric vehicles charging station. *Sustainability, 14*, 1217836.

Iqbal, K. M. J., Khan, M. I., Mikhaylov, A., Shah, A. A., Yadykin, V., Filho, W. L., Tariq, M. A. U. R., & Ullah, W. (2023). Modeling principles, criteria and indicators to assess water sector governance for climate compatibility and sustainability. *Frontiers in Environmental Science, 11*, 989930.

Irshad, A. S., Ludin, G. A., Masrur, H., Ahmadi, M., Yona, A., Mikhaylov, A., Krishnan, N., & Senjyu, T. (2023). Optimization of grid-photovoltaic and battery hybrid system with most technically efficient PV technology after the performance analysis. *Renewable Energy, 207*, 714–730.

Ivleva, S., Hajimajeed, M., Kartashova, O., & Mikhaylov, A. (2023). A comparative study of regional variations in common visual impairments among school-age children. *Electronic Journal of General Medicine, 20*(5), em504. https://doi.org/10.29333/ejgm/13264

Kalinina, O. V., Barykin, S. E., Sergeev, S. M., Semenova, G. N., Fatkullina, A., Mikhaylov, A., & De La Poza, E. (2022). Smart city perspectives in post-pandemic governance: Externalities reduction policy. *F1000Research, 11*, 1032. https://doi.org/10.12688/f1000research.123195.1

Khan, A. A., Laghari, A. A., Gadekallu, T. R., Shaikh, Z. A., Javed, A. R., Rashid, M., Estrela, V. V., & Mikhaylov, A. (2022). A drone-based data management and optimization using metaheuristic algorithms and blockchain smart contracts in a secure fog environment. *Computers and Electrical Engineering, 102*, 108234. https://doi.org/10.1016/j.compeleceng.2022.108234

Khan, A. A., Shaikh, Z. A., Baitenova, L., Mutaliyeva, L., Moiseev, N., Mikhaylov, A., Laghari, A. A., Idris, S. A., & Alshazly, H. (2021). QoS-ledger: Smart contracts and metaheuristic for secure quality-of-service and cost-efficient scheduling of medical-data processing. *Electronics, 10*, 3083.

Kou, G., Pamucar, D., Dinçer, H., & Yüksel, S. (2023a). From risks to rewards: A comprehensive guide to sustainable investment decisions in renewable energy using a hybrid facial expression-based fuzzy decision-making approach. *Applied Soft Computing, 142*, 110365.

Kou, G., Yuksel, S., & Dincer, H. (2023b). A facial expression and expert recommendation fuzzy decision-making approach for sustainable business investments within the metaverse world. *Applied Soft Computing*, 110849.

Li, J., Dinçer, H., Yüksel, S., Mikhaylov, A., & Barykin, S. E. (2022). Bipolar q-ROF hybrid decision making model with golden cut for analyzing the levelized cost of renewable energy alternatives. *IEEE Access, 10*, 35900. https://doi.org/10.1109/ACCESS.2022.3168315

Liu, L., Kato, T., Mandal, P., Mikhaylov, A., Hemeida, A. M., & Senjyu, T. (2021a). Multi-area power system load frequency control by using model predictive control by introducing renewable energy sources. In *2021 International conference on science & contemporary technologies (ICSCT)* (pp. 1–6). https://doi.org/10.1109/ICSCT53883.2021.9642521

Liu, L., Kato, T., Mandal, P., Mikhaylov, A., Hemeida, A. M., Takahashi, H., & Senjyu, T. (2021b). Two cases studies of model predictive control approach for hybrid renewable energy systems. *AIMS Energy, 9*(6), 1241–1259.

Liu, Z.-J., Panfilova, E., Mikhaylov, A., & Kurilova, A. (2022a). COVID-19 crisis impact on the stability between parties in crowdfunding and crowdsourcing. *Wireless Personal Communications, 122*, 915–930.

Liu, Z., Panfilova, E., Mikhaylov, A., & Kurilova, A. (2022b). Assessing stability in the relationship between parties in crowdfunding and crowdsourcing projects during the COVID-19 crisis. *Journal of Global Information Management (JGIM), 30*(4), 1–18. https://doi.org/10.4018/JGIM.297905

Martínez, L., Dinçer, H., & Yüksel, S. (2023). A hybrid decision making approach for new service development process of renewable energy investment. *Applied Soft Computing, 133*, 109897.

Mehta, R., Verghese, J., Mahajan, S., Barykin, S., Bozhuk, S., Kozlova, N., Kapustina, I., Mikhaylov, A., Naumova, E., & Dedyukhina, N. (2022). Consumers' behavior in conversational commerce marketing based on messenger chatbots. *F1000Research, 11*(647). https://doi.org/10.12688/f1000research.122037.1

Mikhaylov, A. (2018). The yield of the carry trade strategy. *Finance: Theory and Practice, 22*(3), 52–63. https://doi.org/10.26794/2587-5671-2018-22-3-52-63

Mikhaylov, A. (2020a). Cryptocurrency market analysis from the open innovation perspective. *Journal of Open Innovation: Technology, Market, and Complexity, 6*(4), 197.

Mikhaylov, A. (2020b). Geothermal energy development in Iceland. *International Journal of Energy Economics and Policy, 10*(4), 31–35.

Mikhaylov, A. (2020c). Cryptocurrency market development: Hurst method. *Finance: Theory and Practice, 24*(3), 81–91.

Mikhaylov, A. (2021a). Development of Friedrich von Hayek's theory of private money and economic implications for digital currencies. *Terra Economicus, 19*(1), 53–62.

Mikhaylov, A. (2021b). Lichens as indicators of atmospheric pollution in urban ecosystems. *Israel Journal of Ecology & Evolution, 67*(1–2), 60–68. https://doi.org/10.1163/22244662-bja10016

Mikhaylov, A. (2022a). Efficiency of renewable energy plants in Russia. *Anais da Academia Brasileira de Ciências, 94*(4), e20191226. https://doi.org/10.1590/0001-3765202220191226

Mikhaylov, A. (2022b). Sustainable development and renewable energy: A new view to a global problem. *Energies, 15*, 1397. https://doi.org/10.3390/en15041397

Mikhaylov, A. (2023). Understanding the risks associated with wallets, depository services, trading, lending, and borrowing in the crypto space. *Journal of Infrastructure, Policy and Development, 6*, 2.

Mikhaylov, A., Bhatti, I. M., Dinçer, H., & Yüksel, S. (2022a). Integrated decision recommendation system using iteration-enhanced collaborative filtering, golden cut bipolar for analyzing the risk-based oil market spillovers. *Computational Economics*. https://doi.org/10.1007/s10614-022-10341-8

Mikhaylov, A., Yumashev, A., & Kolpak, E. (2022b). Quality of life, anxiety and depressive disorders in patients with extrasystolic arrhythmia. *Archives of Medical Science, 18*(2). https://doi.org/10.5114/aoms.2020.101359

Mikhaylov, A., Burakov, D., & Didenko, V. (2019). Relationship between oil price and macroeconomic indicators in Russia. *Finance: Theory and Practice, 23*(2), 105–116. https://doi.org/10.26794/2587-5671-2019-23-2-105-116

Mikhaylov, A., Dinçer, H., Yüksel, S., Pinter, G., & Shaikh, Z. A. (2023a). Bitcoin mempool growth and trading volumes: Integrated approach based on QROF multi-SWARA and aggregation operators. *Journal of Innovation & Knowledge, 8*(3), 100378.

Mikhaylov, A., Dinçer, H., & Yüksel, S. (2023b). Analysis of financial development and open innovation oriented fintech potential for emerging economies using an integrated decision-making approach of MF-X-DMA and golden cut bipolar q-ROFSs. *Financial Innovation, 9*(1), 10. https://doi.org/10.1186/s40854-022-00234-w

Mikhaylov, A., & Grilli, M. L. (2022). Machine learning methods and sustainable development: Metal oxides and multilayer metal-oxides. *Metals, 12*(4).

Mikhaylov, A., Sokolinskaya, N., & Nyangarika, A. (2018). Optimal carry trade strategy based on currencies of energy and developed economies. *Journal of Reviews on Global Economics, 7*, 582–592.

Mikhaylov, A., & Tarakanov, S. (2020). Development of Levenberg-Marquardt theoretical approach for electric network. *Journal of Physics: Conference Series, 1515*, 052006.

Moiseev, N., Mikhaylov, A., Dinçer, H., & Yüksel, S. (2023a). Effects of market capitalization shocks for open innovation models in e-commerce industry: Golden cut QROF MCDM analysis. *Financial Innovation, 9*(1), 32.

Moiseev, N., Mikhaylov, A., Dinçer, H., & Yüksel, S. (2023b). Market capitalization shock effects on open innovation models in e-commerce: Golden cut q-rung orthopair fuzzy multicriteria decision-making analysis. *Financial Innovation, 9*(1), 55.

Moiseev, N., Mikhaylov, A., Varyash, I., & Saqib, A. (2020). Investigating the relation of GDP per capita and corruption index. *Entrepreneurship and Sustainability Issues, 8*(1), 780–794.

Moiseev, N., Sorokin, A., Zvezdina, N., Mikhaylov, A., Khomyakova, L., & Danish, M. S. S. (2021). Credit risk theoretical model on the base of DCC-GARCH in time-varying parameters framework. *Mathematics, 9*, 2423. https://doi.org/10.3390/math9192423

Mukhametov, A., Bekhorashvili, A., Avdeenko, A., & Mikhaylov, A. (2021). The impact of growing legume plants under conditions of biologization and soil cultivation on Chernozem fertility and productivity of rotation crops. *Legume Research, 44*, 12.

Mutalimov, V., Kovaleva, I., Mikhaylov, A., & Stepanova, D. (2021). Assessing regional growth of small business in Russia. *Entrepreneurial Business and Economics Review, 9*(3), 119–133. https://doi.org/10.15678/EBER.2021.090308

Nie, D., Panfilova, E., Samusenkov, V., & Mikhaylov, A. (2020). E-learning financing models in Russia for sustainable development. *Sustainability, 12*(11), 4412.

Niu, X., Yüksel, S., & Dinçer, H. (2023). Emission strategy selection for the circular economy-based production investments with the enhanced decision support system. *Energy, 274*, 127446.

Nyangarika, A., Mikhaylov, A., Muyeen, S. M., Yadykin, V., Mottaeva, A., Pryadko, I., Barykin, S., Fomenko, N., Rykov, G., & Shvandar, K. (2022). Energy stability and decarbonization in developing countries: Random Forest approach for forecasting of crude oil trade flows and macro indicators. *Frontiers in Environmental Science, 10*, 1031343. https://doi.org/10.3389/fenvs.2022.1031343

Nyangarika, A., Mikhaylov, A., & Richter, U. (2019a). Influence oil price towards economic indicators in Russia. *International Journal of Energy Economics and Policy, 9*(1), 123–129.

Nyangarika, A., Mikhaylov, A., & Richter, U. (2019b). Oil price factors: Forecasting on the base of modified auto-regressive integrated moving average model. *International Journal of Energy Economics and Policy, 9*(1), 149–159.

Nyangarika, A., Mikhaylov, A., & Tang, B.-J. (2018). Correlation of oil prices and gross domestic product in oil producing countries. *International Journal of Energy Economics and Policy, 8*(5), 42–48.

Saqib, A., Chan, T.-H., Mikhaylov, A., & Lean, H. H. (2021). Are the responses of sectoral energy imports asymmetric to exchange rate volatilities in Pakistan? Evidence from recent foreign exchange regime. *Frontiers in Energy Research, 9*, 614463. https://doi.org/10.3389/fenrg.2021.614463

Sediqi, M. M., Nakadomari, A., Mikhaylov, A., Krishnan, N., Lotfy, M. E., Yona, A., & Senjyu, T. (2022). Impact of time-of-use demand response program on optimal operation of Afghanistan real power system. *Energies, 15*, 296.

Shaikh, Z. A., Kraikin, A., Mikhaylov, A., & Pinter, G. (2022). Forecasting stock prices of companies producing solar panels using machine learning methods. *Complexity, 2022*, 9186265. https://doi.org/10.1155/2022/9186265

Shaikh, Z. A., Moiseev, N., Mikhaylov, A., & Yüksel, S. (2021). Facile synthesis of copper oxide-cobalt oxide/nitrogen-doped carbon (Cu2O-Co3O4/CN) composite for efficient water splitting. *Applied Sciences, 11*, 9974. https://doi.org/10.3390/app11219974

Tamashiro, K., Alharbi, T., Mikhaylov, A., Hemeida, A. M., Krishnan, N., Lotfy, M. E., & Senjyu, T. (2021). Investigation of home energy management with advanced direct load control and optimal scheduling of controllable loads. *Energies, 14*, 7314. https://doi.org/10.3390/en14217314

Tamashiro, K., Omine, E., Krishnan, N., Mikhaylov, A., Hemeida, A. M., & Senjyu, T. (2023). Optimal components capacity based multi-objective optimization and optimal scheduling based MPC-optimization algorithm in smart apartment buildings. *Energy & Buildings, 278*, 112616. https://doi.org/10.1016/j.enbuild.2022.112616

Uyeh, D. D., Asem-Hiablie, S., Park, T., Kim, K.-M., Mikhaylov, A., Woo, S., & Ha, Y. (2021). Could Japonica rice be an alternative variety for increased global food security and climate change mitigation? *Food, 10*, 1869. https://doi.org/10.3390/foods10081869

Varyash, I., Mikhaylov, A., Moiseev, N., & Aleshin, K. (2020). Triple bottom line and corporate social responsibility performance indicators for Russian companies. *Entrepreneurship and Sustainability Issues, 8*(1), 313–331.

Wan, Q., Miao, X., Wang, C., Dinçer, H., & Yüksel, S. (2023). A hybrid decision support system with golden cut and bipolar q-ROFSs for evaluating the risk-based strategic priorities of fintech lending for clean energy projects. *Financial Innovation, 9*(1), 1–25.

Wang, Q., Guo, J., Li, R., Mikhaylov, A., & Moiseev, N. (2023). Does technical assistance alleviate energy poverty in sub-Saharan African countries? A new perspective on spatial spillover effects of technical assistance. *Energy Strategy Reviews, 45*, 101047. https://doi.org/10.1016/j.esr.2022.101047

Xu, X., Yüksel, S., & Dinçer, H. (2023). An integrated decision-making approach with golden cut and bipolar q-ROFSs to renewable energy storage investments. *International Journal of Fuzzy Systems, 25*(1), 168–181.

Yüksel, S., & Dinçer, H. (2023). Sustainability analysis of digital transformation and circular industrialization with quantum spherical fuzzy modeling and golden cuts. *Applied Soft Computing, 138*, 110192.

Yüksel, S., Dinçer, H., & Uluer, G. S. (2023). Socio-economic indicators of nuclear energy investments: A study on countries. In *Inclusive developments through socio-economic indicators: New theoretical and empirical insights* (pp. 33–44). Emerald Publishing Limited.

Yumashev, A., & Mikhaylov, A. (2020). Development of polymer film coatings with high adhesion to steel alloys and high wear resistance. *Polymer Composites, 41*(7), 2875–2880.

Yumashev, A., Ślusarczyk, B., Kondrashev, S., & Mikhaylov, A. (2020). Global indicators of sustainable development: Evaluation of the influence of the human development index on consumption and quality of energy. *Energies, 13*, 2768.

Zhao, Q., Cherkasov, S., Avdeenko, A., Kondratenko, L., & Mikhaylov, A. (2021). Integral estimate of the added value of agricultural crops in the synergetic agriculture on the example of Vicia faba (Vicia faba L.). *Polish Journal of Environmental Studies, 30*(5), 1–9. https://doi.org/10.15244/pjoes/131949

Chapter 12
Conceptual and Methodological Foundations of Multi-Criteria Selection of Priority Projects Energy-Efficient Projects for the Development of Large Industrial Complexes

Kanato Tamashiro, Raya Karlibaeva, and Diana Stepanova

12.1 Introduction

In practice, there is often a situation where there are several alternative options for achieving the required level of energy efficiency of PC operations. At the same time, all alternative options are equally economical (from the point of view of financial and economic indicators, it is impossible to give preference to any project) and financially feasible (there is a possibility of financing the project and achieving positive financial results) (Aksoy et al., 2023). As a result, in such conditions, the main criterion for the effectiveness of implemented projects is their impact on the comprehensive assessment of PC energy efficiency in terms of achieving the maximum positive effect expressed by the energy efficiency indicators of PC activities in the forecast period (Eti et al., 2022). Therefore, the ultimate goal of the proposed multi-criteria selection of priority energy-efficient projects is to determine a comprehensive assessment of the energy efficiency of the PC in the forecast period for each of the alternative options of the implemented projects and select the one that is characterized by the best values of such an assessment (Niu et al., 2023).

If there are no alternative options, then the forecast value of the integrated energy efficiency assessment of the PC is used to judge the feasibility of implementing the

K. Tamashiro (✉)
University of the Ryukyus, Nishihara, Japan
e-mail: k218483@eve.u-ryukyu.ac.jp

R. Karlibaeva
Tashkent State University of Economics, Tashkent, Uzbekistan

D. Stepanova
Plekhanov Russian University of Economics, Moscow, Russian Federation

H. Dinçer et al. (eds.), *Decision Making in Interdisciplinary Renewable Energy Projects*, Contributions to Management Science,
https://doi.org/10.1007/978-3-031-51532-3_12

submitted project, comparing this value with the value of the assessment at the current stage as well as its dynamics in the forecast period (Ai et al., 2023).

12.2 Literature Review

Variant assessment of the impact of the implementation of PEP on PC energy efficiency indicators. The assessment takes into account the impact on the PC energy efficiency indicators of each of the implemented projects as well as the possible effect of their complex impact. At the same time, to obtain the initial characteristics of such an influence, the following can be used (Wan et al., 2023):

- Direct assessment of impact indicators based on the technical and economic characteristics of individual projects and their complex (Bhuiyan et al., 2021, 2022b; Dong et al., 2021, Mikhaylov, 2021; Barykin et al., 2022a; Liu et al., 2022a, b; Danish et al., 2022a, b; Saqib et al., 2021; Mukhametov et al., 2021; Candila et al., 2021; Mikhaylov & Grilli, 2022, Li et al., 2022).
- Use of domestic and foreign experience in implementing similar projects and the results of their impact on the energy efficiency indicators of PC activities using the methods of similarity theory and expert analysis (Khan et al., 2022; Dinçer et al., 2022a, b; Badr et al., 2022; Barykin et al., 2022b; Mehta et al., 2022; Kalinina et al. 2022; Shaikh et al., 2022; Mikhaylov et al., 2022a, b; Nyangarika et al., 2022).
- In some cases, when it is difficult to obtain direct estimates of the impact of the implementation of a particular project (or the mutual influence of a group of projects) on the energy efficiency indicators of PC activities, methods of economic and statistical modeling can be used, which in this situation are almost the only way to solve the required problem (Denisova et al., 2019; Nyangarika et al., 2019a, b; Huang et al., 2021a, b; Mikhaylov, 2018, 2022; Mikhaylov et al., 2019; Conteh et al., 2021; Sediqi et al., 2022, Khan et al., 2021; Bhuiyan et al., 2022a; Liu et al., 2021a, b; Daniali et al., 2021; Moiseev et al., 2023a, b).

At the same time, the algorithm for using these methods, as well as the composition of modeled indicators and influencing factors for each specific case of implementing certain projects, is built individually, and it is difficult to offer any unified approach (Yüksel & Dinçer, 2023). The final result of the stage is a predictive assessment of the PC energy efficiency for various variants of energy-efficient projects (Dayong et al., 2020; Mikhaylov et al., 2018, 2023; Nyangarika et al., 2018; Danish et al., 2020, 2021; An et al., 2021; Uyeh et al., 2021; Tamashiro et al., 2021, 2023; Shaikh et al., 2021; Wang et al. 2023; He et al., 2023; Irshad et al., 2023; Iqbal et al., 2023; Aksoy et al., 2023).

Final selection of the most optimal option for implementing priority energy-efficient PC development projects (Moiseev et al., 2023a, b). If different options have approximately the same impact on the value of the integrated energy efficiency assessment of the PC, then choose the option that improves the energy efficiency of the PC to a greater extent. In addition, it is important to consider the social impact of implementing a particular variant of energy-efficient projects (An et al., 2019a, b, 2020a, b, c, 2022; Mikhaylov & Tarakanov, 2020; Moiseev et al., 2020, 2021; Gura et al., 2022; Dooyum et al., 2020; Mikhaylov et al., 2020; Mikhaylov, 2020a, b, c).

Also, when solving the problem of selecting priority energy-efficient PC development projects, it is proposed to actively use the methodology for selecting the optimal portfolio of projects for developing and improving the competitiveness of the production complex in conditions of limited resources (Kou et al., 2023a, b).

12.3 Results

A separate big task in conducting an energy efficiency assessment is to determine the basic (reference) values of PC energy efficiency indicators (Martínez et al., 2023). It is proposed to use various methods and approaches to solve this problem. In particular, the following approaches and techniques can be used (Han et al., 2023):

- Use of the best indicators achieved in certain types of economic activity, production, and technological processes
- Indicators of the main competitors (Eti et al., 2023)
- Methods of expert analysis and expert evaluations
- The method of analogies
- Use of approved standards and generally accepted requirements and standards
- Methods of pattern recognition theory, etc.

Creating a database on PC energy efficiency indicators, the database that is being created is linked to two main elements:

The methodology for assessing the energy efficiency of a PC is discussed in Clause 5.3. The database for this element is usually filled in from the following main information sources:

- Statistical forms and indicators of the Federal Statistics Service (Rosstat) as well as reporting indicators of federal and regional ministries and departments.
- Reporting indicators for the PC and its constituent enterprises. PC reporting is the main information resource. In most cases, this source covers more than 80% of the required information.
- Reporting indicators of large energy companies and enterprises involved in the supply of energy resources for the PC under study. First of all, such companies include PJSC Gazprom, oil companies, coal companies, PJSC Rosseti, JSC System Operator of the Unified Energy System (JSC SO UES), regional energy companies, etc..
- Results of expert analysis and model calculations, which are necessary in cases where the required data is not directly supported by various statistical reporting systems.

- Reference and regulatory documentation.

The parameters and technical, economic, and financial characteristics of possible or implemented priority energy-efficient PC development projects.

Enterprises and companies involved in their implementation are responsible for preparing primary project documentation. In some cases (especially when implementing projects with state participation), these may include federal, regional, and municipal executive authorities, housing and utilities management companies, and others. At the same time, for each of the projects, the relevant responsible persons prepare information on a unified form of indicators. Among these indicators, we note the following:

- Technical and economic indicators of the project (production capacity, indicators of specific costs of fuel and energy resources produced or consumed, etc.) (Dinçer et al., 2023a, b, c, d, e, f).
- Indicators of rational consumption of fuel and energy resources and efficiency of their use.
- Indicators of financial feasibility of implementing energy-efficient projects. The ultimate goal of adopting and implementing a set of priority energy-efficient PC development projects is to eliminate "bottlenecks" and increase the energy efficiency of the corresponding PC's activities. If it is not possible to achieve this goal during the implementation of individual projects, such projects should be excluded from the priority list.

Assessment of PC energy efficiency at the present stage and retrospective analysis of its dynamics. Key tasks are the following:

- Current assessment of energy efficiency indicators for PC operations. At the same time, "ideally" such an assessment is made at three levels—E1, E2, and E3. However, in many cases, it is sufficient to limit yourself only to the E1 level.
- Obtaining comprehensive assessments for indicative blocks and levels of assessment and identifying priority areas for implementing energy-efficient PC development projects.
- Determination of a comprehensive assessment of PC energy efficiency and dynamics of its change. Such an assessment is the main outcome indicator of the methodology, combining estimates for the entire set of indicators and assessment levels and reflecting the complex impact of implemented projects on PC energy efficiency indicators. 4. Variant (post-scenario) formation of a list of priority energy-efficient PC development projects recommended for implementation, taking into account economic feasibility and financial constraints. It is based on the following:

1. Assessment of PC energy efficiency at the current moment and for previous periods of time. Analysis of the indicator dynamics allows determining the priority of implementation of energy-efficient PC development projects.
2. List of possible energy-efficient PC development projects. As a rule, projects are evaluated in a comprehensive manner, since it is the cumulative effect of their

implementation that allows achieving the planned energy efficiency indicators of pKa.

In the climate agenda, technological development implies a transition to "clean energy," to the production and consumption of environmentally friendly goods, the reduction and recycling of waste, and environmental protection. Financial support for investing in green technologies is based on the principles of sustainable development outlined in the Declaration on Environment and Development, the main one being the principle of manufacturability.

Russia does not change its position on the climate agenda and the green economy, despite the changing geopolitical situation in the world, but is not among the leading countries and participates in international climate agreements. For Russia, the support of green investments is of particular importance today, due to the tightening of sanctions against the import of technologies that ensure the functioning of the sector of hydrocarbon production, processing and transportation, as well as the urgent problems of technological modernization of the energy industry (Dinçer et al., 2023a, b; Ivleva et al., 2023; Aissou et al., 2023; Mikhaylov, 2023).

The main directions of green investments of Russian enterprises are reduction of emissions of pollutants into the environment (atmospheric air, land, water resources), more than 70%, and reduction of energy costs for production—about 50%. And only 30% is the share of enterprises implementing environmental innovations to replace raw materials with safe or less dangerous ones. The latter direction of green innovation and investment seems to be the most significant.

Stimulating the development of green technologies, reflected in the agendas of many countries, is considered as an instrument of technological modernization of Russia. It was announced by the President of the Russian Federation V. V. Putin at the plenary session of the Eastern Economic Forum (September 5–9, 2022) and was developed in the discussions of the energy track of the forum program. In this regard, the generalization of the accumulated experience in tax incentives for investments in green technologies and the development of proposals for building Russia's own tax incentive system that meets national interests seem timely.

This article systematizes the tax policy instruments used by countries to stimulate investment in green innovation and identifies approaches that are recognized as key in order to stimulate green investment within the borders of individual countries.

Charging adequate fees for environmental damage will also contribute to the innovative development of environmentally friendly products and production processes.

The transition to "clean energy" involves both the use of renewable energy sources and zero greenhouse gas emissions in the production of energy resources from fossil energy sources and the use of technologies that prevent the release of greenhouse gases into the atmosphere and capture them when they enter the atmosphere.

12.4 Conclusion

The estimated period was 2012–2020 (in contrast to the competitiveness assessment, where the period 2010–2020 was taken). Since the beginning of this period, most of the companies reviewed in their reports provide publicly available information necessary for evaluating energy efficiency indicators. From our point of view, this period is representative and allows us to identify the main trends and problems in the field of energy efficiency of the companies under consideration.

To ensure comparability of companies, all their cost indicators were converted to a single currency. The base currency is the US dollar (USD), which is the main one in the reporting of most of the companies under consideration (Xu et al., 2023).

In accordance with this approach, the assessment of energy efficiency for individual indicators is carried out in the same way as the assessment of competitiveness indicators, i.e., by comparing their current and baseline values (Yüksel et al., 2023). It should also be noted that, in most cases, the determination of the basic values of energy efficiency indicators of companies was based on the best indicators achieved by leading competitors (Carayannis et al., 2023).

References

Ai, R., Zheng, Y., Yüksel, S., & Dinçer, H. (2023). Investigating the components of fintech ecosystem for distributed energy investments with an integrated quantum spherical decision support system. *Financial Innovation, 9*(1), 27.

Aissou, A., Ahmed, H. S., Hussein, M. E., Hassan, M. R., Mikhaylov, A., & Mahmoud, E. (2023). Optimizing reliability of flow networks using RWGA subject to total capacity constraint. *Applied Mathematics and Information Sciences, 7*(4), 559–565. https://doi.org/10.18576/amis/170404

Aksoy, T., Yüksel, S., Dınçer, H., Hacioglu, U., Mikhaylov, A., & Muyeen, S. M. (2023). Assessing the electricity production capacities of emerging markets for the sustainable investment. *IEEE Access, 11*, 24574–24590.

An, J., Mikhaylov, A., & Moiseev, N. (2019a). Oil price predictors: Machine learning approach. *International Journal of Energy Economics and Policy, 9*(5), 1–6.

An, J., Mikhaylov, A., & Sokolinskaya, N. (2019b). Machine learning in economic planning: Ensembles of algorithms. *Journal of Physics: Conference Series, 1353*, 012126.

An, J., Mikhaylov, A., & Jung, S.-U. (2020a). The strategy of South Korea in the global oil market. *Energies, 13*(10), 2491.

An, J., Mikhaylov, A., & Kim, K. (2020b). Machine learning approach in heterogeneous group of algorithms for transport safety-critical system. *Applied Sciences, 10*(8), 2670.

An, J., Mikhaylov, A., & Richter, U. H. (2020c). Trade war effects: Evidence from sectors of energy and resources in Africa. *Heliyon, 6*, e05693. https://doi.org/10.1016/j.heliyon.2020.e05693

An, J., Mikhaylov, A., & Jung, S.-U. (2021). A linear programming approach for robust network revenue management in the airline industry. *Journal of Air Transport Management, 91*, 101979.

An, J., Mikhaylov, A., Dinçer, H., & Yüksel, S. (2022). Economic modelling of electricity generation: Long short-term memory and Q-rung orthopair fuzzy sets. *Heliyon, 8*, e12345.

Badr, A. M. N. A. A., Mohamed, T. I., Osman, N. A. Q., & Mikhaylov, A. (2022). A review of social media website users' interaction paths with governmental accounts during the COVID-19 pandemic. *Informatics, 9*, 50. https://doi.org/10.3390/informatics9030050

Barykin, S., Mikheev, A., Kiseleva, E., Putikhin, Y., Alekseeva, N., & Mikhaylov, A. (2022a). An empirical analysis of Russian regions' debt sustainability. *Economies, 10*(4), 296.
Barykin, S. E., Kapustina, I. V., Sergeev, S. M., Daniali, S. M., Kopteva, L. A., Semenova, G. N., Pryadko, I. P., Mikhaylov, A., Baboshkin, P., Datsyuk, P., & Senjyu, T. (2022b). Financial logistics models based on systematic approach improving management solutions. *F1000Research, 11*, 570. https://doi.org/10.12688/f1000research.111252.1
Bhuiyan, M. A., An, J., Mikhaylov, A., Moiseev, N., & Danish, M. S. S. (2021). Renewable energy deployment and COVID-19 measures for sustainable development. *Sustainability, 13*, 4418.
Bhuiyan, M. A., Dinçer, H., Yüksel, S., Mikhaylov, A., Danish, M. S. S., Pinter, G., Uyeh, D. D., & Stepanova, D. (2022a). Economic indicators and bioenergy supply in developed economies: QROF-DEMATEL and random forest models. *Energy Reports, 8*(2022), 561–570.
Bhuiyan, M. A., Zhang, Q., Khare, V., Mikhaylov, A., Pinter, G., & Huang, X. (2022b). Renewable energy consumption and economic growth nexus—A systematic literature review. *Frontiers in Environmental Science, 10*, 878394. https://doi.org/10.3389/fenvs.2022.878394
Candila, V., Maximov, D., Mikhaylov, A., Moiseev, N., Senjyu, T., & Tryndina, N. (2021). On the relationship between oil and exchange rates of oil-exporting and oil-importing countries: From the great recession period to the Covid-19 era. *Energies, 14*, 8065.
Carayannis, E., Kostis, P., Dinçer, H., & Yüksel, S. (2023). Quality function deployment-oriented strategic outlook to sustainable energy policies based on quintuple innovation helix. *Journal of the Knowledge Economy*, 1–19.
Conteh, F., Takahashi, H., Hemeida, A. M., Krishnan, N., Mikhaylov, A., & Senjyu, T. (2021). Hybrid grid-connected renewable power generation for sustainable electricity supply in Sierra Leone: Case study Lungi Town. *Sustainability, 13*(18).
Daniali, S. M., Barykin, S. E., Kapustina, I. V., Khortabi, F. M., Sergeev, S. M., Kalinina, O. V., Mikhaylov, A., Veynberg, R., Zasova, L., & Senjyu, T. (2021). Predicting volatility index according to technical index and economic indicators on the basis of deep learning algorithm. *Sustainability, 13*, 14011.
Danish, M. S. S., Bhattacharya, A., Stepanova, D., Mikhaylov, A., Grilli, M. L., Khosravy, M., & Senjyu, T. (2020). A systematic review of metal oxide applications for energy and environmental sustainability. *Metals, 10*(12), 1604.
Danish, M. S. S., Senjyu, T., Sabory, N. R., Khosravy, M., Grilli, M. L., Mikhaylov, A., & Majidi, H. (2021). A forefront framework for sustainable aquaponics modeling and design. *Sustainability, 13*, 9313.
Danish, M. S., Senjyu, T. S., Sabory, N. R., & Mikhaylov, A. (2022a). Lessons from COVID-19 conferring environmental re-engineering opportunity. In M. Danish & T. Senjyu (Eds.), *Eco-friendly and agile energy strategies and policy development* (pp. 99–118). IGI Global. https://doi.org/10.4018/978-1-7998-9502-2.ch005
Danish, M. S., Senjyu, T. S., Sabory, N. R., & Mikhaylov, A. (2022b). Interplay consequences of COVID-19 on global environmental sustainability. In M. Danish & T. Senjyu (Eds.), *Eco-friendly and agile energy strategies and policy development*. IGI Global. https://doi.org/10.4018/978-1-7998-9502-2.ch003
Dayong, N., Mikhaylov, A., Bratanovsky, S., Shaikh, Z. A., & Stepanova, D. (2020). Mathematical modeling of the technological processes of catering products production. *Journal of Food Process Engineering, 43*(2), 1–9.
Denisova, V., Mikhaylov, A., & Lopatin, E. (2019). Blockchain infrastructure and growth of global power consumption. *International Journal of Energy Economics and Policy, 9*(4), 22–29.
Dinçer, H., Yüksel, S., Mikhaylov, A., Pinter, G., & Shaikh, Z. A. (2022a). Analysis of renewable-friendly smart grid technologies for the distributed energy investment projects using a hybrid picture fuzzy rough decision-making approach. *Energy Reports, 8*, 11466–11477. https://doi.org/10.1016/j.egyr.2022.08.275
Dinçer, H., Yüksel, S., Mikhaylov, A., Barykin, S. E., Aksoy, T., & Hacioğlu, U. (2022b). Analysis of environmental priorities for green project investments using an integrated

q-Rung orthopair fuzzy modelling. *IEEE Access, 10*, 50996–51007. https://doi.org/10.1109/ACCESS.2022.3174058

Dinçer, H., Yüksel, S., Hacioglu, U., Yilmaz, M. K., & Delen, D. (2023a). Development of a sustainable corporate social responsibility index for performance evaluation of the energy industry: A hybrid decision-making methodology. *Resources Policy, 85*, 103940.

Dinçer, H., Eti, S., Gökalp, Y., Yüksel, S., & Çelebi, B. (2023b). The role of small-scale wind turbines in increasing the competitiveness of hospitals. In *Handbook of research on quality and competitiveness in the healthcare services sector* (pp. 336–352). IGI Global.

Dinçer, H., Yüksel, S., Mikhaylov, A., Muyeen, S. M., Chang, T., Barykin, S., & Kalinina, O. (2023c). CO2 emissions integrated fuzzy model: A case of seven emerging economies. *Energy Reports, 9*, 5741–5751.

Dinçer, H., Yüksel, S., Aksoy, T., Hacıoğlu, Ü., Mikhaylov, A., & Pinter, G. (2023d). Analysis of solar module alternatives for efficiency-based energy investments with hybrid 2-tuple IVIF modeling. *Energy Reports, 10*, 61–71.

Dinçer, H., Eti, S., Yüksel, S., Özdemir, S., Yílmaz, A. E., & Ergün, E. (2023e). Integrating data mining and fuzzy decision-making techniques for analyzing the key minimizing factors of carbon emissions. *Journal of Intelligent & Fuzzy Systems, 45*, 1–17.

Dinçer, H., Eti, S., Aksoy, T., Yüksel, S., Hacioglu, U., Mikhaylov, A., & Muyeen, S. M. (2023f). Analysis of environmental impact for material production investments using a novel soft computing methodology. *IEEE Access, 11*, 37987–38001. https://doi.org/10.1109/ACCESS.2023.3266524

Dong, B., Ikonnikova, I., Rogulin, R., Sakulyeva, T., & Mikhaylov, A. (2021). Environmental economic approach to optimization of transport communication in megacities. *Journal of Environmental Science and Health, Part A, 56*(6), 660–666.

Dooyum, U. D., Mikhaylov, A., & Varyash, I. (2020). Energy security concept in Russia and South Korea. *International Journal of Energy Economics and Policy, 10*(4), 102–107.

Eti, S., Dinçer, H., Yüksel, S., & Gökalp, Y. (2022). Analysis of the suitability of the solar panels for hospitals: A new fuzzy decision-making model proposal with the T-spherical TOP-DEMATEL method. *Journal of Intelligent & Fuzzy Systems, 44*, 1–13.

Eti, S., Dinçer, H., Gökalp, Y., Yüksel, S., & Kararoğlu, D. (2023). Identifying key issues to handle the inflation problem in the healthcare industry caused by energy prices: An evaluation with decision-making models. In *Managing inflation and supply chain disruptions in the global economy* (pp. 162–178). Springer.

Gura, D., Mikhaylov, A., Glushkov, S., Zaikov, M., & Shaikh, Z. A. (2022). Model for estimating power dissipation along the interconnect length in single on-chip topology. *Evolutionary Intelligence, 15*, 2369–2373. https://doi.org/10.1007/s12065-020-00407-7

Han, Z., Gong, L., Chen, H., & Yüksel, S. (2023). The role of mineral and oil resources in achieving a clean environment in emerging economies: An asymmetric analysis. *Resources Policy, 86*, 104099.

He, H., Huang, Y., Nakadomari, A., Masrur, H., Krishnan, N., Hemeida, A. M., Mikhaylov, A., Omine, E., & Senjyu, T. (2023). Potential and economic viability of green hydrogen production from seawater electrolysis using renewable energy in remote Japanese islands. *Renewable Energy, 202*, 1436–1447. https://doi.org/10.1016/j.renene.2022.12.046

Huang, Y., Masrur, H., Shigenobu, R., Hemeida, A. M., Mikhaylov, A., & Senjyu, T. (2021a). A comparative design of a campus microgrid considering a multi-scenario and multi-objective approach. *Energies, 14*, 2853.

Huang, Y., Yona, A., Takahashi, H., Hemeida, A. M., Mandal, P., Mikhaylov, A., Senjyu, T., & Lotfy, M. E. (2021b). Energy management system optimization of drug store operate electric vehicles charging station. *Sustainability, 14*, 1217836.

Iqbal, K. M. J., Khan, M. I., Mikhaylov, A., Shah, A. A., Yadykin, V., Filho, W. L., Tariq, M. A. U. R., & Ullah, W. (2023). Modeling principles, criteria and indicators to assess water sector governance for climate compatibility and sustainability. *Frontiers in Environmental Science, 11*, 989930.

Irshad, A. S., Ludin, G. A., Masrur, H., Ahmadi, M., Yona, A., Mikhaylov, A., Krishnan, N., & Senjyu, T. (2023). Optimization of grid-photovoltaic and battery hybrid system with most technically efficient PV technology after the performance analysis. *Renewable Energy, 207*, 714–730.

Ivleva, S., Hajimajeed, M., Kartashova, O., & Mikhaylov, A. (2023). A comparative study of regional variations in common visual impairments among school-age children. *Electronic Journal of General Medicine, 20*(5), em504. https://doi.org/10.29333/ejgm/13264

Kalinina, O. V., Barykin, S. E., Sergeev, S. M., Semenova, G. N., Fatkullina, A., Mikhaylov, A., & De La Poza, E. (2022). Smart city perspectives in post-pandemic governance: Externalities reduction policy. *F1000Research, 11*, 1032. https://doi.org/10.12688/f1000research.123195.1

Khan, A. A., Shaikh, Z. A., Baitenova, L., Mutaliyeva, L., Moiseev, N., Mikhaylov, A., Laghari, A. A., Idris, S. A., & Alshazly, H. (2021). QoS-ledger: Smart contracts and metaheuristic for secure quality-of-service and cost-efficient scheduling of medical-data processing. *Electronics, 10*, 3083.

Khan, A. A., Laghari, A. A., Gadekallu, T. R., Shaikh, Z. A., Javed, A. R., Rashid, M., Estrela, V. V., & Mikhaylov, A. (2022). A drone-based data management and optimization using metaheuristic algorithms and blockchain smart contracts in a secure fog environment. *Computers and Electrical Engineering, 102*, 108234. https://doi.org/10.1016/j.compeleceng.2022.108234

Kou, G., Pamucar, D., Dinçer, H., & Yüksel, S. (2023a). From risks to rewards: A comprehensive guide to sustainable investment decisions in renewable energy using a hybrid facial expression-based fuzzy decision-making approach. *Applied Soft Computing, 142*, 110365.

Kou, G., Yuksel, S., & Dincer, H. (2023b). A facial expression and expert recommendation fuzzy decision-making approach for sustainable business investments within the metaverse world. *Applied Soft Computing, 148*, 110849.

Li, J., Dinçer, H., Yüksel, S., Mikhaylov, A., & Barykin, S. E. (2022). Bipolar q-ROF hybrid decision making model with golden cut for analyzing the levelized cost of renewable energy alternatives. *IEEE Access, 10*, 35900. https://doi.org/10.1109/ACCESS.2022.3168315

Liu, L., Kato, T., Mandal, P., Mikhaylov, A, Hemeida, A. M., & Senjyu T. (2021a). Multi-area power system load frequency control by using model predictive control by introducing renewable energy sources. In *2021 International Conference on Science & Contemporary Technologies (ICSCT)*, pp. 1–6. https://doi.org/10.1109/ICSCT53883.2021.9642521

Liu, L., Kato, T., Mandal, P., Mikhaylov, A., Hemeida, A. M., Takahashi, H., & Senjyu, T. (2021b). Two cases studies of model predictive control approach for hybrid renewable energy systems. *AIMS Energy, 9*(6), 1241–1259.

Liu, Z., Panfilova, E., Mikhaylov, A., & Kurilova, A. (2022a). Assessing stability in the relationship between parties in crowdfunding and crowdsourcing projects during the COVID-19 crisis. *Journal of Global Information Management (JGIM), 30*(4), 1–18. https://doi.org/10.4018/JGIM.297905

Liu, Z.-J., Panfilova, E., Mikhaylov, A., & Kurilova, A. (2022b). COVID-19 crisis impact on the stability between parties in crowdfunding and crowdsourcing. *Wireless Personal Communications, 122*, 915–930.

Martínez, L., Dinçer, H., & Yüksel, S. (2023). A hybrid decision making approach for new service development process of renewable energy investment. *Applied Soft Computing, 133*, 109897.

Mehta, R., Verghese, J., Mahajan, S., Barykin, S., Bozhuk, S., Kozlova, N., Kapustina, I., Mikhaylov, A., Naumova, E., & Dedyukhina, N. (2022). Consumers' behavior in conversational commerce marketing based on messenger chatbots. *F1000Research, 11*(647), 10.12688/f1000research.122037.1.

Mikhaylov, A. (2018). The yield of the carry trade strategy. *Finance: Theory and Practice, 22*(3), 52–63. https://doi.org/10.26794/2587-5671-2018-22-3-52-63

Mikhaylov, A. (2020a). Cryptocurrency market analysis from the open innovation perspective. *Journal of Open Innovation: Technology, Market, and Complexity, 6*(4), 197.

Mikhaylov, A. (2020b). Geothermal energy development in Iceland. *International Journal of Energy Economics and Policy, 10*(4), 31–35.

Mikhaylov, A. (2020c). Cryptocurrency market development: Hurst method. *Finance: Theory and Practice, 24*(3), 81–91.
Mikhaylov, A. (2021). Lichens as indicators of atmospheric pollution in urban ecosystems. *Israel Journal of Ecology & Evolution, 67*(1–2), 60–68. https://doi.org/10.1163/22244662-bja10016
Mikhaylov, A. (2022). Sustainable development and renewable energy: A new view to a global problem. *Energies, 15*, 1397. https://doi.org/10.3390/en15041397
Mikhaylov, A. (2023). Understanding the risks associated with wallets, depository services, trading, lending, and borrowing in the crypto space. *Journal of Infrastructure, Policy and Development, 6*, 2.
Mikhaylov, A., & Grilli, M. L. (2022). Machine learning methods and sustainable development: Metal oxides and multilayer metal-oxides. *Metals, 12*(4).
Mikhaylov, A., & Tarakanov, S. (2020). Development of Levenberg-Marquardt theoretical approach for electric network. *Journal of Physics: Conference Series, 1515*, 052006.
Mikhaylov, A., Sokolinskaya, N., & Nyangarika, A. (2018). Optimal carry trade strategy based on currencies of energy and developed economies. *Journal of Reviews on Global Economics, 7*, 582–592.
Mikhaylov, A., Burakov, D., & Didenko, V. (2019). Relationship between oil price and macroeconomic indicators in Russia. *Finance: Theory and Practice, 23*(2), 105–116. https://doi.org/10.26794/2587-5671-2019-23-2-105-116
Mikhaylov, A., Moiseev, N., Aleshin, K., & Burkhardt, T. (2020). Global climate change and greenhouse effect. *Entrepreneurship and Sustainability Issues, 7*(4), 2897–2913. https://doi.org/10.9770/jesi.2020.7.4(21)
Mikhaylov, A., Bhatti, I. M., Dinçer, H., & Yüksel, S. (2022a). Integrated decision recommendation system using iteration-enhanced collaborative filtering, golden cut bipolar for analyzing the risk-based oil market spillovers. *Computational Economics*, 1. https://doi.org/10.1007/s10614-022-10341-8
Mikhaylov, A., Yumashev, A., & Kolpak, E. (2022b). Quality of life, anxiety and depressive disorders in patients with extrasystolic arrhythmia. *Archives of Medical Science, 18*(2), 328. https://doi.org/10.5114/aoms.2020.101359
Mikhaylov, A., Dinçer, H., Yüksel, S., Pinter, G., & Shaikh, Z. A. (2023). Bitcoin mempool growth and trading volumes: Integrated approach based on QROF multi-SWARA and aggregation operators. *Journal of Innovation & Knowledge, 8*(3), 100378.
Moiseev, N., Mikhaylov, A., Varyash, I., & Saqib, A. (2020). Investigating the relation of GDP per capita and corruption index. *Entrepreneurship and Sustainability Issues, 8*(1), 780–794.
Moiseev, N., Sorokin, A., Zvezdina, N., Mikhaylov, A., Khomyakova, L., & Danish, M. S. S. (2021). Credit risk theoretical model on the base of DCC-GARCH in time-varying parameters framework. *Mathematics, 9*, 2423. https://doi.org/10.3390/math9192423
Moiseev, N., Mikhaylov, A., Dinçer, H., & Yüksel, S. (2023a). Effects of market capitalization shocks for open innovation models in e-commerce industry: Golden cut QROF MCDM analysis. *Financial Innovation, 9*(1), 32.
Moiseev, N., Mikhaylov, A., Dinçer, H., & Yüksel, S. (2023b). Market capitalization shock effects on open innovation models in e-commerce: Golden cut q-Rung orthopair fuzzy multicriteria decision-making analysis. *Financial Innovation, 9*(1), 55.
Mukhametov, A., Bekhorashvili, A., Avdeenko, A., & Mikhaylov, A. (2021). The impact of growing legume plants under conditions of biologization and soil cultivation on Chernozem fertility and productivity of rotation crops. *Legume Research, 44*, 12.
Niu, X., Yüksel, S., & Dinçer, H. (2023). Emission strategy selection for the circular economy-based production investments with the enhanced decision support system. *Energy, 274*, 127446.
Nyangarika, A., Mikhaylov, A., & Tang, B.-J. (2018). Correlation of oil prices and gross domestic product in oil producing countries. *International Journal of Energy Economics and Policy, 8*(5), 42–48.
Nyangarika, A., Mikhaylov, A., & Richter, U. (2019a). Influence oil price towards economic indicators in Russia. *International Journal of Energy Economics and Policy, 9*(1), 123–129.

Nyangarika, A., Mikhaylov, A., & Richter, U. (2019b). Oil price factors: Forecasting on the base of modified auto-regressive integrated moving average model. *International Journal of Energy Economics and Policy, 9*(1), 149–159.

Nyangarika, A., Mikhaylov, A., Muyeen, S. M., Yadykin, V., Mottaeva, A., Pryadko, I., Barykin, S., Fomenko, N., Rykov, G., & Shvandar, K. (2022). Energy stability and decarbonization in developing countries: Random forest approach for forecasting of crude oil trade flows and macro indicators. *Frontiers in Environmental Science, 10*, 1031343. https://doi.org/10.3389/fenvs.2022.1031343

Saqib, A., Chan, T.-H., Mikhaylov, A., & Lean, H. H. (2021). Are the responses of sectoral energy imports asymmetric to exchange rate volatilities in Pakistan? Evidence from recent foreign exchange regime. *Frontiers in Energy Research, 9*, 614463. https://doi.org/10.3389/fenrg.2021.614463

Sediqi, M. M., Nakadomari, A., Mikhaylov, A., Krishnan, N., Lotfy, M. E., Yona, A., & Senjyu, T. (2022). Impact of time-of-use demand response program on optimal operation of Afghanistan real power system. *Energies, 15*, 296.

Shaikh, Z. A., Moiseev, N., Mikhaylov, A., & Yüksel, S. (2021). Facile synthesis of copper oxide-cobalt oxide/nitrogen-doped carbon (Cu2O-Co3O4/CN) composite for efficient water splitting. *Applied Sciences, 11*, 9974. https://doi.org/10.3390/app11219974

Shaikh, Z. A., Kraikin, A., Mikhaylov, A., & Pinter, G. (2022). Forecasting stock prices of companies producing solar panels using machine learning methods. *Complexity, 2022*, 9186265. https://doi.org/10.1155/2022/9186265

Tamashiro, K., Alharbi, T., Mikhaylov, A., Hemeida, A. M., Krishnan, N., Lotfy, M. E., & Senjyu, T. (2021). Investigation of home energy management with advanced direct load control and optimal scheduling of controllable loads. *Energies, 14*, 7314. https://doi.org/10.3390/en14217314

Tamashiro, K., Omine, E., Krishnan, N., Mikhaylov, A., Hemeida, A. M., & Senjyu, T. (2023). Optimal components capacity based multi-objective optimization and optimal scheduling based MPC-optimization algorithm in smart apartment buildings. *Energy & Buildings, 278*, 112616. https://doi.org/10.1016/j.enbuild.2022.112616

Uyeh, D. D., Asem-Hiablie, S., Park, T., Kim, K.-M., Mikhaylov, A., Woo, S., & Ha, Y. (2021). Could Japonica rice be an alternative variety for increased global food security and climate change mitigation? *Food, 10*, 1869. https://doi.org/10.3390/foods10081869

Wan, Q., Miao, X., Wang, C., Dinçer, H., & Yüksel, S. (2023). A hybrid decision support system with golden cut and bipolar q-ROFSs for evaluating the risk-based strategic priorities of fintech lending for clean energy projects. *Financial Innovation, 9*(1), 1–25.

Wang, Q., Guo, J., Li, R., Mikhaylov, A., & Moiseev, N. (2023). Does technical assistance alleviate energy poverty in sub-Saharan African countries? A new perspective on spatial spillover effects of technical assistance. *Energy Strategy Reviews, 45*, 101047. https://doi.org/10.1016/j.esr.2022.101047

Xu, X., Yüksel, S., & Dinçer, H. (2023). An integrated decision-making approach with golden cut and bipolar q-ROFSs to renewable energy storage investments. *International Journal of Fuzzy Systems, 25*(1), 168–181.

Yüksel, S., & Dinçer, H. (2023). Sustainability analysis of digital transformation and circular industrialization with quantum spherical fuzzy modeling and golden cuts. *Applied Soft Computing, 138*, 110192.

Yüksel, S., Dinçer, H., & Uluer, G. S. (2023). Socio-economic indicators of nuclear energy investments: A study on countries. In *Inclusive developments through socio-economic indicators: New theoretical and empirical insights* (pp. 33–44). Emerald Publishing.

Chapter 13
Navigating the Energy Transition: How R&D Investment and Governance Quality Drive Clean Energy in the MENA Region

Ibrahim Nandom Yakubu, Ayhan Kapusuzoglu, and Nildag Basak Ceylan

13.1 Introduction

The emission of carbon dioxide (CO_2) into the atmosphere due to the use of traditional energy sources, mainly fossil fuels, substantially drives climate change, which is presently regarded as one of the most critical environmental challenges (Alam et al., 2021). The broad reliance on fossil fuels as a key energy source has heightened global awareness of the need for sustainability that guarantees adequate access to energy while limiting ecological harm (Murshed, 2020). It is widely acknowledged that addressing and mitigating the effects of climate change need immediate action, which includes shifting away from non-REN sources and toward REN sources. This transition is regarded as a viable option for lowering carbon emissions and achieving a more sustainable global economy (Droege, 2011).

In the past few years, nations across the globe, including those in the MENA area, have begun to recognize how crucial it is to switch to environmentally friendly sources of energy to battle climate change and advance sustainable development. Although the MENA area is well recognized for having large fossil fuel reserves, it also holds tremendous potential for REN, especially in the form of solar and wind resources (Middle East Institute, 2021). Given the region's wide deserts and strong sun irradiation, solar energy has huge prospects. Countries in the region including Saudi Arabia, the United Arab Emirates, and Morocco have started zealous renewable energy plans to exploit their potential for solar power and branch out their energy mix. Saudi Arabia, for example, has established the objective of achieving a

I. N. Yakubu
School of Business, University for Development Studies, Tamale, Ghana

A. Kapusuzoglu (✉) · N. B. Ceylan
Faculty of Business, Ankara Yildirim Beyazit University, Ankara, Turkey
e-mail: akapusuzoglu@aybu.edu.tr; nbceylan@aybu.edu.tr

H. Dinçer et al. (eds.), *Decision Making in Interdisciplinary Renewable Energy Projects*, Contributions to Management Science,
https://doi.org/10.1007/978-3-031-51532-3_13

50% REN production target by the year 2030, with a strong focus on solar energy (International Trade Administration, 2021). Along with solar energy, wind power is increasingly gaining prominence in the area. To benefit from their adept wind conditions, nations including Egypt, Jordan, and Morocco have made investments in wind farms. These initiatives not only help to lower greenhouse gas emissions, but they also open up exciting possibilities for economic development and job creation.

The MENA region has also seen the emergence of a number of projects and initiatives aimed at expediting the switch to renewable energy. The Arab Renewable Energy Initiative, for instance, intends to expand the Arab region's proportion of renewable energy to 30% by 2030. Likewise, the Gulf Cooperation Council countries have formed the GCC Interconnection Grid, an initiative aimed at connecting member states' energy grids, allowing for the exchange of REN resources and improving the stability of the grid (International Renewable Energy Agency, 2014).

The adoption of renewable energy not only contributes to reducing greenhouse gas emissions but also provides other advantages, including improved energy security by diversifying energy sources and lowering reliance on imported fossil fuels (Owusu & Asumadu-Sarkodie, 2016). Renewable energy projects have the potential to drive economic growth, attract investment, and generate new job opportunities (Ibrahim et al., 2021).

The United Nations (UN) has underscored the need for nations to collaborate to guarantee access to low-cost, reliable, and efficient sources of energy globally in acknowledgment of the necessity of ensuring a sustainable energy supply and environmental conservation (Murshed, 2020). This is manifested in the 7th Sustainable Development Goal (SDG) of the UN's 2030 agenda, which seeks to considerably boost the fraction of energy from renewable sources in the overall global energy utilization and increase the rates of energy-use efficiency worldwide (United Nations, 2015). The UN aims to create a more sustainable and ecologically responsible approach to the production and use of energy by giving priority to these goals. Nevertheless, due to some obstacles, including a lack of technological expertise and ineffective resource management techniques, reaching sustainable energy objectives may be daunting for emerging economies including countries in the MENA region.

The researchers contend, in light of the preceding discussions, that effective research and development (R&D) efforts assume a vital role in resolving the challenges pertaining to shifting to renewable energy sources. By boosting R&D, countries may overcome obstacles including restricted access to technology and expertise deficiency. Consequently, this could mean unlocking the door for a more seamless transition to a sustainable energy economy, which might ultimately restore ecological balance and lessen dependency on fossil fuels. Aside from R&D spending, governance quality is also an important aspect of advancing renewable energy. Governance quality substantially influences energy consumption because it fosters investment and innovation in the energy sector. That is, good governance may play an important role in tackling issues like corruption, bureaucratic red tape, and political instability, which are foremost disablements to the advancement of the energy industry. This study, therefore, seeks to explore how investment in R&D and governance quality drives REN adoption in the MENA region.

Although previous studies have looked at the individual impact of R&D spending and institutional quality on REN (see Adedoyin et al., 2020; Li & Ullah, 2022; Wang et al., 2022; Zhang et al., 2023), the contribution and novelty of this study lie in its focus on the MENA region, which continues to be a key player in the global energy landscape. The researchers have no knowledge of any prior research that has comprehensively modelled the impact of R&D spending and governance quality on REN adoption in the MENA region. Thus, the research addresses a notable void in the existing body of knowledge. Additionally, by exploring the dynamic link between R&D investment, institutional quality, and REN utilization, the study may offer fruitful insights for policymakers on how they can accelerate the energy evolution process in the MENA states. The study further augments the broader literature on the drivers of clean energy adoption, particularly in the context of emerging economies.

The rest of the chapter is organized as follows: Section 13.2 reviews the literature, while the methodology is discussed in Sect. 13.3. In Sect. 13.4, the results are presented, and the final section concludes with implications.

13.2 Literature Review

13.2.1 Theoretical Background

Although there is no direct theory to study the nexus between R&D investment and REN adoption, the institutional theory provides important insights into how the effectiveness of governance affects the use of REN. The decision-making of organizations is greatly influenced by institutions, which include formal rules, conventions, and informal practices (Scott, 2013). Institutions are therefore essential in the study of renewable energy since they help foster the adoption and use of green energy sources. The institutional environment for the growth of REN is significantly influenced by the quality of governance, which is demonstrated by clear and effective regulatory frameworks, supporting policies, and strong enforcement mechanisms (Newell & Mulvaney, 2013). Effective governance creates a favorable environment that promotes the development of renewable energy. While supporting policies give incentives and financial tools to encourage renewable energy projects, transparent rules offer investors clarity and direction. Strong enforcement measures also guarantee compliance and inspire trust in stakeholders, promoting a stable environment for investments in renewable energy projects. In contrast, poor governance may stymie renewable energy usage by creating uncertainty and impediments for potential consumers. Inadequate regulatory frameworks, inconsistent regulations, and ineffective enforcement hinder and dissuade investment in REN initiatives, hampering the deployment of REN technologies. In a nutshell, creating a favorable institutional climate is critical for promoting renewable energy and accomplishing a sustainable energy transition, according to the assumption of institutional theory.

13.2.2 *Empirical Literature*

The surge in demand for renewable energy has sparked extensive research to understand the fundamental drivers behind the utilization of REN sources. In this section, we present a comprehensive review of the prevailing literature, focusing on the factors that influence REN consumption and the role of research and development (R&D) as well as governance quality in shaping the adoption of REN.

Using data from African countries, Ergun et al. (2019) examined the key drivers of REN consumption. The study found that while GDP per capita and human development decrease renewable energy utilization, foreign direct investment (FDI) encourages the demand for REN. Akintande et al. (2020) evidenced that the growth in REN usage is driven by human capital, population growth, and the use of electricity. Employing a sample of OECD nations, Li et al. (2020) revealed that eco-innovation and the productivity of energy resources predict the consumption of renewable energy. Lau et al. (2018) evidenced that FDI and the level of economic growth positively influence the adoption of renewable electricity, while trade decreases it in the long term. Papież et al. (2018) discovered that the distribution of REN sources is essential in REN development in the European Union. Somoye et al. (2022) explored the factors explaining REN usage in Nigeria. The authors discover that GDP and financial development boost REN utilization, whereas reliance on fossil fuels inhibits it. They also showed that trade openness has a detrimental influence on REN usage. In the case of China, Zhao et al. (2020) reported that the development of the financial sector and per capita income increase the use of both REN and non-REN sources, whereas the effect of trade openness on renewable consumption is negative. According to Dogan et al. (2023), energy tariffs and environmental taxes inimically affect the use of REN in the European Union (EU), whereas the impact of economic growth and oil prices is positive. Applying panel regression methods, Sachan et al. (2023) assessed the key factors driving the use of REN in the BRICS economies. The findings demonstrated that human development and environmental policy stringency significantly spur REN demand. The study, however, noted that CO_2 emissions reduce the share of REN in the bloc. Yakubu et al. (2022) recounted that the boom in the trade of ICT goods stimulates the use of REN in the BRICS economies. Chu et al. (2023) found that the development of REN is directly impacted by energy security risks in high-energy consumption economies. In divergence to trade openness, Bunyaminu and Yakubu (2022) found that financial growth and FDI reduce REN consumption in Africa. Lawal (2023) discovered that whereas economic indicators positively drive the use of renewable energy, environmental-related factors inhibit it. In Malaysia, the conclusions of Mohamed Yusoff et al. (2023) indicated that economic growth and urbanization improve the utilization of REN. The authors, conversely, provided evidence that the growth in FDI and trade liberalization diminish its usage.

Regarding the influence of R&D spending, revealed that an intensification in R&D is concomitant with enhanced levels of REN usage in China and Japan in both the short and long term. Wang et al. (2020) found that R&D is a primary factor contributing to promoting REN growth in some selected countries in the G20. Churchill et al. (2021) documented that R&D expenditure is positively and substantially related with renewable energy uptake in OECD nations until the year 1996 when the impact reverses. Chmielewski et al. (2023) noted that the reliance on REN in the EU is significantly influenced by R&D expenditure. In the Asia-Pacific countries, Khezri et al. (2021) empirically noted that R&D significantly moderates the effect of financial development and GDP per capita on the development of renewable energy sources.

Turning to the nexus between institutional quality and REN consumption, the findings are quite mixed. While some researchers revealed that the quality of institutions enhances the adoption of REN sources (see Uzar, 2020; Wang et al., 2022; Rahman & Sultana, 2022; Saadaoui & Chtourou, 2022; Vatamanu & Zugravu, 2023), the empirical work of Asongu and Odhiambo (2022) revealed that renewable energy development is negatively influenced by governance quality.

From the review of the empirical literature, it can be deduced that several factors have been identified as determinants of REN consumption. Regarding the impact of R&D, the majority of studies provide strong support for its positive effect on the adoption of REN. However, the impact of institutions on REN development shows mixed findings. Besides, it is worth noting that research on the dynamic link between R&D spending, institutional quality, and the use of REN in individual countries within the MENA area is deficient with no study holistically examining this interplay in the region. Therefore, this study aims to bridge this empirical literature gap.

13.3 Research Methodology

13.3.1 Data and Variables

This study utilized data spanning from 1995 to 2021 and focuses on ten countries within the MENA region. The selected countries, listed in alphabetical order, are "Algeria, Egypt, Iran, Jordan, Kuwait, Morocco, Oman, Qatar, Saudi Arabia, and the United Arab Emirates." While the study initially considered all countries in the MENA region, some had to be omitted due to incomplete data for the variables of relevance. The primary dependent variable in this study is renewable energy consumption (RENC), which is measured as a percentage of total final energy consumption. This variable serves as a proxy for clean energy, drawing upon the approach adopted by Usman et al. (2021). The study also examines two key independent variables: research and development investment (R&D) and governance quality (GOVQ). To provide a comprehensive analysis, several control factors are considered, including financial development (FINDEV), foreign direct investment (FDI), and trade openness (TRA). R&D investment is measured by the expenditure on research and development as a percentage of the gross domestic product (GDP). In line with the work of Yakubu et al. (2021), governance quality is an index derived from various sub-governance indicators. These sub-measures include "government effectiveness, rule of law, regulatory quality, voice and accountability, political stability and absence of violence/terrorism, and control of corruption." Financial development is approximated by the proportion of domestic credit allocated to the private sector relative to the GDP. Additionally, FDI is captured by the net inflows of foreign direct investment, while trade is measured by the combined value of exports and imports of goods and services as a percentage of GDP. To ensure data consistency and reliability, all the variables used were obtained from the World Bank database.

13.3.2 Empirical Model and Estimation Technique

Based on the variables, the basic empirical models for the study are as follows:

$$\text{RENC}_{\text{it}} = \beta_0 + \beta_1 \text{R\&D}_{\text{it}} + \beta_2 \text{FINDEV}_{\text{it}} + \beta_3 \text{FDI}_{\text{it}} + \beta_4 \text{TRA}_{\text{it}} + \varepsilon_{\text{it}} \quad (13.1)$$

$$\text{RENC}_{\text{it}} = \beta_0 + \beta_1 \text{GOVQ}_{\text{it}} + \beta_2 \text{FINDEV}_{\text{it}} + \beta_3 \text{FDI}_{\text{it}} + \beta_4 \text{TRA}_{\text{it}} + \varepsilon_{\text{it}} \quad (13.2)$$

In the models, β_0 connotes the intercept, and β_1, β_2, β_3, and β_4 are the coefficients associated with their respective independent variables. The error term, ε, captures any unexplained variation in the dependent variable. The countries are denoted by "i," and the time period is denoted by "t."

In the model estimation, the study employs three distinct techniques: pooled OLS, fixed effects, and random effects. To determine the appropriate model for the results' interpretation, the Hausman test (Hausman, 1978) is conducted. The Hausman test serves as a criterion, where random effects estimates are utilized unless the test rejects this assumption. A rejection of the Hausman test implies that the underlying suppositions of random effects are violated, leading to the preference for fixed-effects estimates (Wooldridge, 2009). Consequently, in the event that the test statistic from the Hausman test demonstrates statistical significance at a 5% significance level, the fixed-effects estimator is deemed preferable, indicating an inconsistency in the random effects model.

13.4 Empirical Results

13.4.1 Descriptive Statistics

The descriptive statistics presented in Table 13.1 provide a summary of the key variables in the dataset. The average renewable energy consumption (RENC) in the MENA region is approximately 2.580%, indicating a moderate level of adoption.

Table 13.1 Descriptive statistics results

Statistics/variables	RENC	R&D	GOVQ	FINDEV	FDI	TRA
Mean	2.580	0.301	9.19E−16	48.852	2.178	80.980
Maximum	23.000	0.715	3.810	138.858	23.537	172.803
Minimum	0.000	0.042	−4.324	3.907	−2.760	29.228
Std. dev.	4.672	0.207	2.129	25.662	2.887	30.739
Skewness	2.132	0.595	−0.116	0.580	2.890	0.753
Kurtosis	6.857	2.026	2.025	3.006	16.775	3.526
Jarque-Bera	344.306	6.792	9.627	12.580	2501.108	27.696
Probability	0.000	0.034	0.008	0.002	0.000	0.000

The maximum value of 23.000 suggests that there are countries in the MENA region where renewable energy consumption is relatively high. Research and development (R&D) investment accounts for an average of 0.301% of GDP, suggesting a relatively low emphasis on innovation. Governance quality (GOVQ) is represented by a mean value of 9.19E–16, which is close to zero. This indicates that, on average, governance quality in the MENA region is relatively neutral. The skewness of −0.116 portrays a slightly left-skewed distribution, indicative of a slight imbalance in governance quality scores. The kurtosis value of 2.025 indicates a relatively moderate level of peakedness in the distribution. Financial development (FINDEV) is relatively higher, with domestic credit to the private sector constituting around 48.852% of GDP. FDI has a mean value of 2.178. The skewness of 2.890 suggests a highly right-skewed distribution, indicating a concentration of FDI in a few countries. The kurtosis value of 16.775 indicates a highly peaked distribution with heavy tails, suggesting extreme values and potential outliers. Trade openness (TRA) is relatively high, with trade accounting for approximately 80.980% of GDP. The skewness of 0.753 suggests a slightly right-skewed distribution, indicating relatively balanced trade openness. The Jarque-Bera test reveals significant deviations from normality for all variables, suggesting non-normal distributions potentially influenced by outliers or inherent data characteristics.

13.4.2 Multicollinearity Analysis

Table 13.2 presents the correlation and variance inflation factor (VIF) analyses among the variables. Analyzing the correlation coefficients, it is observed that R&D exhibits a moderate negative correlation with GOVQ and TRA while showing weak negative correlations with FINDEV and FDI. GOVQ demonstrates a positive correlation with TRA, and FINDEV shows weak positive correlations with GOVQ and TRA. Meanwhile, FDI displays weak positive correlations with FINDEV and TRA. The VIF values range from 1.317 to 3.815, which are below the threshold of 10, indicating no significant multicollinearity issues (Gujarati, 2003; Yakubu, 2019). Overall, the correlation and VIF analyses suggest that the variables in the model are comparatively independent and do not exhibit high multicollinearity.

Table 13.2 Correlation and variance inflation factor (VIF) analysis results

Variables	R&D	GOVQ	FINDEV	FDI	TRA	VIF	Tolerance
R&D	1.000					1.477	0.677
GOVQ	−0.509	1.000				3.815	0.262
FINDEV	−0.075	0.468	1.000			1.396	0.716
FDI	−0.055	−0.077	0.088	1.000		1.317	0.759
TRA	−0.489	0.794	0.440	0.228	1.000	3.587	0.279

13.4.3 Regression Results

Table 13.3 displays the regression estimates derived from the pooled OLS, fixed effects, and random effects estimators. The outcomes of the Hausman test postulate statistical significance with a value of less than 0.05, indicating that the fixed-effects model is more appropriate. Consequently, the interpretations of the findings are primarily based on the estimates derived from the fixed-effects model.

In Model 1, the coefficient for R&D investment (R&D) is −2.576, and it is significant at 1% level. This suggests that an intensification in R&D investment is associated with a significant decrease in renewable energy consumption. This result might seem counterintuitive, as higher R&D investment is typically expected to lead to greater adoption of renewable energy as evidenced in prior studies (Wang et al., 2020; Churchill et al., 2021; Chmielewski et al., 2023). However, there could be several explanations for this finding. It is possible that the MENA countries in the study have primarily focused their R&D efforts on nonrenewable energy sources or have faced challenges in effectively translating R&D investments into renewable energy projects.

Moving on to the governance quality variable (GOVQ), the coefficient is 0.256, and it is significant at 5% level as shown in Model 2. This indicates that better governance quality is positively associated with higher renewable energy consumption. This aligns with previous research that emphasizes the importance of strong institutional mechanisms in creating an encouraging environment for renewable energy development (Rahman & Sultana, 2022; Saadaoui & Chtourou, 2022; Vatamanu & Zugravu, 2023). MENA countries with more favorable governance structures may have implemented effective policies and regulations that encourage renewable energy adoption.

Regarding the control variables, the coefficient for FINDEV is positive and significant in Model 2. This suggests that the level of financial development significantly promotes renewable energy adoption in MENA countries. The coefficient for foreign direct investment (FDI) is not statistically significant in both models, indicating that FDI does not exert a weighty impact on renewable energy consumption in the MENA countries under investigation. Trade openness influences renewable energy consumption negatively and the effect is significant in Model 2. The finding suggests that increased trade openness is associated with lower levels of renewable energy consumption in MENA countries. One possible explanation is that trade

Table 13.3 Regression analysis results

	Pooled OLS		Random effects		Fixed effects	
Variables/models	Model 1	Model 2	Model 1	Model 2	Model 1	Model 2
R&D	1.625 (1.860)		−2.137*** (0.741)		−2.576*** (0.757)	
GOVQ		−0.080 (0.149)		0.237** (0.096)		0.256** (0.098)
FINDEV	0.082*** (0.021)	0.082*** (0.009)	0.029** (0.011)	0.011*** (0.004)	0.021 (0.012)	0.010** (0.004)
FDI	0.381*** (0.120)	0.236*** (0.067)	−0.074 (0.052)	−0.032 (0.023)	−0.080 (0.053)	−0.035 (0.023)
TRA	−0.095*** (0.017)	−0.070*** (0.012)	−0.007 (0.014)	−0.026*** (0.006)	−0.003 (0.014)	−0.025*** (0.006)
C	4.335*** (1.430)	3.032*** (0.953)	2.833 (1.721)	4.030*** (1.003)	2.362** (1.135)	3.591*** (0.494)
R^2	0.491	0.387	0.350	0.189	0.977	0.958
Adjusted R^2	0.455	0.372	0.304	0.170	0.972	0.954
F-statistic	13.721	27.117	7.663	10.016	177.265	283.267
Prob (F-statistic)	0.000	0.000	0.000	0.000	0.000	0.000
Hausman test χ^2 [Prob. > χ^2]					9.943 [0.041]	10.516 [0.0326]

Notes: Standard errors are in ( ), ** and *** are significance at 5% and 1%, respectively

openness may lead to increased dependence on conventional energy sources, such as fossil fuels, which could reduce the incentive for adopting renewable energy technologies.

Explaining the diagnostic statistics, the R2 values in both Models 1 and 2 suggest a high level of explanatory power, indicating that the models capture a substantial portion of the variation in renewable energy utilization. The significant F-statistics in the models further support the conclusion that the models, as a whole, are statistically significant in explaining the linkage between R&D investment, governance quality, the control variables, and renewable energy usage in the MENA countries.

13.5 Conclusion and Discussion

In conclusion, this study investigated the effects of R&D investment and governance quality on the adoption of renewable energy (REN) in MENA countries from 1995 to 2021. The analysis also considered the influence of financial development, FDI, and trade openness as control factors. By utilizing the fixed-effects technique, the regression results revealed several noteworthy findings. Firstly, there was a negative relationship observed between R&D investment and REN consumption, indicating that higher R&D investment is associated with lower adoption of renewable energy. Secondly, the study reported a positive link between governance quality and REN consumption, suggesting that countries with better governance structures exhibited higher levels of renewable energy adoption. Furthermore, the analysis demonstrated that financial development had a positive influence on REN consumption, while trade openness had a negative impact. However, the effect of FDI on REN consumption was found to be insignificant.

Based on the findings of this study, several policy implications can be derived to promote the adoption of renewable energy in MENA countries. Firstly, policymakers should prioritize enhancing R&D investment in the renewable energy sector. Despite the negative relationship observed between R&D investment and REN consumption, it is crucial to recognize the importance of targeted R&D efforts. Governments should allocate resources to support R&D initiatives specifically focused on renewable energy technologies. This includes funding research projects, incentivizing collaboration between academia and industry, and providing grants or tax incentives for innovative clean energy solutions. By promoting R&D investment, countries can drive technological advancements and overcome barriers hindering renewable energy adoption.

Secondly, the positive association between governance quality and REN consumption highlights the significance of strong institutional frameworks and transparent regulatory systems. Policymakers should focus on improving governance quality to create an encouraging environment for REN adoption. This entails the development and implementation of effective policies, regulations, and supportive frameworks that facilitate the deployment and integration of renewable energy sources. Streamlining administrative procedures, ensuring legal clarity, and establishing mechanisms for monitoring and enforcing renewable energy targets are essential steps in this regard. Strengthening governance quality encourages investment, fosters public-private partnerships, and instills trust in the renewable energy sector.

Besides, policymakers should promote financial development to support renewable energy projects. The positive influence of financial development on REN consumption suggests the need for robust financial mechanisms in the sector. Policymakers should collaborate with financial institutions to develop innovative financing solutions tailored to renewable energy investments. This can involve establishing green financing mechanisms, such as green bonds or specialized loan programs, to provide affordable capital for renewable energy projects. Additionally, governments can offer incentives like tax credits or grants to attract private sector investments in the renewable energy sector, stimulating economic growth and job creation.

Managing trade openness is also crucial for promoting REN adoption. The negative influence of trade openness on REN consumption highlights the potential challenges posed by international trade dynamics. Policymakers should carefully assess the trade policies and agreements in place to ensure they do not hinder domestic renewable energy development. It is important to strike a balance between promoting trade and protecting the domestic renewable energy industry. Governments can consider implementing measures like import tariffs or preferential market access for domestically produced renewable energy equipment to support the local renewable energy sector and foster its competitiveness.

Notwithstanding the fact that this study provides valuable insights into the factors influencing renewable energy adoption in MENA countries, there are still avenues for further research that can contribute to a deeper understanding of this topic. For instance, future studies could explore the specific mechanisms through which R&D investment impacts REN consumption. By examining the relationship in more detail, including the types of R&D activities and their effects on different renewable energy technologies, researchers can provide more nuanced insights into how R&D can be effectively utilized to drive renewable energy adoption. Likewise, investigating the role of specific governance factors and their differential impacts on renewable energy consumption could shed light on the most critical aspects of governance quality. This could involve examining the influence of individual governance indicators and analyzing their relative importance in facilitating or hindering renewable energy development.

References

Adedoyin, F. F., Bekun, F. V., & Alola, A. A. (2020). Growth impact of transition from non-renewable to renewable energy in the EU: The role of research and development expenditure. *Renewable Energy, 159*, 1139–1145.

Akintande, O. J., Olubusoye, O. E., Adenikinju, A. F., & Olanrewaju, B. T. (2020). Modeling the determinants of renewable energy consumption: Evidence from the five most populous nations in Africa. *Energy, 206*, 117992.

Alam, M. S., Apergis, N., Paramati, S. R., & Fang, J. (2021). The impacts of R&D investment and stock markets on clean-energy consumption and CO2 emissions in OECD economies. *International Journal of Finance & Economics, 26*(4), 4979–4992.

Asongu, S., & Odhiambo, N. M. (2022). Governance and renewable energy consumption in sub-Saharan Africa. *International Journal of Energy Sector Management, 16*(2), 209–223.

Bunyaminu, A., & Yakubu, I. N. (2022). Green energy demand and financial development: Evidence from Africa. In *Finance for sustainability in a turbulent economy* (pp. 26–48). IGI Global.

Chmielewski, W., Postuła, M., & Dubel, P. (2023). The impact of expenditure on research and development on selected energy factors in the European Union. *Energies, 16*(8), 3554.

Chu, L. K., Ghosh, S., Doğan, B., Nguyen, N. H., & Shahbaz, M. (2023). Energy security as new determinant of renewable energy: The role of economic complexity in top energy users. *Energy, 263*, 125799.

Churchill, S. A., Inekwe, J., & Ivanovski, K. (2021). R&D expenditure and energy consumption in OECD nations. *Energy Economics, 100*, 105376.

Dogan, E., Hodžić, S., & Šikić, T. F. (2023). Do energy and environmental taxes stimulate or inhibit renewable energy deployment in the European Union? *Renewable Energy, 202*, 1138–1145.

Droege, P. (2011). *Urban energy transition: From fossil fuels to renewable power*. Elsevier.

Ergun, S. J., Owusu, P. A., & Rivas, M. F. (2019). Determinants of renewable energy consumption in Africa. *Environmental Science and Pollution Research, 26*(15), 15390–15405.

Gujarati, D. (2003). *Basic econometrics* (4th ed.). McGraw Hill.

Hausman, J. A. (1978). Specification tests in econometrics. *Econometrica: Journal of the Econometric Society, 46*, 1251–1271.

Ibrahim, I. D., Hamam, Y., Alayli, Y., Jamiru, T., Sadiku, E. R., Kupolati, W. K., Ndambuki, J. M., & Eze, A. A. (2021). A review on Africa energy supply through renewable energy production: Nigeria, Cameroon, Ghana and South Africa as a case study. *Energy Strategy Reviews, 38*, 100740.

International Renewable Energy Agency. (2014). *Pan-Arab clean energy initiative*. Retrieved from https://www.irena.org/Energy-Transition/Country-engagement/Regional-Initiatives/Pan-Arab-Clean-Energy-Initiative

International Trade Administration. (2021). *Saudi Arabia renewable energy*. Retrieved from https://www.trade.gov/market-intelligence/saudi-arabia-renewable-energy

Khezri, M., Heshmati, A., & Khodaei, M. (2021). The role of R&D in the effectiveness of renewable energy determinants: A spatial econometric analysis. *Energy Economics, 99*, 105287.

Lau, L. S., Yii, K. J., Lee, C. Y., Chong, Y. L., & Lee, E. H. (2018). Investigating the determinants of renewable energy consumption in Malaysia: An ARDL approach. *International Journal of Business and Society, 19*(3), 886–903.

Lawal, A. I. (2023). Determinants of renewable energy consumption in Africa: Evidence from system GMM. *Energies, 16*(5), 2136.

Li, W., & Ullah, S. (2022). Research and development intensity and its influence on renewable energy consumption: Evidence from selected Asian economies. *Environmental Science and Pollution Research, 29*(36), 54448–54455.

Li, J., Zhang, X., Ali, S., & Khan, Z. (2020). Eco-innovation and energy productivity: New determinants of renewable energy consumption. *Journal of Environmental Management, 271*, 111028.

Middle East Institute. (2021). *Fossil fuel subsidies and renewable energies in MENA: An oxymoron?* Retrieved from https://www.mei.edu/publications/fossil-fuel-subsidies-and-renewable-energies-mena-oxymoron

Mohamed Yusoff, N. Y., Ridzuan, A. R., Soseco, T., Wahjoedi, Narmaditya, B. S., & Ann, L. C. (2023). Comprehensive outlook on macroeconomic determinants for renewable energy in Malaysia. *Sustainability, 15*(5), 3891.

Murshed, M. (2020). *Does ICT-trade openness ensure energy and environmental sustainability? Empirical evidence from selected South Asian economies*. Retrieved from https://mpra.ub.uni-muenchen.de/97944/1/MPRA_paper_97944.pdf

Newell, P., & Mulvaney, D. (2013). The political economy of the 'just transition'. *The Geographical Journal, 179*(2), 132–140.

Owusu, P. A., & Asumadu-Sarkodie, S. (2016). A review of renewable energy sources, sustainability issues and climate change mitigation. *Cogent Engineering, 3*(1), 1167990.

Papież, M., Śmiech, S., & Frodyma, K. (2018). Determinants of renewable energy development in the EU countries. A 20-year perspective. *Renewable and Sustainable Energy Reviews, 91*, 918–934.

Rahman, M. M., & Sultana, N. (2022). Impacts of institutional quality, economic growth, and exports on renewable energy: Emerging countries perspective. *Renewable Energy, 189*, 938–951.

Saadaoui, H., & Chtourou, N. (2022). Do institutional quality, financial development, and economic growth improve renewable energy transition? Some evidence from Tunisia. *Journal of the Knowledge Economy, 14*, 2927–2958.

Sachan, A., Sahu, U. K., Pradhan, A. K., & Thomas, R. (2023). Examining the drivers of renewable energy consumption: Evidence from BRICS nations. *Renewable Energy, 202*, 1402–1411.

Scott, W. R. (2013). *Institutions and organizations: Ideas, interests, and identities*. Sage.

Somoye, O. A., Ozdeser, H., & Seraj, M. (2022). Modeling the determinants of renewable energy consumption in Nigeria: Evidence from autoregressive distributed lagged in error correction approach. *Renewable Energy, 190*, 606–616.

United Nations. (2015). *Ensure access to affordable, reliable, sustainable and modern energy*. Retrieved from https://www.un.org/sustainabledevelopment/energy/#:~:text=Goal%207%20is%20about%20ensuring,hinders%20economic%20and%20human%20development

Usman, M., Ma, Z., Zafar, M. W., Waheed, A., & Li, M. (2021). Analyzing the determinants of clean energy consumption in a sustainability strategy: Evidence from EU-28 countries. *Environmental Science and Pollution Research, 28*(39), 54551–54564.

Uzar, U. (2020). Political economy of renewable energy: Does institutional quality make a difference in renewable energy consumption? *Renewable Energy, 155*, 591–603.

Vatamanu, A. F., & Zugravu, B. G. (2023). Financial development, institutional quality and renewable energy consumption. A panel data approach. *Economic Analysis and Policy, 78*, 765–775.

Wang, Q., Li, S., & Pisarenko, Z. (2020). Heterogeneous effects of energy efficiency, oil price, environmental pressure, R&D investment, and policy on renewable energy -- evidence from the G20 countries. *Energy, 209*, 118322.

Wang, E., Gozgor, G., Mahalik, M. K., Patel, G., & Hu, G. (2022). Effects of institutional quality and political risk on the renewable energy consumption in the OECD countries. *Resources Policy, 79*, 103041.

Wooldridge, J. M. (2009). *Introductory econometrics: A modern approach* (4th ed.). South-Western College Publishing.

Yakubu, I. N. (2019). Does corruption grease or sand the wheels of bank profitability in Ghana? *Cogent Economics & Finance, 7*(1), 1701909.

Yakubu, I. N., Kapusuzoğlu, A., & Ceylan, N. B. (2021). Examining the nexus between institutional quality and stock market development: Evidence from Ghana. *The Journal of Financial Researches and Studies, 13*(25), 864–878.

Yakubu, I. N., Kapusuzoglu, A., & Ceylan, N. B. (2022). ICT trade and energy transition in the BRICS economies. In *Sustainability in energy business and finance* (pp. 13–24). Springer.

Zhang, P., Li, Z., Ghardallou, W., Yan, X., & Jie, C. (2023). Nexus of institutional quality and technological innovation on renewable energy development: Moderating role of green finance. *Renewable Energy, 214*, 233–241.

Zhao, P., Lu, Z., Fang, J., Paramati, S. R., & Jiang, K. (2020). Determinants of renewable and non-renewable energy demand in China. *Structural Change and Economic Dynamics, 54*, 202–209.

Chapter 14
Assessment and Improvement of Energy Efficiency in the Context of Competitiveness

Mahmoud Zadehbagheri and Ashraf Hemeida

14.1 Introduction

Similar to the task of assessing competitiveness, UMMC's energy efficiency assessment was carried out for the company's copper division. The company's main specialization is a closed technological cycle for the production of copper products, including the extraction of raw materials and the final production of finished copper products (Eti et al., 2022). Such production is extremely energy intensive, with particular emphasis on the use of energy in the electrolytic process. The demand for energy production in a number of processes using copper concentrate is in the range of 14–20 GJ/t of copper cathode. The amount of energy consumed at the electrolytic copper purification stage is 300–400 kWh per ton of copper. This is once again emphasized by the extreme urgency of the task of evaluating and improving the energy efficiency of the production complex (Eti et al., 2023).

When forming the initial information on energy consumption indicators, as in the task of assessing competitiveness, the summary information was formed as a result indicator for the main enterprises of the company operating in the following main complexes operating in the field of copper production and products made from it: mineral resources, metallurgy, processing, and electrical and machine building (Martínez et al., 2023).

M. Zadehbagheri (✉)
Department of Electrical Engineering, Yasuj Branch, Islamic Azad University, Yasuj, Iran

A. Hemeida
Department of Electrical Engineering, Faculty of Energy Engineering, Aswan University, Aswan, Egypt
e-mail: ashraf@aswu.edu.eg

H. Dinçer et al. (eds.), *Decision Making in Interdisciplinary Renewable Energy Projects*, Contributions to Management Science,
https://doi.org/10.1007/978-3-031-51532-3_14

14.2 Literature Review

The assessment of energy efficiency in the block of environmental efficiency of fuel and energy resources consumption showed that the ratio CO of CO2 emissions to primary energy consumption in Russia is the best in comparison with all competing companies (Dinçer et al., 2023a, b, c, d, e, f). Thus, at the end of 2020CO, CO2 emissions in relation to primary consumption amounted to 2.64 g of CO2/g of cu. t, slightly inferior to the best value—2.42 g of CO2/g of cu.t (Freeport company).

At the same time, similar trends were observed for most other companies. Unfortunately, the company lags significantly behind its leading competitors Anglo American and CODELCO (in 2020, the value of the indicator is 435.7 g of CO2/USD against 156.1 and 319.8 g of CO2/USD, respectively) (An et al., 2019a, b, c, 2020a, b, c, d, 2022; Mikhaylov & Tarakanov, 2020; Moiseev et al., 2020, 2021; Gura et al., 2022; Dooyum et al., 2020; Mikhaylov et al., 2020; Mikhaylov, 2020a, b, c).

It should be noted that the ratio CO of CO2 emissions to gross revenue, similar to the indicators of energy and electricity consumption, is strongly influenced by market factors (Carayannis et al., 2023). This is confirmed by estimates for 2015–2016, where the value for almost all companies was below the baseline level (Dayong et al., 2020; Mikhaylov et al., 2018, 2023; Nyangarika et al., 2018; Danish et al., 2020, 2021; An et al., 2021; Uyeh et al., 2021; Tamashiro et al., 2021, 2023; Shaikh et al., 2021; Wang et al. 2023; He et al., 2023; Irshad et al., 2023; Iqbal et al., 2023; Aksoy et al., 2023).

Thus, this indicator exceeded 1 g of CO2/kWh (it was in the range of 1.091–1.141 g of CO2/kWh), which is higher than that of the main competitors: at the end of 2020, this indicator for competing companies was less than 0.7 g of CO2/kWh (Denisova et al., 2019; Nyangarika et al., 2019a, b; Huang et al., 2021a, b; Mikhaylov, 2018, 2022; Mikhaylov et al., 2019; Conteh et al., 2021; Sediqi et al., 2022, Khan et al., 2021; Bhuiyan et al., 2022a; Liu et al., 2021a, b; Daniali et al., 2021; Moiseev et al., 2023a, b).

For example, by the end of 2020, this indicator for the company (CO2/t) was two or more times higher than the carbon intensity of copper products produced by its main competitors, while UMMC lagged behind the best companies (Khan et al., 2022; Dinçer et al., 2022a, b; Badr et al., 2022; Barykin et al., 2022b; Mehta et al., 2022; Kalinina et al. 2022; Shaikh et al., 2022; Mikhaylov et al., 2022a, b; Nyangarika et al., 2022).

At the same time, for most of the competitors considered, a significant part of the copper products sold is copper concentrate, i.e., raw materials, the production of which requires less energy consumption than copper products of a high degree of readiness (Bhuiyan et al., 2021; Dong et al., 2021, Mikhaylov, 2021b; Barykin et al., 2022a; Liu et al., 2022a; b; Bhuiyan et al., 2022b; Danish et al., 2022a, b; Saqib et al., 2021; Mukhametov et al., 2021, Candila et al., 2021; Mikhaylov & Grilli, 2022, Li et al., 2022).

14.3 Results

Due to the nature of the problem being solved (assessment of the energy efficiency of the largest companies, which include many enterprises and divisions, often located on different continents), as well as information limitations (as a rule, data on indicators of consumption of various energy resources by most companies and enterprises are publicly available in very limited volumes), the energy efficiency assessment was carried out at production complex as a whole, according to the following set of indicators (Yüksel et al., 2023):

1. Block 1. Efficiency of energy-consuming systems and energy saving
   (a) Energy intensity of the gross revenue of the production complex by primary energy consumption in comparable prices
   (b) Electrical capacity of the gross revenue of the production complex by primary energy consumption in comparable prices
   (c) Specific energy consumption of the production complex for the production of copper products
2. Block 2. Economic efficiency of energy consumption and efficiency of use of fixed assets
   (a) Coefficient of advance of electric energy consumption in relation to the growth of the production volume of the production complex
   (b) Coefficient of advance of primary energy resources consumption in relation to the growth of the production volume of the production complex (Xu et al., 2023)
   (c) Energy intensity of fixed production assets of the production complex
   (d) Electrical capacity of the main production assets of the production complex
3. Block 3. Ecological efficiency of fuel and energy resources consumption (Han et al., 2023)
   (a) The ratio CO of CO2 emissions to primary consumption of energy resources in the entire production complex
   (b) The ratio CO of CO2 emissions to the gross revenue of the production complex in comparable prices
   (c) The ratio CO of CO2 emissions to primary electricity consumption in the entire production complex
   (d) Carbon intensity of copper products of the production complex

In contrast to the assessment of competitiveness, when assessing energy efficiency, due to the specifics of copper production and products made from it in terms of energy consumption, not all the companies identified in Clause 3.3 were considered as objects for comparative analysis, but only those that specialize mainly in the production of copper and products made from it or whose reporting is possible separately highlight the copper sector among all the company's activities (Kou et al.,

2023a, b). As a result, the energy efficiency assessment was carried out in comparison with the following companies (Moiseev et al., 2023a, b):

- CODELCO
- Freeport-McMoRan (Freeport)
- BHP Billiton (BHP)
- KGHM Polska Miedz S.A. (KGHM)
- Antofagasta plc (Antofagasta)
- Anglo American plc (Anglo American)

Below is a brief description and analysis of the results obtained.

When evaluating the energy intensity of UMMC's gross revenue by primary energy consumption, we note that UMMC generally has a good position among its competitors. According to the results of calculations in 2020, UMMC occupied an intermediate position among all companies with an energy intensity of 164.9 g/t/USD, significantly second only to Anglo American, BHP Billiton and KGHM. This is the result of a planned development program implemented over the past decade and aimed at updating the production and technological base, taking into account environmental and energy requirements (Yüksel & Dinçer, 2023).

This was primarily due to the devaluation of the Russian ruble against the US dollar. Naturally, during this period, the company's revenue volumes in dollar terms significantly decreased. In this regard, the direct consumption of energy resources in the copper sector in the period 2012–2020 not only did not increase but, on the contrary, decreased (1150 thousand tons of cu.t in 2012 against 1114 thousand tons of cu.t in 2020).

Now, when more than 80% of gas supplies to Europe are tied to prices on the markets, there are much fewer such incentives. The income of suppliers and so on fluctuate in accordance with price dynamics. So, when prices rise, Gazprom can take a wait-and-see position, limiting itself to fulfilling its long-term contractual obligations. From an economic point of view, the situation when gas is traded as a classic commodity, the price of which is formed due to the current balance and supply, is more correct; however, the market becomes volatile and unstable, because there are few large gas suppliers to Europe. Gas prices will remain high for a long time. It's basically connected with quotas for greenhouse gas emissions; as mentioned above, they have risen in price several times. Many manufacturers use raw fuel oil, gas, coal, etc. (Wan et al., 2023). Thus, the increase in emission quotas affects the products produced. Compared to oil and coal, gas is a less "dirty" fuel, so this factor also plays a role in raising gas prices.

Huge investments are being directed to the global energy sector; in 2019, they amounted to $ 2000 billion. About a quarter of these investments is invested in renewable energy and energy efficiency. But even these investments are considered insufficient goals set by the UN Sustainable Development Goals and the Paris Climate Agreement. It can be assumed that, despite multimillion-dollar investments and various legislative instruments adopted to increase the share of renewable energy in the total energy balance of the world, the demand for oil and gas will always be, and it tends to grow in the near future (Ai et al., 2023).

In the BP Statistical Review of World Energy, CO2 reduction due to the contraction of the economy, it is estimated at 6.3%. To achieve the goals according to the Paris Agreement, the corporation proposes to reduce carbon emissions for 30 years at the pace of 2020.

By the end of 2020, global energy consumption decreased by 4.7%, which led to an adjustment in gas prices (in Europe in 2020, $113 per cubic meter compared to $159) and, in 2019, then sharply increased several times.

Oil and gas prices will change, but despite this, the development of global energy in the near future is impossible to imagine without oil and gas sectors (Niu et al., 2023). Due to various sanctions and bills in support of renewable energy and against non-"green energy," criticism of renewable energy has intensified. In the long term, the green transition will intensify, which in turn will have an impact on the conditions for financing oil and gas projects (Dınçer et al., 2023a, b; Ivleva et al., 2023; Aissou et al., 2023; Mikhaylov, 2023).

Recorded gas prices in Europe were helped by the rapid economic recovery and the problems of green energy. The short-term effect of the price crisis may be some softening of European requirements for Gazprom, but in the long run, expensive gas will stimulate the EU to reduce dependence on hydrocarbon imports (Aksoy et al., 2023).

Grouping approaches:

1. Tax benefits in the field of green technologies
2. Tax credit and (or) investment tax credit (for the production of energy from renewable sources, for energy projects in clean energy, for the production of biofuels)
3. Environmental taxes (carbon tax) and energy taxes (indirect taxes, including VAT and excise taxes on fuel, motor oils), which are considered as a source of formation of special climate funds and financial support for investments in green technologies; grant support for investments in green technologies (energy equipment)
4. Voluntary agreements with energy-intensive companies on preferential environmental taxation for innovation developers
5. A combination of tax incentives for investments in green technologies and tax incentives environmental pollution destimulation
6. Tax administration and measures to counteract evasion of environmental taxes and fees and illegal application of preferential taxation of investment activities in the field of green technologies

14.4 Conclusion

The disadvantage of the energy intensity indicator of the gross revenue of the production complex for primary energy consumption is its significant dependence on world copper prices. In 2015–2016, when copper prices were low, this indicator was the highest due to lower revenue volumes, while in 2012–2013, when copper prices were low, the energy intensity of companies' gross revenue was extremely low.

The electric capacity indicator of the company's gross revenue for primary energy consumption was even better than the previous indicator. On the other hand, the value of electrical capacity depends very much on the technologies used and, in this respect, may differ significantly from company to company. Therefore, this indicator is less informative than the previous one and should be used only with adaptation to the conditions of specific companies, which was difficult to perform in the framework of this study. Perhaps, the most significant criterion for evaluating the energy efficiency of the production of competing companies is the specific energy consumption consumed in the production of products in physical terms. It is not influenced by market factors; as a result, its value reflects the real level of energy efficiency of the company.

References

Ai, R., Zheng, Y., Yüksel, S., & Dinçer, H. (2023). Investigating the components of fintech ecosystem for distributed energy investments with an integrated quantum spherical decision support system. *Financial Innovation, 9*(1), 27.

Aissou, A., Ahmed, H. S., Hussein, M. E., Hassan, M. R., Mikhaylov, A., & Mahmoud, E. (2023). Optimizing reliability of flow networks using RWGA subject to total capacity constraint. *Applied Mathematics and Information Sciences, 7*(4), 559–565. https://doi.org/10.18576/amis/170404

Aksoy, T., Yüksel, S., Dınçer, H., Hacioglu, U., Mikhaylov, A., & Muyeen, S. M. (2023). Assessing the electricity production capacities of emerging markets for the sustainable investment. *IEEE Access, 11*, 24574–24590

An, J., Mikhaylov, A., & Moiseev, N. (2019a). Oil price predictors: Machine learning approach. *International Journal of Energy Economics and Policy, 9*(5), 1–6.

An, J., Mikhaylov, A., & Sokolinskaya, N. (2019b). Machine learning in economic planning: Ensembles of algorithms. *Journal of Physics: Conference Series, 1353*, 012126.

An, J., Mikhaylov, A., & Sokolinskaya, N. (2019c). Oil incomes spending in sovereign fund of Norway (GPFG). *Investment Management and Financial Innovations*, 16(3), 10–17.

An, J., Mikhaylov, A., & Jung, S.-U. (2020a). The strategy of South Korea in the global oil market. *Energies, 13*(10), 2491.

An, J., Mikhaylov, A., & Kim, K. (2020b). Machine learning approach in heterogeneous group of algorithms for transport safety-critical system. *Applied Sciences, 10*(8), 2670.

An, J., Mikhaylov, A., & Richter, U. H. (2020c). Trade war effects: Evidence from sectors of energy and resources in Africa. *Heliyon, 6*, e05693. https://doi.org/10.1016/j.heliyon.2020.e05693

An, J., Dorofeev, M., & Zhu, S. (2020d). Development of energy cooperation between Russia and China. *International Journal of Energy Economics and Policy*, 10(1), 134–139.

An, J., Mikhaylov, A., & Jung, S.-U. (2021). A linear programming approach for robust network revenue management in the airline industry. *Journal of Air Transport Management, 91*, 101979.

An, J., Mikhaylov, A., Dinçer, H., & Yüksel, S. (2022). Economic modelling of electricity generation: Long short-term memory and Q-rung Orthopair fuzzy sets. *Heliyon, 8*, e12345.

Badr, A. M. N. A. A., Mohamed, T. I., Osman, N. A. Q., & Mikhaylov, A. (2022). A review of social media website users' interaction paths with governmental accounts during the COVID-19 pandemic. *Informatics, 9*, 50. https://doi.org/10.3390/informatics9030050

Barykin, S., Mikheev, A., Kiseleva, E., Putikhin, Y., Alekseeva, N., & Mikhaylov, A. (2022a). An empirical analysis of Russian regions' debt sustainability. *Economies, 10*(4), 296.

Barykin, S. E., Kapustina, I. V., Sergeev, S. M., Daniali, S. M., Kopteva, L. A., Semenova, G. N., Pryadko, I. P., Mikhaylov, A., Baboshkin, P., Datsyuk, P., & Senjyu, T. (2022b). Financial logistics models based on systematic approach improving management solutions. *F1000Research, 11, 570*, 10.12688/f1000research.111252.1.

Bhuiyan, M. A., An, J., Mikhaylov, A., Moiseev, N., & Danish, M. S. S. (2021). Renewable energy deployment and COVID-19 measures for sustainable development. *Sustainability, 13*, 4418.

Bhuiyan, M. A., Dinçer, H., Yüksel, S., Mikhaylov, A., Danish, M. S. S., Pinter, G., Uyeh, D. D., & Stepanova, D. (2022a). Economic indicators and bioenergy supply in developed economies: QROF-DEMATEL and random forest models. *Energy Reports, 8*(2022), 561–570.

Bhuiyan, M. A., Zhang, Q., Khare, V., Mikhaylov, A., Pinter, G., & Huang, X. (2022b). Renewable energy consumption and economic growth nexus—A systematic literature review. *Frontiers in Environmental Science, 10*, 878394. https://doi.org/10.3389/fenvs.2022.878394

Candila, V., Maximov, D., Mikhaylov, A., Moiseev, N., Senjyu, T., & Tryndina, N. (2021). On the relationship between oil and exchange rates of oil-exporting and oil-importing countries: From the great recession period to the Covid-19 era. *Energies, 14*, 8065.

Carayannis, E., Kostis, P., Dinçer, H., & Yüksel, S. (2023). Quality function deployment-oriented strategic outlook to sustainable energy policies based on quintuple innovation helix. *Journal of the Knowledge Economy*, 1–19.

Conteh, F., Takahashi, H., Hemeida, A. M., Krishnan, N., Mikhaylov, A., & Senjyu, T. (2021). Hybrid grid-connected renewable power generation for sustainable electricity supply in Sierra Leone: Case study Lungi Town. *Sustainability, 13*(18).

Daniali, S. M., Barykin, S. E., Kapustina, I. V., Khortabi, F. M., Sergeev, S. M., Kalinina, O. V., Mikhaylov, A., Veynberg, R., Zasova, L., & Senjyu, T. (2021). Predicting volatility index according to technical index and economic indicators on the basis of deep learning algorithm. *Sustainability, 13*, 14011.

Danish, M. S. S., Bhattacharya, A., Stepanova, D., Mikhaylov, A., Grilli, M. L., Khosravy, M., & Senjyu, T. (2020). A systematic review of metal oxide applications for energy and environmental sustainability. *Metals, 10*(12), 1604.

Danish, M. S. S., Senjyu, T., Sabory, N. R., Khosravy, M., Grilli, M. L., Mikhaylov, A., & Majidi, H. (2021). A forefront framework for sustainable aquaponics modeling and design. *Sustainability, 13*, 9313.

Danish, M. S., Senjyu, T. S., Sabory, N. R., & Mikhaylov, A. (2022a). Lessons from COVID-19 conferring environmental re-engineering opportunity. In M. Danish & T. Senjyu (Eds.), *Eco-friendly and agile energy strategies and policy development* (pp. 99–118). IGI Global. https://doi.org/10.4018/978-1-7998-9502-2.ch005

Danish, M. S., Senjyu, T. S., Sabory, N. R., & Mikhaylov, A. (2022b). Interplay consequences of COVID-19 on global environmental sustainability. In M. Danish & T. Senjyu (Eds.), *Eco-friendly and agile energy strategies and policy development*. IGI Global. https://doi.org/10.4018/978-1-7998-9502-2.ch003

Dayong, N., Mikhaylov, A., Bratanovsky, S., Shaikh, Z. A., & Stepanova, D. (2020). Mathematical modeling of the technological processes of catering products production. *Journal of Food Process Engineering, 43*(2), 1–9.

Denisova, V., Mikhaylov, A., & Lopatin, E. (2019). Blockchain infrastructure and growth of global power consumption. *International Journal of Energy Economics and Policy, 9*(4), 22–29.

Dinçer, H., Yüksel, S., Mikhaylov, A., Pinter, G., & Shaikh, Z. A. (2022a). Analysis of renewable-friendly smart grid technologies for the distributed energy investment projects using a hybrid picture fuzzy rough decision-making approach. *Energy Reports, 8*, 11466–11477. https://doi.org/10.1016/j.egyr.2022.08.275

Dinçer, H., Yüksel, S., Mikhaylov, A., Barykin, S. E., Aksoy, T., & Hacioğlu, U. (2022b). Analysis of environmental priorities for green project investments using an integrated

q-rung orthopair fuzzy modelling. *IEEE Access, 10*, 50996–51007. https://doi.org/10.1109/ACCESS.2022.3174058

Dinçer, H., Yüksel, S., Hacioglu, U., Yilmaz, M. K., & Delen, D. (2023a). Development of a sustainable corporate social responsibility index for performance evaluation of the energy industry: A hybrid decision-making methodology. *Resources Policy, 85*, 103940.

Dinçer, H., Eti, S., Gökalp, Y., Yüksel, S., & Çelebi, B. (2023b). The role of small-scale wind turbines in increasing the competitiveness of hospitals. In *Handbook of research on quality and competitiveness in the healthcare services sector* (pp. 336–352). IGI Global.

Dinçer, H., Yüksel, S., Mikhaylov, A., Muyeen, S. M., Chang, T., Barykin, S., & Kalinina, O. (2023c). CO2 emissions integrated fuzzy model: A case of seven emerging economies. *Energy Reports, 9*, 5741–5751.

Dinçer, H., Yüksel, S., Aksoy, T., Hacıoğlu, Ü., Mikhaylov, A., & Pinter, G. (2023d). Analysis of solar module alternatives for efficiency-based energy investments with hybrid 2-tuple IVIF modeling. *Energy Reports, 10*, 61–71.

Dinçer, H., Eti, S., Yüksel, S., Özdemir, S., Yílmaz, A. E., & Ergün, E. (2023e). Integrating data mining and fuzzy decision-making techniques for analyzing the key minimizing factors of carbon emissions. *Journal of Intelligent & Fuzzy Systems, 45*, 1–17.

Dinçer, H., Eti, S., Aksoy, T., Yüksel, S., Hacioglu, U., Mikhaylov, A., & Muyeen, S. M. (2023f). Analysis of environmental impact for material production investments using a novel soft computing methodology. *IEEE Access, 11*, 37987–38001. https://doi.org/10.1109/ACCESS.2023.3266524

Dong, B., Ikonnikova, I., Rogulin, R., Sakulyeva, T., & Mikhaylov, A. (2021). Environmental economic approach to optimization of transport communication in megacities. *Journal of Environmental Science and Health, Part A, 56*(6), 660–666.

Dooyum, U. D., Mikhaylov, A., & Varyash, I. (2020). Energy security concept in Russia and South Korea. *International Journal of Energy Economics and Policy, 10*(4), 102–107.

Eti, S., Dinçer, H., Yüksel, S., & Gökalp, Y. (2022). Analysis of the suitability of the solar panels for hospitals: A new fuzzy decision-making model proposal with the T-spherical TOP-DEMATEL method. *Journal of Intelligent & Fuzzy Systems, Preprint*, 1–13.

Eti, S., Dinçer, H., Gökalp, Y., Yüksel, S., & Kararoğlu, D. (2023). Identifying key issues to handle the inflation problem in the healthcare industry caused by energy prices: An evaluation with decision-making models. In *Managing inflation and supply chain disruptions in the global economy* (pp. 162–178). Springer.

Gura, D., Mikhaylov, A., Glushkov, S., Zaikov, M., & Shaikh, Z. A. (2022). Model for estimating power dissipation along the interconnect length in single on-chip topology. *Evolutionary Intelligence, 15*, 2369–2373. https://doi.org/10.1007/s12065-020-00407-7

Han, Z., Gong, L., Chen, H., & Yüksel, S. (2023). The role of mineral and oil resources in achieving a clean environment in emerging economies: An asymmetric analysis. *Resources Policy, 86*, 104099.

He, H., Huang, Y., Nakadomari, A., Masrur, H., Krishnan, N., Hemeida, A. M., Mikhaylov, A., Omine, E., & Senjyu, T. (2023). Potential and economic viability of green hydrogen production from seawater electrolysis using renewable energy in remote Japanese islands. *Renewable Energy, 202*, 1436–1447. https://doi.org/10.1016/j.renene.2022.12.046

Huang, Y., Masrur, H., Shigenobu, R., Hemeida, A. M., Mikhaylov, A., & Senjyu, T. (2021a). A comparative design of a campus microgrid considering a multi-scenario and multi-objective approach. *Energies, 14*, 2853.

Huang, Y., Yona, A., Takahashi, H., Hemeida, A. M., Mandal, P., Mikhaylov, A., Senjyu, T., & Lotfy, M. E. (2021b). Energy management system optimization of drug store operate electric vehicles charging station. *Sustainability, 14*, 1217836.

Iqbal, K. M. J., Khan, M. I., Mikhaylov, A., Shah, A. A., Yadykin, V., Filho, W. L., Tariq, M. A. U. R., & Ullah, W. (2023). Modeling principles, criteria and indicators to assess water sector governance for climate compatibility and sustainability. *Frontiers in Environmental Science, 11*, 989930.

Irshad, A. S., Ludin, G. A., Masrur, H., Ahmadi, M., Yona, A., Mikhaylov, A., Krishnan, N., & Senjyu, T. (2023). Optimization of grid-photovoltaic and battery hybrid system with most technically efficient PV technology after the performance analysis. *Renewable Energy, 207*, 714–730.

Ivleva, S., Hajimajeed, M., Kartashova, O., & Mikhaylov, A. (2023). A comparative study of regional variations in common visual impairments among school-age children. *Electronic Journal of General Medicine, 20*(5), em504. https://doi.org/10.29333/ejgm/13264

Kalinina, O. V., Barykin, S. E., Sergeev, S. M., Semenova, G. N., Fatkullina, A., Mikhaylov, A., & De La Poza, E. (2022). Smart city perspectives in post-pandemic governance: Externalities reduction policy. *F1000Research, 11*, 1032. https://doi.org/10.12688/f1000research.123195.1

Khan, A. A., Shaikh, Z. A., Baitenova, L., Mutaliyeva, L., Moiseev, N., Mikhaylov, A., Laghari, A. A., Idris, S. A., & Alshazly, H. (2021). QoS-ledger: Smart contracts and metaheuristic for secure quality-of-service and cost-efficient scheduling of medical-data processing. *Electronics, 10*, 3083.

Khan, A. A., Laghari, A. A., Gadekallu, T. R., Shaikh, Z. A., Javed, A. R., Rashid, M., Estrela, V. V., & Mikhaylov, A. (2022). A drone-based data management and optimization using metaheuristic algorithms and blockchain smart contracts in a secure fog environment. *Computers and Electrical Engineering, 102*, 108234. https://doi.org/10.1016/j.compeleceng.2022.108234

Kou, G., Pamucar, D., Dinçer, H., & Yüksel, S. (2023a). From risks to rewards: A comprehensive guide to sustainable investment decisions in renewable energy using a hybrid facial expression-based fuzzy decision-making approach. *Applied Soft Computing, 142*, 110365.

Kou, G., Yuksel, S., & Dincer, H. (2023b). A facial expression and expert recommendation fuzzy decision-making approach for sustainable business investments within the metaverse world. *Applied Soft Computing, 148*, 110849.

Li, J., Dinçer, H., Yüksel, S., Mikhaylov, A., & Barykin, S. E. (2022). Bipolar q-ROF hybrid decision making model with golden cut for analyzing the levelized cost of renewable energy alternatives. *IEEE Access, 10*, 35900. https://doi.org/10.1109/ACCESS.2022.3168315

Liu, L., Kato, T., Mandal, P., Mikhaylov, A., & Hemeida, A. M., & Senjyu, T. (2021a). Multi-area power system load frequency control by using model predictive control by introducing renewable energy sources. In *2021 International Conference on Science & Contemporary Technologies (ICSCT)*, pp. 1–6. https://doi.org/10.1109/ICSCT53883.2021.9642521.

Liu, L., Kato, T., Mandal, P., Mikhaylov, A., Hemeida, A. M., Takahashi, H., & Senjyu, T. (2021b). Two cases studies of model predictive control approach for hybrid renewable energy systems. *AIMS Energy, 9*(6), 1241–1259.

Liu, Z., Panfilova, E., Mikhaylov, A., & Kurilova, A. (2022a). Assessing stability in the relationship between parties in crowdfunding and crowdsourcing projects during the COVID-19 crisis. *Journal of Global Information Management (JGIM), 30*(4), 1–18. https://doi.org/10.4018/JGIM.297905

Liu, Z.-J., Panfilova, E., Mikhaylov, A., & Kurilova, A. (2022b). COVID-19 crisis impact on the stability between parties in crowdfunding and crowdsourcing. *Wireless Personal Communications, 122*, 915–930.

Martínez, L., Dinçer, H., & Yüksel, S. (2023). A hybrid decision making approach for new service development process of renewable energy investment. *Applied Soft Computing, 133*, 109897.

Mehta, R., Verghese, J., Mahajan, S., Barykin, S., Bozhuk, S., Kozlova, N., Kapustina, I., Mikhaylov, A., Naumova, E., & Dedyukhina, N. (2022). Consumers' behavior in conversational commerce marketing based on messenger chatbots. *F1000Research, 11*(647). https://doi.org/10.12688/f1000research.122037.1

Mikhaylov, A. (2018). The yield of the carry trade strategy. *Finance: Theory and Practice, 22*(3), 52–63. https://doi.org/10.26794/2587-5671-2018-22-3-52-63

Mikhaylov, A. (2020a). Cryptocurrency market analysis from the open innovation perspective. *Journal of Open Innovation: Technology, Market, and Complexity, 6*(4), 197.

Mikhaylov, A. (2020b). Geothermal energy development in Iceland. *International Journal of Energy Economics and Policy, 10*(4), 31–35.

Mikhaylov, A. (2020c). Cryptocurrency market development: Hurst method. *Finance: Theory and Practice, 24*(3), 81–91.

Mikhaylov, A. (2021). Lichens as indicators of atmospheric pollution in urban ecosystems. *Israel Journal of Ecology & Evolution, 67*(1–2), 60–68. https://doi.org/10.1163/22244662-bja10016

Mikhaylov, A. (2022). Sustainable development and renewable energy: A new view to a global problem. *Energies, 15*, 1397. https://doi.org/10.3390/en15041397

Mikhaylov, A. (2023). Understanding the risks associated with wallets, depository services, trading, lending, and borrowing in the crypto space. *Journal of Infrastructure, Policy and Development, 6*, 2.

Mikhaylov, A., & Grilli, M. L. (2022). Machine learning methods and sustainable development: Metal oxides and multilayer metal-oxides. *Metals, 12*(4).

Mikhaylov, A., & Tarakanov, S. (2020). Development of Levenberg-Marquardt theoretical approach for electric network. *Journal of Physics: Conference Series, 1515*, 052006.

Mikhaylov, A., Sokolinskaya, N., & Nyangarika, A. (2018). Optimal carry trade strategy based on currencies of energy and developed economies. *Journal of Reviews on Global Economics, 7*, 582–592.

Mikhaylov, A., Burakov, D., & Didenko, V. (2019). Relationship between oil price and macroeconomic indicators in Russia. *Finance: Theory and Practice, 23*(2), 105–116. https://doi.org/10.26794/2587-5671-2019-23-2-105-116

Mikhaylov, A., Moiseev, N., Aleshin, K., & Burkhardt, T. (2020). Global climate change and greenhouse effect. *Entrepreneurship and Sustainability Issues, 7*(4), 2897–2913. https://doi.org/10.9770/jesi.2020.7.4(21)

Mikhaylov, A., Bhatti, I. M., Dinçer, H., & Yüksel, S. (2022a). Integrated decision recommendation system using iteration-enhanced collaborative filtering, golden cut bipolar for analyzing the risk-based oil market spillovers. *Computational Economics*. https://doi.org/10.1007/s10614-022-10341-8

Mikhaylov, A., Yumashev, A., & Kolpak, E. (2022b). Quality of life, anxiety and depressive disorders in patients with extrasystolic arrhythmia. *Archives of Medical Science, 18*(2), 328. https://doi.org/10.5114/aoms.2020.101359

Mikhaylov, A., Dinçer, H., Yüksel, S., Pinter, G., & Shaikh, Z. A. (2023). Bitcoin mempool growth and trading volumes: Integrated approach based on QROF multi-SWARA and aggregation operators. *Journal of Innovation & Knowledge, 8*(3), 100378.

Moiseev, N., Mikhaylov, A., Varyash, I., & Saqib, A. (2020). Investigating the relation of GDP per capita and corruption index. *Entrepreneurship and Sustainability Issues, 8*(1), 780–794.

Moiseev, N., Sorokin, A., Zvezdina, N., Mikhaylov, A., Khomyakova, L., & Danish, M. S. S. (2021). Credit risk theoretical model on the base of DCC-GARCH in time-varying parameters framework. *Mathematics, 9*, 2423. https://doi.org/10.3390/math9192423

Moiseev, N., Mikhaylov, A., Dinçer, H., & Yüksel, S. (2023a). Effects of market capitalization shocks for open innovation models in e-commerce industry: Golden cut QROF MCDM analysis. *Financial Innovation, 9*(1), 32.

Moiseev, N., Mikhaylov, A., Dinçer, H., & Yüksel, S. (2023b). Market capitalization shock effects on open innovation models in e-commerce: Golden cut q-rung orthopair fuzzy multicriteria decision-making analysis. *Financial Innovation, 9*(1), 55.

Mukhametov, A., Bekhorashvili, A., Avdeenko, A., & Mikhaylov, A. (2021). The impact of growing legume plants under conditions of biologization and soil cultivation on Chernozem fertility and productivity of rotation crops. *Legume Research, 44*, 12.

Niu, X., Yüksel, S., & Dinçer, H. (2023). Emission strategy selection for the circular economy-based production investments with the enhanced decision support system. *Energy, 274*, 127446.

Nyangarika, A., Mikhaylov, A., & Tang, B.-J. (2018). Correlation of oil prices and gross domestic product in oil producing countries. *International Journal of Energy Economics and Policy, 8*(5), 42–48.

Nyangarika, A., Mikhaylov, A., & Richter, U. (2019a). Influence oil price towards economic indicators in Russia. *International Journal of Energy Economics and Policy, 9*(1), 123–129.

Nyangarika, A., Mikhaylov, A., & Richter, U. (2019b). Oil price factors: Forecasting on the base of modified auto-regressive integrated moving average model. *International Journal of Energy Economics and Policy, 9*(1), 149–159.

Nyangarika, A., Mikhaylov, A., Muyeen, S. M., Yadykin, V., Mottaeva, A., Pryadko, I., Barykin, S., Fomenko, N., Rykov, G., & Shvandar, K. (2022). Energy stability and decarbonization in developing countries: Random Forest approach for forecasting of crude oil trade flows and macro indicators. *Frontiers in Environmental Science, 10*, 1031343. https://doi.org/10.3389/fenvs.2022.1031343

Saqib, A., Chan, T.-H., Mikhaylov, A., & Lean, H. H. (2021). Are the responses of sectoral energy imports asymmetric to exchange rate volatilities in Pakistan? Evidence from recent foreign exchange regime. *Frontiers in Energy Research, 9*, 614463. https://doi.org/10.3389/fenrg.2021.614463

Sediqi, M. M., Nakadomari, A., Mikhaylov, A., Krishnan, N., Lotfy, M. E., Yona, A., & Senjyu, T. (2022). Impact of time-of-use demand response program on optimal operation of Afghanistan real power system. *Energies, 15*, 296.

Shaikh, Z. A., Moiseev, N., Mikhaylov, A., & Yüksel, S. (2021). Facile synthesis of copper oxide-cobalt oxide/nitrogen-doped carbon (Cu2O-Co3O4/CN) composite for efficient water splitting. *Applied Sciences, 11*, 9974. https://doi.org/10.3390/app11219974

Shaikh, Z. A., Kraikin, A., Mikhaylov, A., & Pinter, G. (2022). Forecasting stock prices of companies producing solar panels using machine learning methods. *Complexity, 2022*, 9186265. https://doi.org/10.1155/2022/9186265

Tamashiro, K., Alharbi, T., Mikhaylov, A., Hemeida, A. M., Krishnan, N., Lotfy, M. E., & Senjyu, T. (2021). Investigation of home energy management with advanced direct load control and optimal scheduling of controllable loads. *Energies, 14*, 7314. https://doi.org/10.3390/en14217314

Tamashiro, K., Omine, E., Krishnan, N., Mikhaylov, A., Hemeida, A. M., & Senjyu, T. (2023). Optimal components capacity based multi-objective optimization and optimal scheduling based MPC-optimization algorithm in smart apartment buildings. *Energy & Buildings, 278*, 112616. https://doi.org/10.1016/j.enbuild.2022.112616

Uyeh, D. D., Asem-Hiablie, S., Park, T., Kim, K.-M., Mikhaylov, A., Woo, S., & Ha, Y. (2021). Could Japonica rice be an alternative variety for increased global food security and climate change mitigation? *Food, 10*, 1869. https://doi.org/10.3390/foods10081869

Wan, Q., Miao, X., Wang, C., Dinçer, H., & Yüksel, S. (2023). A hybrid decision support system with golden cut and bipolar q-ROFSs for evaluating the risk-based strategic priorities of fintech lending for clean energy projects. *Financial Innovation, 9*(1), 1–25.

Wang, Q., Guo, J., Li, R., Mikhaylov, A., & Moiseev, N. (2023). Does technical assistance alleviate energy poverty in sub-Saharan African countries? A new perspective on spatial spillover effects of technical assistance. *Energy Strategy Reviews, 45*, 101047. https://doi.org/10.1016/j.esr.2022.101047

Xu, X., Yüksel, S., & Dinçer, H. (2023). An integrated decision-making approach with golden cut and bipolar q-ROFSs to renewable energy storage investments. *International Journal of Fuzzy Systems, 25*(1), 168–181.

Yüksel, S., & Dinçer, H. (2023). Sustainability analysis of digital transformation and circular industrialization with quantum spherical fuzzy modeling and golden cuts. *Applied Soft Computing, 138*, 110192.

Yüksel, S., Dinçer, H., & Uluer, G. S. (2023). Socio-economic indicators of nuclear energy investments: A study on countries. In *Inclusive developments through socio-economic indicators: New theoretical and empirical insights* (pp. 33–44). Emerald Publishing.

Chapter 15
Charismatic Leadership in Work Organizations: An Evaluation for Energy Industry

Gergely Czukor and Cemre Çınar

15.1 Introduction

Leadership occupies a key role in motivating employees to achieve organizational goals. One significant source of such leadership influence is charisma. Charismatic leadership is a widespread application in leader development and practice, and charismatic leaders positively affect organizational outcomes. However, a dark side of its application includes overemphasis on charismatic communication skills and relationship building while neglecting the achievement-oriented aspects of charisma. Emphasizing soft skills while understating accomplishment encourages superficiality in leadership development and practice. Prior theories view charisma as an attribution of the followers to the leader (Conger & Kanungo, 1998; Yukl, 1999) as a reflection of the relationship between the leader and the followers (Howell & Shamir, 2005) or as a communication strategy (Antonakis et al., 2016). We argue that these perspectives can be enriched by adopting a social/cultural anthropology viewpoint, proposing that individuals earn charisma via achievement through adversity (Lindholm, 2002; Tambiah, 1984). We propose a model integrating current views within the framework of earned charisma, aiming to establish cross-disciplinary consistency in the charismatic leadership literature and emphasize the achievement aspect of charisma in leadership practice in work organizations.

In the model, we address leader influence from a social identity perspective concerning the leader's representation of the struggling follower prototype. The social identity approach suggests that in order to emerge and be perceived as effective,

G. Czukor (✉)
I/O Psychology Graduate Program, Istanbul Bilgi University, Istanbul, Turkey
e-mail: gergely.czukor@bilgi.edu.tr

C. Çınar
School of Management, Sabanci University, Istanbul, Turkey

H. Dinçer et al. (eds.), *Decision Making in Interdisciplinary Renewable Energy Projects*, Contributions to Management Science,
https://doi.org/10.1007/978-3-031-51532-3_15

leaders must demonstrate that they are prototypical members of their in-group (Hogg, 2001; Reicher et al., 2005). However, to correspond with charismatic leaders often considered unusually or even supernaturally gifted, we propose that these leaders identify the attributes of the struggling follower prototype and act as their guiders and saviors. The charismatic leader condenses the collective norms, values, needs, and goals into the prototype to simplify diversity, enabling the leader to develop a powerful emotional bond with the followers.

Organizations in the energy sector, including fossil fuel and utility companies, face hardships resulting from the lack of leadership to lead employees, build teams, and manage employee careers (Center for Creative Leadership, 2020). Unfortunately, infamous leaders, including Jeffrey Skilling of Enron, have damaged the reputation of leadership in the sector for the past twenty years. At Enron, self-interested top executives abused their power and put pressure on their employees to engage in unethical business and accounting practices (Cruver, 2003; Tourish & Vatcha, 2005). Eventually, customers lost their wealth, and Enron filed bankruptcy, leaving thousands of people jobless. The sector needs leaders who act in the long-term, constructive interest of their employees and customers. Additionally, energy firms have the responsibility to ensure sustainability. The amendments proposed by this chapter to charismatic leadership as applied to organizations can provide insights into how leaders can influence their employees constructively. Leadership in the energy sector needs to embrace the needs of the followers by leaders who demonstrate outstanding achievement and are driven by universal ethics.

15.2 Defining Charismatic Leadership

One of the inconsistencies in the literature on charisma and charismatic leadership is the abundance of often conflicting definitions. Although having many approaches enhances construct validity, in this chapter, we attempt to advance the literature by aligning and clarifying existing definitions. As charismatic leadership is a subset of the broader field of leadership studies, it is helpful to examine how the construct of leadership is defined. The most widely accepted definition of leadership posits that it is the influence of one individual on a group to achieve a common goal (Northouse, 2012). That is, when scholars use the term leadership, they refer to the existing influence of a leader on the followers. The influence process refers to convincing the group to accept a vision and then taking action to achieve the goals embedded in that vision (Sy et al., 2018). Social influence occurs because the leader inspires the followers via his/her communication and exemplary behavior (Conger & Kanungo, 1998). Thus, leadership is distinguished from management, in which coerciveness and leading by exception are some of the ways to influence subordinates (Kotter, 1990).

Leadership theories attempt to explain the underlying mechanism by which leaders influence the followers. For example, the trait approaches argue that leaders possess certain traits, such as determination and intelligence, which help to persuade others of a vision (e.g., Bryman, 1992). The behaviorist view suggests that leaders engage in relationship and task-oriented behaviors directed toward their followers to achieve a goal (Stogdill, 1974). In line with these approaches, charismatic leadership refers to how a leader's charismatic character influences the followers. Accordingly, the definition of charismatic leadership may be proposed as the influence of a leader on followers using charisma to achieve a common goal.

15.3 Approaches to Charisma

Intriguingly, the term, charisma, can be more challenging to define despite being used in everyday language. Research on generic charisma maintains that charisma is a multidimensional competency attribute that individuals can develop, and accordingly, anyone may have some levels of charisma (Verčič & Verčič, 2011). Research specifically on leader charisma has approached charisma as a personal attribute of extreme self-confidence, dominance, and conviction of moral righteousness (House, 1976). Because perceivers can differ in the extent to which they perceive a leader as charismatic, researchers have proposed attribution and relationship explanations. According to the attribution explanation, charisma is an attribution made by perceivers based on certain observable leader behaviors (Conger & Kanungo, 1998). Thus, the followers bestow charisma (i.e., perceiving a leader as admirable and a role model) onto their leaders. Hence, in this view, charisma does not constitute a distinct individual attribute. A closely related approach argues that charisma exists in the relationship between the leader and followers and refers to the energetic and enthusiastic leader-follower relationship (Howell & Shamir, 2005). Theorists, in this view, speak of charismatic leader-follower relations, which develop based on the followers' positive perceptions and emotional responses to the leaders' vision and specific behaviors, such as being decisive and a risk-taker.

More recently, researchers have disputed the attribution and relationship approaches because these views need clarification for the purposes of scientific study. Critics argue that in these approaches the cause (i.e., charisma) and its effects (e.g., follower enthusiasm) are not distinct factors but instead overlap or even refer to the same construct (Antonakis et al., 2016). That is, charismatic-leader-follower relations and attributions seem more of outcomes, while the focal predictor, which is supposed to be charisma, remains obscure. To overcome these theoretical and methodological limitations, charisma should be clearly defined as a distinct, measurable construct so researchers can study its effects on various outcome variables, such as attribution or charismatic leader-follower relations.

Because leaders influence followers through communication, researchers studied leader speeches and derived nine charismatic leadership techniques that can be used to distinguish charismatic and noncharismatic leaders (Antonakis et al., 2021). In this view, charisma refers to charismatic communication, in which the leader engages in signaling using emotions, values, and symbols relying on charismatic leadership tactics. These include verbal tactics such as metaphors, anecdotes and rhetorical questions, and nonverbal techniques such as pauses and gestures. Charismatic leaders are argued to use such tactics, which help them to trigger enthusiasm in their followers, motivating them to follow a vision and engage in collective action.

The communication approach has implications for leadership development. Leaders can be trained to develop verbal and nonverbal charismatic leadership tactics to increase their followers' motivation to achieve organizational goals. Experimental research shows that when a leader used the nine charismatic leadership tactics in delivering a motivational speech, subordinates showed enhanced performance compared to a speech not using those tactics (Antonakis et al., 2021). However, communicating charismatically does not necessarily transfer to influence the followers. That is, *charismatic leadership* (referring to the occurring influence of a charismatic leader on the followers) is not the same construct as being a *charismatic leader*. A leader can use a charismatic communication style yet not influence the followers. Not every follower will respond with the same sort of enthusiasm to the same charismatic leader. Some followers may feel inspired in response to the leader's charisma, while others may not be affected. Hence, leader charisma may not necessarily be associated with charismatic leadership.

15.4 Earned Charisma

The reviewed approaches to charisma and charismatic leadership have provided insights into the psychological mechanisms of how charismatic leaders influence their followers. However, these current approaches deviate from views of other disciplines, specifically social/cultural anthropology; hence, the literature includes a cross-disciplinary inconsistency. Anthropologists maintain that charisma develops by individuals demonstrating endurance, resilience, ability, and skills to overcome austerity (Lindholm, 2002; Tambiah, 1984). Accordingly, cultural/social anthropology provides a perspective to embrace the achievement aspect of charisma, which we believe, is missing in the attribution, relations, and communication views. Further, due to this shortcoming, the legitimacy behind the charismatic effect is understated. For example, approaching charisma solely from a communication point of view implies that communication tactics weigh more than task-oriented achievement skills. As applied to the workplace, we believe that charismatic leadership should incorporate achievement-oriented charisma to avoid shallowness in leadership development and practice. We incorporate the recent approaches into a more achievement-focused model using earned charisma to create consistency among different disciplines and approaches in the charismatic leadership literature and to make the organizational application of charismatic leadership theoretically better founded.

Weber was the first to provide a synthesis of charismatic influence. In his view, the ideal type of charisma was personal charisma, in which the leader inserts emotional domination over the followers who perceive their leader as of divine origin

(Weber, 1978). Weber developed his approach mainly in the context of religious prophets and tribal leaders. Social/cultural anthropology further added that charisma, as a source of influence, was earned by some unique individuals who developed insight by going through austerity. Examples include Sufi leaders, Christian monks, and other tribal general or religious leaders, such as shamans. In those groups, a common practice included individual isolation from the community for an extended period or exposure to physically and psychologically challenging rituals to foster inner growth and wisdom. For example, in the Sioux Native American tribe, young males performed the Sun Dance to earn legitimacy to be considered as warriors. A ten feet leather string hanged from a secret pole was attached to a small piece of bone inserted into the performers' skin underneath their chests. They performed the Sun Dance to tear the skin off and let go of the hold. Many would lose consciousness in this painful ritual, lasting for many hours. Upon completing the task, the performers earned the charisma to influence others.

The social/cultural anthropology literature further states that charismatics respond to their communities' needs to influence members' attitudes and behaviors. These acts include but are not limited to healing people and recommending health and dietary practices, interacting with supernatural forces in hopes of communal benefits such as praying for rain, psychologically preparing warriors for war, and helping members cope with the loss of loved ones. Charismatic religious leaders provided their communities with insights into how to live their lives to be religiously appreciated. In sum, the literature on charisma from the anthropological perspective maintains that individuals earn charisma through austerity and use charisma to benefit their community. Hence, in this view, manipulators and liars cannot be charismatic – accordingly, charisma is associated with constructiveness and integrity (Horvath, 2013).

15.5 Earned Charisma in Organizations

We attempt to draw a parallel between the cultural/social anthropologist perspective and charisma in organizations concerning austerity and demonstrating outstanding achievement, respectively. In modern organizational and institutionalized contexts, individuals earn charisma with endurance and/or unusual talent. Examples include acquiring a respectable profession after long years of studying, passing a career milestone following years of hard work, showing outstanding achievement, passing advanced ability tests, or winning an achievement award. Accordingly, scientists, medical doctors, lawyers, PhDs, artists, athletes, military leaders, entrepreneurs, and organizational leaders, in general, earn charisma by completing some critical career, life-changing tasks. This achievement is associated with mastering a challenge, at the end of which the achiever gains not only a profession or a rank but also a psychological quality, represented in earned charisma. Often, such achievement is associated with modern rituals inducting the person into his/her new role and arranging the social and physical environment to enhance charisma (Hentrup et al.,

2019). For example, when a high achieving employee receives a promotion, she/he is collectively celebrated and provided with a bigger office to emphasize his/her important role.

Apart from the achievement-related and present-time orientation of earned charisma, personal change can also be a source of charisma and inspiration among followers. Because personal change is perceived as more effortful than stability, improving life from an undesirable past is more inspiring than desirable standings with stability (Klein & O'Brien, 2017). Thus, combined with achievements, personal change can be a source of inspiration and charisma when evaluating the past and present of the leaders. Such change indicates effort through evoking followers' empathy and identification. Earned charisma is a distinct factor because the quantity of mastery is measurable and applicable in a scientific study to research the impact of earned charisma. Earned charisma is role achievement based; hence, it stems from earning a high social status. Simply put, high-status individuals as leaders possess higher levels of earned charisma than individuals with a lower status.

Charismatic leadership was applied to organizational research and practice from the 1970s when American corporations struggled to remain competitive globally (Conger, 2020). US companies were administrative and inflexible to transform and adapt to fierce international competition. Charismatic corporate leadership was introduced to turn employees into enthusiastic followers and transform organizations in novel directions. Charismatic leaders motivated followers to contribute to change regarding goals, work practices, roles, responsibilities, working hours, locations, travel, teams, technology, and learning. These leaders displayed dominance, moral convictions, self-confidence, hard work, and self-sacrifice to influence employees. Intriguingly, job security declined due to organizational efforts to lessen expenditures. Thus, paradoxically, downsizing and increasing commitment were concurrent targets.

From a critical point of view, it remains unclear whether employees would enhance motivation in response to their leaders' charisma or worry that withholding explicit enthusiasm may lead to losing jobs. Apparently, corporate charismatic leaders and anthropological charismatics are in contrast. The corporate model emphasizes peripheral routes to influence, involving emotional domination and displays of confidence to benefit the organization primarily. In a more faithful approach to anthropological charisma, the leader uses earned charisma and enhances the welfare of the employees professionally, psychologically, and even in terms of physiological health. Without these essential attributes, charismatic leadership is only frivolously applied in organizations because the followers remain means instead of the ends of organizational goals and a broader vision.

In the following, we present our application of the social identity approach to charismatic leadership to provide a theoretical framework for earned charisma and follower needs in our proposed model. The social identity approach provides foundations for interpreting leadership influence in groups. The theory outlines how group membership motivates individual attitudes and behavior in response to leaders' attempts to influence.

15.6 The Social Identity Approach

Social identity theory has been applied to leadership (Hogg, 2001; Reicher et al., 2005). However, it has yet to be extensively used in the realm of charismatic leadership. The social identity approach argues that leaders influence by demonstrating that they are prototypical members of their groups. In other words, leaders present themselves as one of their followers, so the followers are more likely to identify and be influenced (Hogg, 2001). Leaders demonstrate additional attributes, such as acting as entrepreneurs of identity, which refers to forming the group's identity in new contexts (Haslam et al., 2011). Social identity theory argues that people aspire to be part of groups that positively reflect on them and gives people sense of "we," rather than "I" as a separate, unique individual by rooting individuals' sense of self into group identity (Ashforth & Mael, 1989). Individuals maintain positive distinctiveness by intergroup comparisons, in which they favorably compare the in-group to out-groups (Tajfel & Turner, 1986). Leaders address follower needs for positive distinctiveness by communicating the group's positive characteristics of values, norms, and goals. The more leaders embody the positively distinct group prototype, the more charismatic they will be perceived, especially by those followers who strongly identify with the group (Steffens et al., 2014).

Group prototypicality also makes followers more susceptible to leaders' influence, especially in groups where social identity is salient, as it indicates leaders' benevolence (i.e., having good intentions to pursue the group's best interests) and induces trust in a leader (Mayer et al., 1995; Giessner & van Knippenberg, 2008; van Knippenberg & Hogg, 2003; Dirks & Ferrin, 2002; Ridgeway, 2003). Leader's group-oriented behaviors, such as self-sacrifice behaviors and considering collective interest, are examples of the association between group prototypicality and group-oriented behavior that is also rooted in the social identity perspective (Shamir et al., 1993; Platow & van Knippenberg, 2001; van Knippenberg & van Knippenberg, 2005). The focus on group-oriented behavior is also consistent with theories of charismatic leadership, which suggest that behaviors such as leader self-sacrifice are a crucial aspect of charisma (Choi & Mai-Dalton, 1998). Research has shown that leaders seen as prototypical group members are likelier to be perceived as charismatic (Van Knippenberg & Hogg, 2003). Additionally, charismatic leaders express trust in their followers' ability, and when goals are achieved, group members feel proud and positively differentiated from out-groups, particularly in intergroup competition or conflict (Eden, 1990).

15.7 The Struggling Follower Prototype

However, group prototypicality does not correspond with the portrayal of charismatic leaders as exceptionally unique, occasionally even argued to have divine origin. Although leadership theorists refrain from suggesting that leaders ought to be above their followers (e.g., Northouse, 2012), charismatic leaders, depicted as saviors, appear to prevail over others in skills, insights, and behaviors. Followers turn to charismatic leaders for guidance, yet charismatic leaders demonstrate that they can precisely identify the urgent collective needs of their followers and communicate their vision to satisfy those needs.

We provide a social identity argument proposing that charismatic leaders address the struggling follower prototype in need of the savior embodied by the charismatic leader. The struggling prototype corresponds with the average group member who struggles to advance in his/her life, and the charismatic leader shows a clear understanding of this struggle and offers a vision of a better future for the followers. For historical examples, Mustafa Kemal Atatürk's Turkish youth speech identified the follower prototype as the brave, independence-loving Turkish young person who aims to defend the homeland from traitors and invaders. Martin Luther King, in his speeches, identified the average black person needing to eliminate racial discrimination in the USA. For a destructive example, Adolf Hitler addressed the average German, humiliated by the loss of World War I and suffering from economic constraints.

15.8 Model of Charismatic Leadership in Work Organizations

Based on the reviewed literature, we propose a charismatic leadership model in work organizations. The model uses earned charisma as a focal predictor. Charismatic communication is the auxiliary predictor and moderator of earned charismatic effects. The outcomes in the model consist of followers' charismatic attributions and charismatic leader-follower relations. The extent to which the leader can address the needs of the struggling follower prototype moderates the impact of earned charisma. Our application of the charismatic leadership approach is specific to organizations. In organizations, managers are assigned, not emerged. Therefore, the classic approach to charismatic leadership, in which the leader signals his/her ability to emerge as a leader, is not clearly transferable. Our focus on earned charisma, associated with the leaders' prior valued achievement, provides a more appropriate application of charismatic leadership to organizations.

In the model, earned charisma is the primary source of charismatic influence on the followers. This effect is indicated by the continuous line toward the outcome variables, charismatic attributions, and charismatic leader-follower relations. That is, a remarkable prior achievement implies that the leader is unique and has a psychological contract with subordinates to be the source of influence. In the model, charismatic communication is an auxiliary predictor of the outcome variables. As indicated by a stripped line, charismatic communication is supposed to have a weaker effect as compared to the effect of the earned charisma. Thus, the model suggests that developing charismatic communication tactics will enhance charismatic attributions and relations. However, in organizations with a culture of meritocracy, communication skills are secondary to achievement. The model includes the leader's address of the struggling follower prototype as a moderator of earned charismatic effect. That is, genuine charismatic leadership takes place when leaders identify their followers' needs, propose visions, and inspire action to act in the best interest of their followers. Followers can have a variety of needs, such as training, autonomy and meaningfulness, limited bureaucracy, workflow, work-life balance, promotion opportunities, and financial benefits. Finally, charismatic communication also moderates the interaction between earned charisma and addressing the struggling prototype. Accordingly, the most substantial charismatic influence takes place when the leaders possess earned charisma, embrace the needs of the struggling follower prototype, and communicate charismatically.

15.9 Limitations

The main limitation of the model is assumed meritocracy despite the prevalence of nepotism in organizational promotions. Earned charisma as a source of influence diminishes when privileges denote acquaintance rather than achievement. Therefore, the model better suits organizations where leaders emerge due to outstanding achievements.

Another limitation of our model is the cross-cultural consistency due to perceiving earned charisma as the focal predictor and charismatic communication as the auxiliary predictor. Although one of the most comprehensive studies of cross-cultural leadership, GLOBE, and other studies found that implicit leadership theories are present and valued in many cultures (House et al., 1999; Carl & Javidan, 2001; Bass & Avolio, 1993), cultural variations in leadership perceptions including information processing of followers to leadership prototypes (Ensari & Murphy, 2003) and the role of primary role of earned charisma or charismatic communication need to be clarified further in the literature (Lord & Emrich, 2000). Based on the individualism and collectivism accounts of attribution styles (dispositional vs. situational) indicated by Markus and Kitayama (1991), cultures may have different social-cognitive processing styles of leadership attributions. As individualistic cultures are more likely to make dispositional attributions for interpreting charismatic leadership, group prototypicality of the leadership may be more effective in Western cultures having individualistic orientation compared to collectivistic cultures with situational attribution styles more likely to focus on collective implications such as company performance outcomes (Ensari & Murphy, 2003). Thus, charismatic communication may be most effective due to the high probability of trait-related, dispositional attributions, whereas earned charisma may be compelling in collectivistic cultures due to outcome-oriented focus on situational factors. Thus, our model needs to be further clarified to address cross-cultural differences about different attribution styles of charisma.

The role of follower expectations in charismatic leadership is also an important consideration. Charismatic leaders are often able to create high expectations among their followers. These expectations can be a source of motivation, but they can also be a source of pressure. If followers cannot meet the leader's expectations, they may become disillusioned or disengaged.

15.10 Conclusion

In conclusion, this chapter has comprehensively reviewed the literature on charismatic leadership. We have proposed a model of charismatic leadership in work organizations that integrate the social identity approach with the earned charisma perspective. The model posits that earned charisma is the primary source of charismatic influence on followers, charismatic communication is an auxiliary predictor of charismatic attributions and relations, and the leader's address of the struggling follower prototype moderates the effect of earned charisma. Despite the limitations, we believe that the model provides a valuable contribution to the study of charismatic leadership by addressing the limitations of previous models by focusing on the role of earned charisma and the leader's address of the struggling follower prototype. The model offers insights about charismatic leadership as more than the embodiment of fixed prototypes of the followers, just reaching achievements or following the party line; it also addresses struggling followers to motivate them to contribute collectively to shared organizational goals. The past twenty years of leadership crisis in the energy sector, characterized by selfishness, fraud, and conformity, may see remedy in this model focusing on high achieving leaders who aim improving the lives of their followers.

References

Antonakis, J., Bastardoz, N., Jaquart, P., & Shamir, B. (2016). Charisma: An ill-defined and ill-measured gift. *Annual Review of Organizational Psychology and Organizational Behavior, 3*(1), 293–319.

Antonakis, J., d'Adda, G., Weber, R., & Zehnder, C. (2021). Just words? Just speeches? On the economic value of charismatic leadership. *Management Science, 9*(68), 6355. https://doi.org/10.1287/mnsc.2021.4219

Ashforth, B. E., & Mael, F. (1989). Social identity theory and the organization. *The Academy of Management Review, 14*(1), 20–39.

Bass, B. M., & Avolio, B. J. (1993). *Transformational leadership: A response to critiques*. Academic Press.

Bryman, A. (1992). *Charisma and leadership in organizations*. Sage.

Carl, D. E., & Javidan, M. (2001, August). Universality of charismatic leadership: A multi-nation study. *Academy of Management Proceedings, 2001*(1), B1–B6.

Center for Creative Leadership. (2020). *The leadership challenge in the energy sector*. Retrieved from https://www.ccl.org/articles/leading-effectively-articles/the-leadership-challenge-in-the-energy-sector.

Choi, Y., & Mai-Dalton, R. R. (1998). On the leadership function of self-sacrifice. *The Leadership Quarterly, 9*(4), 475–501.

Conger, J. A. (2020). Charisma in organizational studies. In J. P. Zúquete (Ed.), *Routledge international handbook of charisma* (pp. 303–312). Taylor & Francis.

Conger, J. A., & Kanungo, R. N. (1998). *Charismatic leadership in organizations*. Sage.

Cruver, B. (2003). *Enron: Anatomy of greed*. Arrow.

Dirks, K. T., & Ferrin, D. L. (2002). Trust in leadership: Meta-analytic findings and implications for research and practice. *Journal of Applied Psychology, 87*(4), 611.

Eden, D. (1990). *Pygmalion in Management*. D. C. Heath and Company.

Ensari, N., & Murphy, S. E. (2003). Cross-cultural variations in leadership perceptions and attribution of charisma to the leader. *Organizational Behavior and Human Decision Processes, 92*(1–2), 52–66.

Giessner, S. R., & van Knippenberg, D. (2008). "License to fail": Goal definition, leader group prototypicality, and perceptions of leadership effectiveness after leader failure. *Organizational Behavior and Human Decision Processes, 105*(1), 14–35.

Haslam, S. A., Reicher, S., & Platow, J. M. (2011). *The new psychology of leadership: Identity, influence and power*. Psychology Press.

Hentrup, L. J., Bastardoz, N., & Menges, J. (2019). Who is seen as a charismatic leader? The role of environmental features in charisma attributions. *Academy of Management Proceedings, 2019*(1), 18504. https://doi.org/10.5465/AMBPP.2019.237

Hogg, M. A. (2001). A social identity theory of leadership. *Personality and Social Psychology Review, 5*, 184–200.

Horvath, A. (2013). *Modernism and charisma*. Palgrave Macmillan.

House, R. (1976). *A 1976 theory of charismatic leadership* (Vol. 77, pp. 1–34). University of Toronto.

House, R. J., Hanges, P. J., Ruiz-Quintanilla, S. A., Dorfman, P. W., Javidan, M., Dickson, M., & Gupta, V. (1999). Cultural influences on leadership and organizations: Project GLOBE. In *Advances in global leadership* (p. 1). JAI Press.

Howell, J. M., & Shamir, B. (2005). The role of followers in the charismatic leadership process: Relationships and their consequences. *The Academy of Management Review, 30*(1), 96–112. https://doi.org/10.2307/20159097

Klein, N., & O'Brien, E. (2017). The power and limits of personal change: When a bad past does (and does not) inspire in the present. *Journal of Personality and Social Psychology, 113*(2), 210–229.

Kotter, J. P. (1990). *A force for change: How leadership differs from management*. Free Press.

Lindholm, C. (2002). *The Islamic Middle East: Tradition and change*. Basil Blackwell.

Lord, R. G., & Emrich, C. G. (2000). Thinking outside the box by looking inside the box: Extending the cognitive revolution in leadership research. *The Leadership Quarterly, 11*(4), 551–579.

Markus, H. R., & Kitayama, S. (1991). Culture and the self: Implications for cognition, emotion, and motivation. *Psychological Review, 98*(2), 224–253. https://doi.org/10.1037/0033-295X.98.2.224

Mayer, R. C., Davis, J. H., & Schoorman, F. D. (1995). An integrative model of organizational trust. *Academy of Management Review, 20*(3), 709–734.

Northouse, P. G. (2012). *Leadership: Theory and practice* (6th ed.). Sage.

Platow, M. J., & van Knippenberg, D. (2001). A social identity analysis of leadership endorsement: The effects of leader ingroup prototypicality and distributive intergroup fairness. *Personality and Social Psychology Bulletin, 27*(11), 1508–1519. https://doi.org/10.1177/01461672012711011

Reicher, S., Haslam, S. A., & Hopkins, N. (2005). Social identity and the dynamics of leadership: Leaders and followers as collaborative agents in the transformation of social reality. *The Leadership Quarterly, 16*(4), 547–568.

Ridgeway, C. (Ed.). (2003). *Status characteristics and leadership*. Sage. https://doi.org/10.4135/9781446216170

Shamir, B., House, R. J., & Arthur, M. B. (1993). The motivational effects of charismatic leadership: A self-concept based theory. *Organization Science, 4*(4), 577–594. https://doi.org/10.1287/orsc.4.4.577

Steffens, N. K., Haslam, S. A., & Reicher, S. D. (2014). Up close and personal: Evidence that shared social identity is a basis for the 'special' relationship that binds followers to leaders. *The Leadership Quarterly, 25*(2), 296–313. https://doi.org/10.1016/j.leaqua.2013.08.008

Stogdill, R. M. (1974). *Handbook of leadership: A survey of theory and research*. Free Press.

Sy, T., Horton, C., & Riggio, R. (2018). Charismatic leadership: Eliciting and channeling follower emotions. *The Leadership Quarterly, 29*(1), 58–69. https://doi.org/10.1016/j.leaqua.2017.12.008

Tajfel, H., & Turner, C. J. (1986). The social identity theory of inter-group behavior. In S. Worchel & L. W. Austin (Eds.), *Psychology of intergroup relations*. Nelson-Hall.

Tambiah, S. (1984). *The Buddhist saints of the Forest and the cult of amulets*. Cambridge University Press.

Tourish, D., & Vatcha, N. (2005). Charismatic leadership and corporate cultism at Enron: The elimination of dissent, the promotion of conformity and organizational collapse. *Leadership, 1*(4), 455–480. https://doi.org/10.1177/1742715005057671

van Knippenberg, B., & van Knippenberg, D. (2005). Leader self-sacrifice and leadership effectiveness: The moderating role of leader prototypicality. *Journal of Applied Psychology, 90*(1), 25–37. https://doi.org/10.1037/0021-9010.90.1.25

Van Knippenberg, D., & Hogg, M. A. (2003). A social identity model of leadership effectiveness in organizations. *Research in Organizational Behavior, 25*, 243–295.

Verčič, A. T., & Verčič, D. (2011). Generic charisma – Conceptualizations and measurement. *Public Relations Review, 37*, 12–19.

Weber, M. (1978 [1922]). *Economy and society*. University of California Press.

Yukl, G. (1999). An evaluation of conceptual weaknesses in transformational and charismatic leadership theories. *Leadership Quarterly, 10*, 285–305. https://doi.org/10.1016/S1048-9843(99)00013

Chapter 16
The Role of Insurance in Promoting Clean Energy Transition

Hasan Meral

16.1 Introduction

Climate change, one of the world's most significant challenges in the twenty-first century, causes environmental and social issues and results in economic losses (Tol, 2009). Over the past ten years, extreme weather events caused 82% of the world's total damage from natural disasters. In 2022, the cost of these events reached USD 211 billion, which is 183% higher than the average cost of the last 30 years, adjusting for inflation (EM-DAT, 2023). According to Swiss Re, by 2050, the size of economic losses due to climate change could reach approximately 18% of the world's GDP (Guo et al., 2021). It presents a significant obstacle, particularly for developing economies that depend primarily on agriculture, have limited social safety nets, and have a high coverage gap (Collier et al., 2009; Tol, 2009).

The Intergovernmental Panel on Climate Change reported that in 2019 64% of global greenhouse gas (GHG) emissions were caused by energy production and industrial activities (IPCC, 2022). To address this issue, it is necessary to increase the use of renewable resources for energy production (Yüksel et al., 2021). However, renewable energy projects usually require a high initial cost, and, as an emerging sector, they are characterized by relatively high uncertainty (Steckel & Jakob, 2018; Dinçer et al., 2022). It leads to an increase in the perceived level of risk and a decrease in the interest of traditional financing sources in such projects. At this point, the insurance industry can play an essential role in promoting the transition to clean energy, thanks to its assets totaling USD 44 trillion (IAIS, 2023).

The insurance industry recognized that climate change amplifies social and environmental risks. According to Swiss Re (2023), insurers suffered substantial

H. Meral (✉)
Institute of Islamic Economics and Finance, Marmara University, Istanbul, Turkey
e-mail: hasan.meral@marmara.edu.tr

H. Dinçer et al. (eds.), *Decision Making in Interdisciplinary Renewable Energy Projects*, Contributions to Management Science,
https://doi.org/10.1007/978-3-031-51532-3_16

financial losses due to environmental risks in 2022. Therefore, promoting the clean energy transition is crucial for the insurance industry, considering it is no longer just a matter of choice but a requirement. The insurance industry can support the transformation of energy resources by investing in clean energy projects and green bonds. Furthermore, they can eliminate uncertainties associated with clean energy investments by leveraging their expertise in risk management and providing financial coverage (Mills, 2007).

Despite the significant relationship between insurance and the clean energy economy, there has been limited academic research on their interaction. This chapter aims to fill this gap in the literature by investigating how the insurance industry can contribute to the shift toward renewable energy. Specifically, the chapter examines how insurance companies can support clean energy projects and minimize the risks. The chapter will begin by examining how climate change affects the insurance industry and how insurers address it. Following that, the chapter will explore different tools and policies the insurance industry can utilize to promote the adoption of clean energy.

16.2 Insurance Industry and Climate Change

Insurers have faced unexpected costs due to the environmental shifts resulting from climate change. In 2022 alone, economic losses from natural disasters, predominantly extreme weather events, amounted to USD 275 billion. Of these losses, insurance companies incurred USD 125 billion. In inflation-adjusted terms, this cost borne by insurers is 54% higher than the last decade's average. Additionally, the 2022 payouts represent the fourth-highest loss recorded over the past half-century. Despite the considerable costs incurred by the insurance industry over the last decade, 61% of total losses were uncovered, indicating that there is still a significant coverage gap in this area (Swiss Re, 2023).

The insurance industry is affected by climate change in numerous ways, both directly and indirectly. Table 16.1 details climate change's impacts and potential consequences on the insurance industry.

Although the exact effects of climate change on natural disasters cannot be measured presently, evaluating the shift in the historical patterns of disasters can give a general idea of its influence. Over the last 30 years, there has been a significant increase in the frequency and severity of weather-related natural disasters. Compared to the previous period, there has been a 216% increase in frequency and a 343% increase in severity. This growth in weather-related disasters is 50% faster than all other natural disasters (EM-DAT, 2023). The costs incurred by these disasters have directly affected insurers' losses, resulting in decreased reserves and profitability. In addition, the shifting trend makes it more difficult to create precise risk models that work effectively for catastrophic events.

The calculation of losses expected to arise from extreme events is based on simulations which are carried out using the results of relevant historical events. However,

Table 16.1 Impacts of climate change on the insurance industry and its possible consequences

Impact	Possible consequences
The increase in frequency and severity of extreme weather events leading an increase in catastrophic losses (EM-DAT, 2023; S&P Global, 2014; Grimaldi et al., 2020)	• Increase in insurance losses • Increase in insurance operational costs • Increase in insurance premiums • Higher capital requirements • Decline in insurance industry profitability
Due to global warming, regional risk profiles are shifting and risk modeling is becoming more challenging (Dlugolecki, 2008; Prudential Regulation Authority, 2015; Wolfrom & Yokoi-Arai, 2016; S&P Global, 2014)	• Difficulty in building risk models based on historical data • Decrease in the success score of risk models • Unforeseen insurance losses • Decrease in risk appetite • Reinsurance capacity problems
Due to rising losses and operational expenses, insurance premiums are increasing (Aerts & Botzen, 2011; Dailey et al., 2009; Kunreuther et al., 1995)	• Decrease in insurance penetration rates • Decreased accessibility to insurance • Increase in insurance coverage gaps (especially in emerging economies) • Increased vulnerability of disadvantaged social groups to natural disaster risks
Due to the indirect losses caused by climate change, diversification of the portfolio is becoming more difficult (Grimaldi et al., 2020)	• Increase in systemic risks in the global insurance market • Decline in sustainable profitability in the insurance industry

using past data in such estimations undermines their predictive power regarding climate change. Misunderstanding the impact of climate change on natural disaster risks can lead to a significant difference between the measured risk and the actual loss incurred (Wolfrom & Yokoi-Arai, 2016). According to S&P Global's analysis (2014), catastrophic losses in the previous decade were 50% greater than reinsurers had initially modelled.

Climate change leads to changes in the global risk landscape. As a result of global warming, losses are now occurring in areas previously considered low risk (Dlugolecki, 2008). However, due to the nature of risks, as the geographic scale decreases, the uncertainty increases, posing a challenge to assessing the impact of climate change at the local level (Prudential Regulation Authority, 2015). Developing countries are particularly vulnerable to climate-related risks due to their limited resilience and low insurance coverage against natural disasters. Countries with large populations living in coastal areas that are close to sea level and depend heavily on agriculture are at the highest risk. Moreover, these countries have limited financial resources and fiscal flexibility to address the impacts of climate change (Wolfrom & Yokoi-Arai, 2016).

The impacts of climate change may interrupt insurance companies' efforts to establish a balanced portfolio and implement effective risk management strategies through risk diversification. Insurers typically seek to mitigate losses from one risk by providing coverage for various risks across different geographies. However, the systemic risks of climate change have increased the likelihood of chain reactions resulting from a single event (Grimaldi et al., 2020). A major disaster can cause damage not only to infrastructure and superstructure but also to human life and other personal and commercial properties. Moreover, it may result in a domino effect, causing business interruptions, increased investment risks, etc.

While climate change presents several challenges for the insurance industry, it also offers opportunities for development in certain areas. First, the increase in global risks also raises the awareness of individuals and organizations about risks. This may be an important opportunity to reduce the global coverage gap. Furthermore, as risks continue to surge, insurance companies emerge as key actors in managing the financial consequences of these risks within the global economic framework. By leveraging big data and data analytics, insurers can effectively mitigate the financial impacts of climate change through innovative product offerings, such as parametric insurance.

Insurers started expressing their concerns about climate change as early as in the 1970s. In the first report prepared by Munich Re (1973), global warming and the increase in GHG emissions were discussed as factors that would increase natural disaster risks in the long term. Since then, the leading players in the global insurance market have been collaborating with international institutions to reduce the effects of climate change (Prudential Regulation Authority, 2015). Despite the difficulties associated with climate change, insurers are taking preventative measures to tackle the issue. Their efforts primarily center on two key aspects: minimizing their carbon footprint and endorsing the transition toward clean energy.

Several insurers, which account for a significant portion of global insurance premium production, have committed to supporting the fight against climate change by reducing their carbon footprint. This approach includes mitigating carbon emissions from insurance operations and investing in low-carbon technologies and renewable energy. For instance, some of the world's largest insurance groups, such as Swiss Re, Munich Re, and Zurich, have committed to achieving net-zero GHG emissions from their operations by 2030. To accomplish this, these companies reduce their energy consumption, increase renewable energy sources' utilization, and buy certified carbon credits (Swiss Re, 2022; Munich Re, 2022; Zurich, 2022).

The insurance industry has taken collective measures in partnership with international organizations to support the transition to clean energy. In 2012, the United Nations Development Programme published the Principles for Sustainable Insurance (PSI), highlighting the environmental, social, and governance (ESG) risks and opportunities the insurance industry encounters. The PSI's efforts led to the Net-Zero Insurance Alliance (NZIA) creation in 2021. This association includes a group of (re)insurance companies that collectively account for a significant portion of global insurance premium production. Members of the NZIA have committed to transforming their underwriting portfolios to achieve net-zero GHG emissions by 2050 (UN, 2023).

In 2016, the insurance industry collaborated with the United Nations, the World Bank, and other international organizations to establish the Insurance Development Forum (IDF) as an initiative to mitigate the effects of climate change. The IDF's goal is to increase global insurance and risk management capabilities to enhance the ability of individuals, communities, and institutions' resilience against natural disasters and associated financial risks. The forum provides guidance and advice to the insurance industry in various areas, including risk modeling, regulatory and legislative frameworks, resilience policies, and customized insurance solutions (IDF, 2023). Although the insurance industry has made efforts to respond to climate change, there is still room for improvement. To help facilitate the transition to clean energy, insurers can utilize the tools and policy recommendations provided in the following section.

16.3 The Role of Insurance in Promoting Clean Energy

The insurance industry is crucial in reducing uncertainty and providing financial protection to individuals and institutions. It has the potential to contribute significantly to the transition to clean energy by supporting renewable energy initiatives. Also, with their expertise in risk management and financial resources, insurance companies can help reduce the impact of climate change.

Table 16.2 outlines the significant policy opportunities available to the insurance industry for promoting clean energy. The next section will comprehensively examine these policies through three principal categories: carbon footprint reduction, clean energy transformation, and clean energy adoption.

Table 16.2 Policies insurers can adopt to promote clean energy

Target	Policy
Reducing carbon footprint	• Net-zero GHG emissions in insurance operations • Incorporating ESG factors into UW processes • Divestment of the fossil fuel industry • Net-zero GHG emissions UW portfolio
Clean energy transition	• Investing in clean energy projects and green bonds • Providing coverage for the development, construction, and operation of clean energy projects • Offering innovative products to increase the predictability and sustainable profitability of clean energy projects • Risk management consultancy for clean energy projects
Clean energy adoption	• Premium discount to encourage the use of clean energy • Sustainable pricing strategy that incentivizes limiting energy consumption

The insurance industry is a key player in the global economy, with the power to influence commercial activities worldwide through its investment and risk management policies. Insurance companies can adopt a policy to support clean energy by achieving carbon neutrality in their operations. By reducing energy consumption and transitioning to renewable energy sources, insurance companies can limit their carbon footprint. While some insurers have set medium-term goals in this area, the entire sector has not yet reached an agreement. Another policy that insurers can use to reduce carbon emissions is divestment from companies that produce energy from fossil fuels. The Global Fossil Fuel Commitments Database (2023) reports that 1591 institutions worldwide, representing USD 40.5 trillion in assets, have committed to divest from fossil fuels. Among these organizations are the world's leading insurance groups, such as Allianz, Aviva, Generali, Mapfre, Munich Re, Swiss Re, Talanx, and Zurich. For instance, Allianz has decreased its GHG emissions by 35% between 2019 and 2022 by investing in low-emission companies (Allianz Trade, 2020). As more insurance companies refrain from funding projects and companies that produce high carbon emissions, financing in this area may become scarce, and these types of investments may lose their appeal.

Insurance companies can also reduce their carbon footprint by incorporating ESG factors into their underwriting processes. As environmental risks increase, ESG factors have gained importance for the insurance industry in recent years. Insurers have realized that ESG-related issues, such as climate change, ecosystem degradation, and pollution, can have major financial consequences. As a result, they are now placing greater emphasis on long-term sustainability when making underwriting decisions (UNEP, 2020). The NZIA member companies, who provide a significant portion of the world's insurance capacity, have pledged not to offer coverage for projects with high carbon emissions. This decision could potentially make it difficult for these projects to maintain their sustainability. A significant adaptation gap remains in the industry despite the potential to incorporate environmental factors into insurance risk assessment procedures. A Capgemini study (2022) reveals that only 45% of insurance firms have integrated ESG factors into their underwriting procedures. Enhancing this ratio in the future would increase insurers' contributions to the clean energy transition.

Supporting clean energy transition through investing in clean energy projects represents another pathway for insurers. Without taking action, it is predicted that energy consumption-related carbon emissions will rise by 70% by 2050. This will result in a more than 50% increase in total greenhouse gas emissions, according to OECD (2012). In order to reach the goal of achieving net-zero GHG emissions by 2050, investments in clean energy must increase significantly. The current average of USD 2 trillion over the past five years needs to rise to USD 5 trillion by 2030 and USD 4.5 trillion by 2050 (IEA, 2021). However, given the long-term and relatively risky nature of these projects, traditional sources of financing for renewable energy face limitations (Kaminker & Stewart, 2012). Insurers, who possess USD 44 trillion in assets, can play a vital role in bridging this financing gap in the transition to clean energy by investing in renewable energy projects and green bonds.

In recent years, the insurance industry has become increasingly interested in investing in clean energy. According to the interviews conducted by Ceres (2019) with major insurance companies in the United States, 52% of the respondents reported having clearly defined strategies and objectives for investing in clean energy. Additionally, 85% stated they considered investing in renewable energy sources. Swiss Re, for instance, has committed to hold at least USD 4 billion in green, social, and sustainability bonds by 2024 and to allocate an extra USD 750 million to renewable energy projects (Swiss Re, 2022). By investing in clean energy, insurers can diversify their investment portfolios and generate long-term sustainable returns while mitigating climate change. Despite these remarkable developments, clean energy investments by insurance companies represent a limited portion of their portfolios. The primary barriers to insurers adopting clean energy investments include inadequate government support, lack of investment experience, and difficulties diversifying investment instruments (Ceres, 2019; Kaminker & Stewart, 2012).

Investing in clean energy projects involves more uncertainty and risk than traditional fossil fuel projects. Since these projects require significant construction investments, they are exposed to property damage, business interruption, and liability risks. In this context, insurance can serve as a valuable tool for reducing uncertainties associated with developing and operating clean energy projects by enabling the transfer of risks from project owners to insurance companies. Traditional insurance products such as construction all risks, industrial risks, liability, business interruption, and loss of profit offer a safeguard against potential financial losses that clean energy projects may incur. These insurance products cover various stages of clean energy projects, from development to financing, construction, and operation. For instance, Munich Re's Photovoltaic Guarantee Insurance, which extends for 30 years, covers losses arising from manufacturer-induced damages in solar power plants (Munich Re, 2023). It facilitates their financing by guaranteeing the long-term profitability of solar energy investments. Another innovative example is Euler Hermes' Green2Green credit insurance, which ensures climate change mitigation projects and invests premiums in green bonds (Allianz Trade, 2020). This credit insurance mechanism not only eases financing for such projects but also invests the premiums earned in green bonds, thereby financing the projects.

Compared to other energy projects, clean energy projects are more sensitive to the impact of environmental factors. For instance, wind power plants depend on the availability and stability of wind levels for energy production. These vulnerabilities contribute to the unpredictability of clean energy projects' financial income and production levels. Parametric insurance, a type of new-generation insurance, offers protection against income loss caused by environmental factors. Unlike traditional insurance, which measures financial loss after the occurrence of a risk, parametric insurance automatically pays out when a predefined event is triggered, measured by a specific parameter or index. For example, Swiss Re (2015) provides parametric insurance for an Australian renewable energy producer's wind farms with more than 500 megawatts of capacity. The insurance product pays the producer a fixed amount per megawatt hour for energy that cannot be produced due to low winds. As of 2021, the global parametric insurance market had a premium production size of USD 11.7

billion and is projected to reach USD 29.3 billion by 2031, with an average annual growth rate of 9.9% (Kanhaiya et al., 2022). Although these forecasts indicate significant growth potential, the penetration of parametric insurance in the global insurance market remains low, given its overall size. There is still much potential to improve the adoption and implementation of parametric insurance.

Insurance companies can act as consultants for clean energy projects by managing their associated risks. They offer extensive risk management services for every project phase, utilizing their expertise and statistical data on various risks in different geographic regions. These services include risk management, avoidance, and mitigation associated with clean energy projects (Kaminker & Stewart, 2012). By accurately measuring and assessing risks associated with clean energy projects, essential strategies can be developed to mitigate these risks and prevent potential losses.

With a global premium production size of USD 7.2 trillion, the insurance industry has a unique opportunity to provide financial incentives to support the adoption of clean energy by individuals and institutions (IAIS, 2023). One direct approach to achieving this objective is offering discounted policies to customers using clean energy sources. For instance, travelers and farmers offer varying discounts on home insurance for green-certified homes. However, these practices are not industry-wide (Roberte, 2021). Similarly, some insurance providers offer reduced premiums for car insurance policies covering electric and hybrid vehicles. Conversely, the increased costs of spare parts and repairs for electric vehicles relative to traditional gas-powered cars result in higher car insurance premiums, ultimately limiting the efficacy of this approach to incentivize the transition to clean energy (Forbes, 2023). As a result, there is limited evidence that insurance companies accelerate the clean energy transition through premium discounts.

A more indirect policy that insurers can implement to support the clean energy transition is incentivizing energy consumption reductions. For instance, Metromile (2023) promotes reduced driving via usage-based car insurance pricing. Under this scheme, policyholders who drive less are charged lower premiums. This initiative reduces carbon emissions, benefiting the environment and lowering the risk of accidents for insurers. Insurance companies may also customize their insurance products to encourage energy efficiency. For instance, following a covered loss, an insurer may offer to rebuild an environmentally friendly and energy-efficient structure by providing additional compensation. Additionally, some insurers offer hybrid vehicle upgrades in commercial vehicle fleets as a replacement option (Capgemini, 2021). Insurance companies increasingly adopt sustainable insurance practices to reduce energy and resource consumption. With growing awareness of the impacts of climate change, these practices are likely to become more widespread and impactful.

16.4 Conclusion

This chapter examined the impacts of climate change on the insurance industry and outlined alternative tools and policies that insurance companies can use to promote the clean energy transition. Insurers should set industry-wide targets to minimize their carbon footprint and to reach net-zero GHG emission UW portfolio. Moreover, they should invest more in clean energy projects and offer financial protection to mitigate uncertainties during their development and operation. Also, companies should encourage the adoption of clean energy by incentivizing environmentally friendly practices among their customers. However, the lack of common strategies for incorporating sustainable insurance policies may create a competitive disadvantage for some companies and slow down the adaptation of the practice. To avoid this, standards and guidelines should be established to assist insurance companies in implementing clean energy strategies effectively.

In order to improve the involvement of the insurance industry in promoting the shift toward clean energy, it is essential for industry stakeholders, policymakers, and researchers to work together and take specific actions:

- Companies should enhance collaboration to increase the transparency of their commitments toward the clean energy transition and make it an industry-wide standard.
- Policymakers should introduce regulatory measures and incentives that encourage the adoption of clean energy initiatives.
- Researchers should propose alternative strategies that insurance can use to mitigate the impacts of climate change and support the clean energy transition.

References

Aerts, J. C., & Botzen, W. W. (2011). Climate change impacts on pricing long-term flood insurance: A comprehensive study for the Netherlands. *Global Environmental Change, 21*(3), 1045–1060.

Allianz Trade. (2020, November 24). *Green2Green single risk: A pioneer insurance product by Euler Hermes to foster green economy*. Retrieved from https://www.allianz-trade.com/en_GB/newsroom/green2green-single-risk-pioneer-insurance-by-euler-hermes.html

Capgemini. (2021). *How P&C insurers can protect and power our journey to a more sustainable world*. Capgemini Research Institute. Retrieved from https://www.capgemini.com/insights/research-library/sustainable-insurance/

Capgemini. (2022). *How insurers can lead climate change resiliency*. Capgemini Research Institute. Retrieved from https://www.capgemini.com/de-de/wp-content/uploads/sites/5/2022/05/2022_World_P__C_Insurance_Report.pdf

Ceres. (2019). *Scaling U.S. insurers' clean energy infrastructure investments*. Ceres. Retrieved from https://www.ceres.org/resources/reports/scaling-us-insurers-clean-energy-infrastructure-investments

Collier, B., Skees, J., & Barnett, B. (2009). Weather index insurance and climate change: Opportunities and challenges in lower income countries. *The Geneva Papers on Risk and Insurance-Issues and Practice, 34*, 401–424.

Dailey, P., Huddleston, M., Brown, S., & Fasking, D. (2009). *The financial risks of climate change. ABI Research Paper, 19*.
Dinçer, H., Yüksel, S., & Martínez, L. (2022). Collaboration enhanced hybrid fuzzy decision-making approach to analyze the renewable energy investment projects. *Energy Reports, 8*, 377–389.
Dlugolecki, A. (2008). Climate change and the insurance sector. *The Geneva Papers on Risk and Insurance-Issues and Practice, 33*, 71–90.
EM-DAT. (2023, April 14). *The international disaster database*. Retrieved from https://www.emdat.be/database
Forbes. (2023, January 4). *Electric car insurance: Why it costs more*. Retrieved from https://www.forbes.com/advisor/car-insurance/electric-vehicle/
Global Divestment Commitments Database. (2023). *The database of fossil fuel divestment commitments made by institutions worldwide*. Retrieved from https://divestmentdatabase.org
Grimaldi, A., Javanmardian, K., Piner, D., Samandari, H., & Strovink, K. (2020, November 19). *Climate change and P&C insurance: The threat and opportunity*. McKinsey & Company, Our Insights. Retrieved from https://www.mckinsey.com/industries/financial-services/our-insights/climate-change-and-p-and-c-insurance-the-threat-and-opportunity
Guo, J., Kubli, D., & Saner, P. (2021). *The economics of climate change: No action not an option*. Swiss Re Institute.
IAIS. (2023). *2022 global insurance market report*. International Association of Insurance Supervisors. Retrieved from https://www.iaisweb.org/uploads/2022/12/GIMAR-2022.pdf
IDF. (2023). *2022 in review a year of influence and implementation*. Insurance Development Forum. Retrieved from https://www.insdevforum.org/4118-2/
IEA. (2021). *Net zero by 2050: A roadmap for the global energy sector*. International Energy Agency. Retrieved from https://www.iea.org/reports/net-zero-by-2050
IPCC. (2022). *Climate change 2022: Mitigation of climate change. Contribution of Working Group III to the Sixth Assessment Report of the Intergovernmental Panel on Climate Change*.
Kaminker, C., & Stewart, F. (2012). The role of institutional investors in financing clean energy. *OECD working papers on finance, insurance and private pensions No. 23*.
Kanhaiya, K. Sourabh, E. & Vineet, K. (2022). *Parametric insurance market research, 2031*. Retrieved from https://www.alliedmarketresearch.com/parametric-insurance-market-A14966
Kunreuther, H., Meszaros, J., Hogarth, R. M., & Spranca, M. (1995). Ambiguity and underwriter decision processes. *Journal of Economic Behavior & Organization, 26*(3), 337–352.
Metromile. (2023). *Pay-per-mile comprehensive car insurance*. Retrieved from https://www.metromile.com/coverage/comprehensive-insurance/
Mills, E. (2007). Synergisms between climate change mitigation and adaptation: An insurance perspective. *Mitigation and Adaptation Strategies for Global Change, 12*, 809–842.
Munich Re. (1973). *Flood – Inundation*. Münchener Rückversicherungs-Gesellschaft.
Munich Re. (2022). *Sustainability report 2022*. Retrieved from https://www.munichre.com/en/company/sustainability.html
Munich Re. (2023). *Photovoltaic warranty insurance*. Retrieved from https://www.munichre.com/en/solutions/for-industry-clients/pv-warranty-insurance-backing-your-solar-investment.html
OECD. (2012). *OECD environmental outlook to 2050: The consequences of inaction*. OECD. https://doi.org/10.1787/9789264122246-en
Prudential Regulation Authority. (2015). *The impact of climate change on the UK insurance sector. A climate change adaptation report*. Retrieved from https://www.bankofengland.co.uk/-/media/boe/files/prudential-regulation/publication/impact-of-climate-change-on-the-uk-insurance-sector.pdf
Roberte, L. (2021, November 16). *Home insurance policies that offer green home discounts*. Retrieved from https://www.thebalancemoney.com/green-home-insurance-discounts-5093053.
S&P Global. (2014, September 3). *Climate change could sting reinsurers that underestimate its impact*. Retrieved from https://www.spglobal.com/ratings/en/research/articles/140903-climate-change-could-sting-reinsurers-that-underestimate-its-impact-8706036

Steckel, J. C., & Jakob, M. (2018). The role of financing cost and de-risking strategies for clean energy investment. *International Economics, 155*, 19–28.

Swiss Re. (2015). *Swiss Re corporate solutions and Infigen energy complete first-in-market wind risk hedge*. Retrieved from https://corporatesolutions.swissre.com/dam/jcr:f63184a3-ea9f-4e51-962b-3fc5deb9a7e6/Swiss-Re-Corporate-Solutions-Infigen-Energy-wind-risk-hedge.pdf

Swiss Re. (2022). *Sustainability report 2022*. Retrieved from https://www.swissre.com/sustainability/approach/reporting.html

Swiss Re. (2023). *Natural catastrophes and inflation in 2022: A perfect storm*. Swiss Re Institute. Retrieved from https://www.swissre.com/institute/research/sigma-research/sigma-2023-01.html

Tol, R. S. J. (2009). The economic effects of climate change. *Journal of Economic Perspectives, 23*(2), 29–51.

UN. (2023). *NZIA target-setting protocol version 1.0*. United Nations. Retrieved from https://www.unepfi.org/industries/insurance/nzia-target-setting-protocol-version-1-0/

UNEP. (2020). *Managing environmental, social and governance risks in non-life insurance business*. United Nations. Retrieved from https://wedocs.unep.org/20.500.11822/33632

Wolfrom, L., & Yokoi-Arai, M. (2016). Financial instruments for managing disaster risks related to climate change. *OECD Journal: Financial Market Trends, 2015*(1), 25–47.

Yüksel, S., Eti, S., Prosekov, S., & Uluer, G. S. (2021). A study on the energy problem in the eastern mediterranean: Text mining and AHP-based strategy recommendations. In *Strategic approaches to energy management: Current trends in energy economics and green investment* (pp. 27–38). Springer.

Zurich. (2022). *Zurich aims for net-zero target in operations by 2030 instead of 2050*. Retrieved from https://www.zurich.com/en/media/news-releases/2022/2022-0331-01

Chapter 17
Investments in Energy Complexes: Evidence from Tajikistan

Kitmo and Md. Mominur Rahman

17.1 Introduction

Amendments will be made to the existing regulatory legal documents, and new preferential conditions will be created for potential investors. From our point of view, the main criteria here are the investment activity of industries and a favorable environment for their improvement. This problem can be solved by improving the investment climate and increasing the volume of investments. In order to solve the issue of attracting investments in the energy complex, ensuring stability and adaptation of the investment activity regime, the necessary conditions and factors for increasing the motivation of the country's energy complex are considered (Aksoy et al., 2023). These relations are the basis for increasing the profitability of the use of investment resources and require the involvement of investors. The volume of investment resources in specific areas depends on the state of the investment environment, the profitability of socioeconomic reconstruction, the level of development of other areas of the economy, production and financial dynamics, and infrastructure for the useful use of natural resources and energy. In order to attract potential investors in the energy sector, the state policy on domestic and foreign investments, the nationalization of foreign property, the conclusion of international treaties, the stability of state institutions, the stability of political power, the priority level of state support for the economic sector, etc. are of great importance (Eti et al., 2023). In modern conditions, in order to solve socioeconomic issues of economic development, the state needs to attract credit or gratuitous resources and expand

Kitmo
Department of Renewable Energy, National Advanced School of Engineering of University of Maroua, Maroua, Cameroon

M. M. Rahman (✉)
Research Wing (Accounting), BIGM, University of Dhaka (Affiliated), Dhaka, Bangladesh

H. Dinçer et al. (eds.), *Decision Making in Interdisciplinary Renewable Energy Projects*, Contributions to Management Science,
https://doi.org/10.1007/978-3-031-51532-3_17

investment activities in the republic and regions through the development and improvement of the level mechanism (Niu et al., 2023). It is necessary to improve the management of investment projects, combining the system of goals of investment development of the country and investment support in the republic (Mikhaylov et al., 2023b; Wang et al., 2023; He et al., 2023; Irshad et al., 2023; Iqbal et al., 2023; Aksoy et al., 2023).

17.2 Literature Review

The evaluation of investment projects is an extremely complex problem and requires various quantitative and qualitative methods and ratios when investing in the country's energy complex, including in relation to foreign investments. When evaluating projects, there are a number of conditions and additional factors that should be taken into account, especially with regard to the profitability of direct and sustainable investments. Therefore, assessing the profitability of any investment project, including in the energy sector, requires a continuous process. It begins from the moment of the feasibility study of the project, is considered as part of its control (monitoring) during the execution of the work, and after its completion presents a summary result (Denisova et al., 2019; Nyangarika et al., 2019a, b; Huang et al., 2021a, b; Mikhaylov, 2018, 2022a, b; Mikhaylov et al., 2019; Conteh et al., 2021; Sediqi et al., 2022, Khan et al., 2021; Bhuiyan et al., 2022a; Liu et al., 2021a, b; Daniali et al., 2021; Moiseev et al., 2023a, b; An et al., 2022).

When developing investment projects, it is important to choose the direction of investment, because where there are better forms of development, this leads to higher profitability (Bhuiyan et al., 2021, 2022b; Dong et al., 2021, Mikhaylov, 2021b; Barykin et al., 2022a; Liu et al., 2022a, b; Danish et al., 2022a, b; Saqib et al., 2021; Mukhametov et al., 2021; Candila et al., 2021; Mikhaylov & Grilli, 2022; Li et al., 2022).

In order to make investment management decisions, it is necessary to accurately determine the volume of investments with the receipt of financial resources and income after the implementation of the investment project (Khan et al., 2022; Dinçer et al., 2022a, b; Badr et al., 2022; Barykin et al., 2022b; Mehta et al., 2022; Kalinina et al. 2022; Shaikh et al., 2022; Mikhaylov et al., 2022b, 2023a; Nyangarika et al., 2022).

To do this, the volume of invested capital and future resources, the time of increase and decrease of funds as a result of the implementation of investment projects in the energy complex should be measured, taking into account the current value and future value of investment projects, it is also necessary to determine the flow of investments (Mikhaylov et al., 2022a; Mikhaylov, 2021a; Varyash et al., 2020; Zhao et al., 2021; An & Mikhaylov, 2020, 2021; Alwaelya et al., 2021; Yumashev & Mikhaylov, 2020; Yumashev et al., 2020; Mutalimov et al., 2021).

The problem of discounted cost of implementation of investment projects in the energy sector (exploitation of new hydropower resources) includes the need to compare current costs and future revenues (Nie et al. 2020; Dayong et al., 2020; Mikhaylov et al., 2018; Nyangarika et al., 2018; Danish et al., 2020, 2021; An et al., 2021; Uyeh et al., 2021; Tamashiro et al., 2021, 2023; Shaikh et al., 2021).

In order to implement investment projects, it is necessary to price attract investments in hydropower (Eti et al., 2022). The methods of project evaluation in this place are mainly based on the collection and processing of information about the discount amount of future financial income and should include the expected income from the completed project and the amount of capital invested (An et al., 2019a, b, 2020a, b, c; Mikhaylov, 2020a, b, c; Mikhaylov & Tarakanov, 2020; Moiseev et al., 2020, 2021; Gura et al., 2022; Dooyum et al., 2020).

17.3 Results

Currently, out of 120 projects proposed for investment in the regions of the Republic of Tatarstan, 22 investment projects (18.3%) belong to the city of Dushanbe, 36 projects (30%) in the Sughd region of projects (11.7%) in the VMKB, 25 projects (20.8%) to the subordinate districts of the Republic (NTR), and 23 projects (19.2%) to the Khatlon Region. Investment projects aimed at implementing the country's strategic development goals in the field of hydropower (52.3%) and investment projects aimed at improving the level and quality of life of the population, such as mining in the industrial, agricultural, and social spheres (36.2%) were involved. If earlier investments were concentrated in such areas as trade and finance, in recent years, rapid development has been noted in a number of other areas (energy, communications, real estate, water supply, various services, etc.). In addition to them, the main factors of ensuring high profitability of using relatively cheap energy resources, raw materials and labor, expanding production volumes, and making a profit from domestic investments are considered. Therefore, when implementing investment projects of the energy complex, methods of assessing the energy complex and expert assessment methods and costs are used to determine the degree of influence of a particular factor on attracting investment and its objectivity. Assessment of the factors of attracting investments in the energy complex of the Republic of Tajikistan requires the creation of favorable conditions for the inflow of investments. In modern conditions, it is planned to intensify the attraction of investments in the energy complex on the basis of improving the level and quality of construction of hydroelectric power plants and, in the future, the life of the population. Among the main factors affecting the return on investment and stimulating investment in the energy complex of the republic, the following factors should also be taken into account (Dinçer et al., 2023a, b; Ivleva et al., 2023; Aissou et al., 2023; Mikhaylov, 2023):

- The level and quality of state support and protection of domestic and foreign investors - The volume of domestic and external demand for electricity
- The level of poverty and social protection of the population

- The level of corruption and the investment environment
- Quality protection of intellectual property rights
- The level of education and the level and quality of life of the population
- The speed of development and decline of bureaucracy

When developing investment projects in the energy complex, it is necessary to take into account the following conditions: trends in the development of industries, the possibility of changing the position of the energy complex in the domestic market and ways to enter the world market, changes in the financial situation of the industry itself and its partners, access to natural materials, and financial and information resources. Based on this information and conditions, various valuation methods are used, which serve to make decisions on the choice of investment strategies in the fields.

We all know that energy is the most important resource of mankind; without energy, it is impossible to imagine modern human life. In this regard, energy is the basis of the country's economy (Martínez et al., 2023).

At the present time, energy issues are the most important and urgent problems of the whole world. We know that in the global energy sector, special attention is paid to renewable energy sources (Dinçer et al., 2023a, b, c, d, e, f), since renewable energy is more sparing of nature (less greenhouse gases emitted into the atmosphere, better ecology, better for humanity). Various bills are being adopted supporting and encouraging renewable energy at the state and interstate levels (Carayannis et al., 2023). But despite this, in 2020, fossil fuels provide 83.1% of the total global energy consumption. By 2020, it still occupies the largest share in the total consumption of primary energy sources of the planet—31.2%. Coal is in second place, 27.2%; gas is third with a share of 24.7%; the contribution of hydropower was 6.9%; renewable energy sources gave 5.7% and nuclear power, 4.3%.

Developing countries account for 61% of the world's total energy consumption. At the same time, China's electricity consumption exceeds the total energy consumption of both the USA and the EU combined (Yüksel et al., 2023).

During the 2020 pandemic, global energy consumption decreased by 4.5% compared to 2019, while wind and solar showed an increase of 9.7%. We explain this amazing fact not by the advantage of renewable energy sources but by their work on nonmarket conditions: guarantees for the purchase of their energy when other generators are blocked and budget subsidies.

Over the ten years (2011–2020), the share of renewable energy in the energy balance of the planet has grown by about 4.0%, in electricity consumption—by 7.6%. At this rate, it will take about 120 years of the current—very high—rates of investments in renewable energy sources for their share in the total primary energy balance to reach at least 50% and about 66 years to achieve the same level in electricity generation. The pandemic and lockdowns led to a 4.3% drop in global GDP in 2020, according to the UN.

17.4 Conclusion

The most common method is to determine the payback period, that is, the number of years required to return the invested capital, which is very convenient for facilitating decision-making, taking into account the payback period and subsequent stages of investment. Another method is to calculate the average return on investment and its initial cost (Xu et al., 2023). This method is determined by dividing the annual net income by the average cost of investments, but it does not take into account temporary financial flows (Han et al., 2023). For foreign companies or industries, the main indicators of investment attractiveness are the timing of covering investment costs (Kou et al., 2023a, b). If we talk about the stages of implementation of investment projects, then, in the process of making investment decisions, three stages are currently used: the amount of capital and the definition of sources of financing, the forecast of money supply growth from the implementation of investment projects, and assessment of the financial condition of the enterprise and its opportunities to participate in investment activities (Ai et al., 2023). At the same time, an important stage in assessing investment attractiveness is the analysis of the state of production facilities, complexes, and the determination of the payback period. To analyze the financial situation of industries, it is most important to estimate fixed costs, that is, costs that do not change when the volume of production changes (Wan et al., 2023). This includes rent, monthly salary of management, etc.; if the volume of electricity sales decreases, then fixed costs will remain at the same level, and as a result, income will also be lower than daily income, but variable costs will also be part of the volume of production changes (Yüksel & Dinçer, 2023). Hence, it can be concluded that the risk of investing in areas with high fixed costs is higher than in areas with high variable costs (Moiseev et al., 2023a, b).

References

Ai, R., Zheng, Y., Yüksel, S., & Dinçer, H. (2023). Investigating the components of fintech ecosystem for distributed energy investments with an integrated quantum spherical decision support system. *Financial Innovation, 9*(1), 27.

Aissou, A., Ahmed, H. S., Hussein, M. E., Hassan, M. R., Mikhaylov, A., & Mahmoud, E. (2023). Optimizing reliability of flow networks using RWGA subject to total capacity constraint. *Applied Mathematics and Information Sciences, 7*(4), 559–565. https://doi.org/10.18576/amis/170404

Aksoy, T., Yüksel, S., Dınçer, H., Hacioglu, U., Mikhaylov, A., & Muyeen, S. M. (2023). Assessing the electricity production capacities of emerging markets for the sustainable investment. *IEEE Access, 11*, 24574–24590.

Alwaelya, S. A., Yousif, N. B. A., & Mikhaylov, A. (2021). Emotional development in preschoolers and socialization. *Early Child Development and Care, 191*, 16.

An, J., & Mikhaylov, A. (2020). Russian energy projects in South Africa. *Journal of Energy in Southern Africa, 31*(3), 58–64.

An, J., & Mikhaylov, A. (2021). Analysis of energy projects financial efficiency and renewable energy generation in Russia. *Finance: Theory and Practice, 25*(5), 79–91.

An, J., Mikhaylov, A., & Moiseev, N. (2019a). Oil price predictors: Machine learning approach. *International Journal of Energy Economics and Policy, 9*(5), 1–6.

An, J., Mikhaylov, A., & Sokolinskaya, N. (2019b). Machine learning in economic planning: Ensembles of algorithms. *Journal of Physics: Conference Series, 1353*, 012126.

An, J., Mikhaylov, A., & Jung, S.-U. (2020a). The strategy of South Korea in the global oil market. *Energies, 13*(10), 2491.

An, J., Mikhaylov, A., & Kim, K. (2020b). Machine learning approach in heterogeneous group of algorithms for transport safety-critical system. *Applied Sciences, 10*(8), 2670.

An, J., Mikhaylov, A., & Richter, U. H. (2020c). Trade war effects: Evidence from sectors of energy and resources in Africa. *Heliyon, 6*, e05693. https://doi.org/10.1016/j.heliyon.2020.e05693

An, J., Mikhaylov, A., & Jung, S.-U. (2021). A linear programming approach for robust network revenue management in the airline industry. *Journal of Air Transport Management, 91*, 101979.

An, J., Mikhaylov, A., Dinçer, H., & Yüksel, S. (2022). Economic modelling of electricity generation: Long short-term memory and Q-rung orthopair fuzzy sets. *Heliyon, 8*, e12345.

Badr, A. M. N. A. A., Mohamed, T. I., Osman, N. A. Q., & Mikhaylov, A. (2022). A review of social media website users' interaction paths with governmental accounts during the COVID-19 pandemic. *Informatics, 9*, 50. https://doi.org/10.3390/informatics9030050

Barykin, S., Mikheev, A., Kiseleva, E., Putikhin, Y., Alekseeva, N., & Mikhaylov, A. (2022a). An empirical analysis of Russian regions' debt sustainability. *Economies, 10*(4), 296.

Barykin, S. E., Kapustina, I. V., Sergeev, S. M., Daniali, S. M., Kopteva, L. A., Semenova, G. N., Pryadko, I. P., Mikhaylov, A., Baboshkin, P., Datsyuk, P., & Senjyu, T. (2022b). Financial logistics models based on systematic approach improving management solutions. *F1000Research, 11*, 570. https://doi.org/10.12688/f1000research.111252.1

Bhuiyan, M. A., An, J., Mikhaylov, A., Moiseev, N., & Danish, M. S. S. (2021). Renewable energy deployment and COVID-19 measures for sustainable development. *Sustainability, 13*, 4418.

Bhuiyan, M. A., Dinçer, H., Yüksel, S., Mikhaylov, A., Danish, M. S. S., Pinter, G., Uyeh, D. D., & Stepanova, D. (2022a). Economic indicators and bioenergy supply in developed economies: QROF-DEMATEL and random forest models. *Energy Reports, 8*(2022), 561–570.

Bhuiyan, M. A., Zhang, Q., Khare, V., Mikhaylov, A., Pinter, G., & Huang, X. (2022b). Renewable energy consumption and economic growth nexus—A systematic literature review. *Frontiers in Environmental Science, 10*, 878394. https://doi.org/10.3389/fenvs.2022.878394

Candila, V., Maximov, D., Mikhaylov, A., Moiseev, N., Senjyu, T., & Tryndina, N. (2021). On the relationship between oil and exchange rates of oil-exporting and oil-importing countries: From the great recession period to the Covid-19 era. *Energies, 14*, 8065.

Carayannis, E., Kostis, P., Dinçer, H., & Yüksel, S. (2023). Quality function deployment-oriented strategic outlook to sustainable energy policies based on quintuple innovation helix. *Journal of the Knowledge Economy*, 1–19.

Conteh, F., Takahashi, H., Hemeida, A. M., Krishnan, N., Mikhaylov, A., & Senjyu, T. (2021). Hybrid grid-connected renewable power generation for sustainable electricity supply in Sierra Leone: Case study Lungi Town. *Sustainability, 13*(18).

Daniali, S. M., Barykin, S. E., Kapustina, I. V., Khortabi, F. M., Sergeev, S. M., Kalinina, O. V., Mikhaylov, A., Veynberg, R., Zasova, L., & Senjyu, T. (2021). Predicting volatility index according to technical index and economic indicators on the basis of deep learning algorithm. *Sustainability, 13*, 14011.

Danish, M. S. S., Bhattacharya, A., Stepanova, D., Mikhaylov, A., Grilli, M. L., Khosravy, M., & Senjyu, T. (2020). A systematic review of metal oxide applications for energy and environmental sustainability. *Metals, 10*(12), 1604.

Danish, M. S. S., Senjyu, T., Sabory, N. R., Khosravy, M., Grilli, M. L., Mikhaylov, A., & Majidi, H. (2021). A forefront framework for sustainable aquaponics modeling and design. *Sustainability, 13*, 9313.

Danish, M. S., Senjyu, T. S., Sabory, N. R., & Mikhaylov, A. (2022a). Lessons from COVID-19 conferring environmental re-engineering opportunity. In M. Danish & T. Senjyu (Eds.),

Eco-friendly and agile energy strategies and policy development (pp. 99–118). IGI Global. https://doi.org/10.4018/978-1-7998-9502-2.ch005

Danish, M. S., Senjyu, T. S., Sabory, N. R., & Mikhaylov, A. (2022b). Interplay consequences of COVID-19 on global environmental sustainability. In M. Danish & T. Senjyu (Eds.), *Eco-friendly and agile energy strategies and policy development*. IGI Global. https://doi.org/10.4018/978-1-7998-9502-2.ch003

Dayong, N., Mikhaylov, A., Bratanovsky, S., Shaikh, Z. A., & Stepanova, D. (2020). Mathematical modeling of the technological processes of catering products production. *Journal of Food Process Engineering, 43*(2), 1–9.

Denisova, V., Mikhaylov, A., & Lopatin, E. (2019). Blockchain infrastructure and growth of global power consumption. *International Journal of Energy Economics and Policy, 9*(4), 22–29.

Dinçer, H., Yüksel, S., Mikhaylov, A., Pinter, G., & Shaikh, Z. A. (2022a). Analysis of renewable-friendly smart grid technologies for the distributed energy investment projects using a hybrid picture fuzzy rough decision-making approach. *Energy Reports, 8*, 11466–11477. https://doi.org/10.1016/j.egyr.2022.08.275

Dinçer, H., Yüksel, S., Mikhaylov, A., Barykin, S. E., Aksoy, T., & Hacioğlu, U. (2022b). Analysis of environmental priorities for green project investments using an integrated q-rung orthopair fuzzy modelling. *IEEE Access, 10*, 50996–51007. https://doi.org/10.1109/ACCESS.2022.3174058

Dinçer, H., Yüksel, S., Hacioglu, U., Yilmaz, M. K., & Delen, D. (2023a). Development of a sustainable corporate social responsibility index for performance evaluation of the energy industry: A hybrid decision-making methodology. *Resources Policy, 85*, 103940.

Dinçer, H., Eti, S., Gökalp, Y., Yüksel, S., & Çelebi, B. (2023b). The role of small-scale wind turbines in increasing the competitiveness of hospitals. In *Handbook of research on quality and competitiveness in the healthcare services sector* (pp. 336–352). IGI Global.

Dinçer, H., Yüksel, S., Mikhaylov, A., Muyeen, S. M., Chang, T., Barykin, S., & Kalinina, O. (2023c). CO2 emissions integrated fuzzy model: A case of seven emerging economies. *Energy Reports, 9*, 5741–5751.

Dinçer, H., Yüksel, S., Aksoy, T., Hacıoğlu, Ü., Mikhaylov, A., & Pinter, G. (2023d). Analysis of solar module alternatives for efficiency-based energy investments with hybrid 2-tuple IVIF modeling. *Energy Reports, 10*, 61–71.

Dinçer, H., Eti, S., Yüksel, S., Özdemir, S., Yflmaz, A. E., & Ergün, E. (2023e). Integrating data mining and fuzzy decision-making techniques for analyzing the key minimizing factors of carbon emissions. *Journal of Intelligent & Fuzzy Systems, 45*, 1–17.

Dinçer, H., Eti, S., Aksoy, T., Yüksel, S., Hacioglu, U., Mikhaylov, A., & Muyeen, S. M. (2023f). Analysis of environmental impact for material production investments using a novel soft computing methodology. *IEEE Access, 11*, 37987–38001. https://doi.org/10.1109/ACCESS.2023.3266524

Dong, B., Ikonnikova, I., Rogulin, R., Sakulyeva, T., & Mikhaylov, A. (2021). Environmental economic approach to optimization of transport communication in megacities. *Journal of Environmental Science and Health, Part A, 56*(6), 660–666.

Dooyum, U. D., Mikhaylov, A., & Varyash, I. (2020). Energy security concept in Russia and South Korea. *International Journal of Energy Economics and Policy, 10*(4), 102–107.

Eti, S., Dinçer, H., Yüksel, S., & Gökalp, Y. (2022). Analysis of the suitability of the solar panels for hospitals: A new fuzzy decision-making model proposal with the T-spherical TOP-DEMATEL method. *Journal of Intelligent & Fuzzy Systems, 44*, 1–13.

Eti, S., Dinçer, H., Gökalp, Y., Yüksel, S., & Kararoğlu, D. (2023). Identifying key issues to handle the inflation problem in the healthcare industry caused by energy prices: An evaluation with decision-making models. In *Managing inflation and supply chain disruptions in the global economy* (pp. 162–178). Springer.

Gura, D., Mikhaylov, A., Glushkov, S., Zaikov, M., & Shaikh, Z. A. (2022). Model for estimating power dissipation along the interconnect length in single on-chip topology. *Evolutionary Intelligence, 15*, 2369–2373. https://doi.org/10.1007/s12065-020-00407-7

Han, Z., Gong, L., Chen, H., & Yüksel, S. (2023). The role of mineral and oil resources in achieving a clean environment in emerging economies: An asymmetric analysis. *Resources Policy, 86*, 104099.

He, H., Huang, Y., Nakadomari, A., Masrur, H., Krishnan, N., Hemeida, A. M., Mikhaylov, A., Omine, E., & Senjyu, T. (2023). Potential and economic viability of green hydrogen production from seawater electrolysis using renewable energy in remote Japanese islands. *Renewable Energy, 202*, 1436–1447. https://doi.org/10.1016/j.renene.2022.12.046

Huang, Y., Masrur, H., Shigenobu, R., Hemeida, A. M., Mikhaylov, A., & Senjyu, T. (2021a). A comparative design of a campus microgrid considering a multi-scenario and multi-objective approach. *Energies, 14*, 2853.

Huang, Y., Yona, A., Takahashi, H., Hemeida, A. M., Mandal, P., Mikhaylov, A., Senjyu, T., & Lotfy, M. E. (2021b). Energy management system optimization of drug store operate electric vehicles charging station. *Sustainability, 14*, 1217836.

Iqbal, K. M. J., Khan, M. I., Mikhaylov, A., Shah, A. A., Yadykin, V., Filho, W. L., Tariq, M. A. U. R., & Ullah, W. (2023). Modeling principles, criteria and indicators to assess water sector governance for climate compatibility and sustainability. *Frontiers in Environmental Science, 11*, 989930.

Irshad, A. S., Ludin, G. A., Masrur, H., Ahmadi, M., Yona, A., Mikhaylov, A., Krishnan, N., & Senjyu, T. (2023). Optimization of grid-photovoltaic and battery hybrid system with most technically efficient PV technology after the performance analysis. *Renewable Energy, 207*, 714–730.

Ivleva, S., Hajimajeed, M., Kartashova, O., & Mikhaylov, A. (2023). A comparative study of regional variations in common visual impairments among school-age children. *Electronic Journal of General Medicine, 20*(5), em504. https://doi.org/10.29333/ejgm/13264

Kalinina, O. V., Barykin, S. E., Sergeev, S. M., Semenova, G. N., Fatkullina, A., Mikhaylov, A., & De La Poza, E. (2022). Smart city perspectives in post-pandemic governance: Externalities reduction policy. *F1000Research, 11*, 1032. https://doi.org/10.12688/f1000research.123195.1

Khan, A. A., Shaikh, Z. A., Baitenova, L., Mutaliyeva, L., Moiseev, N., Mikhaylov, A., Laghari, A. A., Idris, S. A., & Alshazly, H. (2021). QoS-ledger: Smart contracts and metaheuristic for secure quality-of-service and cost-efficient scheduling of medical-data processing. *Electronics, 10*, 3083.

Khan, A. A., Laghari, A. A., Gadekallu, T. R., Shaikh, Z. A., Javed, A. R., Rashid, M., Estrela, V. V., & Mikhaylov, A. (2022). A drone-based data management and optimization using metaheuristic algorithms and blockchain smart contracts in a secure fog environment. *Computers and Electrical Engineering, 102*, 108234. https://doi.org/10.1016/j.compeleceng.2022.108234

Kou, G., Pamucar, D., Dinçer, H., & Yüksel, S. (2023a). From risks to rewards: A comprehensive guide to sustainable investment decisions in renewable energy using a hybrid facial expression-based fuzzy decision-making approach. *Applied Soft Computing, 142*, 110365.

Kou, G., Yuksel, S., & Dincer, H. (2023b). A facial expression and expert recommendation fuzzy decision-making approach for sustainable business investments within the metaverse world. *Applied Soft Computing, 148*, 110849.

Li, J., Dinçer, H., Yüksel, S., Mikhaylov, A., & Barykin, S. E. (2022). Bipolar q-ROF hybrid decision making model with golden cut for analyzing the levelized cost of renewable energy alternatives. *IEEE Access, 10*, 35900–35900. https://doi.org/10.1109/ACCESS.2022.3168315

Liu, L., Kato, T., Mandal, P., Mikhaylov, A, Hemeida, A. M., & Senjyu T. (2021a). Multi-area power system load frequency control by using model predictive control by introducing renewable energy sources. In *2021 International Conference on Science & Contemporary Technologies (ICSCT)*, pp. 1–6. https://doi.org/10.1109/ICSCT53883.2021.9642521

Liu, L., Kato, T., Mandal, P., Mikhaylov, A., Hemeida, A. M., Takahashi, H., & Senjyu, T. (2021b). Two cases studies of model predictive control approach for hybrid renewable energy systems. *AIMS Energy, 9*(6), 1241–1259.

Liu, Z.-J., Panfilova, E., Mikhaylov, A., & Kurilova, A. (2022a). COVID-19 crisis impact on the stability between parties in crowdfunding and crowdsourcing. *Wireless Personal Communications, 122*, 915–930.

Liu, Z., Panfilova, E., Mikhaylov, A., & Kurilova, A. (2022b). Assessing stability in the relationship between parties in crowdfunding and crowdsourcing projects during the COVID-19 crisis. *Journal of Global Information Management (JGIM), 30*(4), 1–18. https://doi.org/10.4018/JGIM.297905

Martínez, L., Dinçer, H., & Yüksel, S. (2023). A hybrid decision making approach for new service development process of renewable energy investment. *Applied Soft Computing, 133*, 109897.

Mehta, R., Verghese, J., Mahajan, S., Barykin, S., Bozhuk, S., Kozlova, N., Kapustina, I., Mikhaylov, A., Naumova, E., & Dedyukhina, N. (2022). Consumers' behavior in conversational commerce marketing based on messenger chatbots. *F1000Research, 11*(647). https://doi.org/10.12688/f1000research.122037.1

Mikhaylov, A. (2018). The yield of the carry trade strategy. *Finance: Theory and Practice, 22*(3), 52–63. https://doi.org/10.26794/2587-5671-2018-22-3-52-63

Mikhaylov, A. (2020a). Cryptocurrency market analysis from the open innovation perspective. *Journal of Open Innovation: Technology, Market, and Complexity, 6*(4), 197.

Mikhaylov, A. (2020b). Geothermal energy development in Iceland. *International Journal of Energy Economics and Policy, 10*(4), 31–35.

Mikhaylov, A. (2020c). Cryptocurrency market development: Hurst method. *Finance: Theory and Practice, 24*(3), 81–91.

Mikhaylov, A. (2021a). Development of Friedrich von Hayek's theory of private money and economic implications for digital currencies. *Terra Economicus, 19*(1), 53–62.

Mikhaylov, A. (2021b). Lichens as indicators of atmospheric pollution in urban ecosystems. *Israel Journal of Ecology & Evolution, 67*(1–2), 60–68. https://doi.org/10.1163/22244662-bja10016

Mikhaylov, A. (2022a). Efficiency of renewable energy plants in Russia. *Anais da Academia Brasileira de Ciências, 94*(4), e20191226. https://doi.org/10.1590/0001-3765202220191226

Mikhaylov, A. (2022b). Sustainable development and renewable energy: A new view to a global problem. *Energies, 15*, 1397. https://doi.org/10.3390/en15041397

Mikhaylov, A. (2023). Understanding the risks associated with wallets, depository services, trading, lending, and borrowing in the crypto space. *Journal of Infrastructure, Policy and Development, 6*, 2.

Mikhaylov, A., & Grilli, M. L. (2022). Machine learning methods and sustainable development: Metal oxides and multilayer metal-oxides. *Metals, 12*(4).

Mikhaylov, A., & Tarakanov, S. (2020). Development of Levenberg-Marquardt theoretical approach for electric network. *Journal of Physics: Conference Series, 1515*, 052006.

Mikhaylov, A., Sokolinskaya, N., & Nyangarika, A. (2018). Optimal carry trade strategy based on currencies of energy and developed economies. *Journal of Reviews on Global Economics, 7*, 582–592.

Mikhaylov, A., Burakov, D., & Didenko, V. (2019). Relationship between oil price and macroeconomic indicators in Russia. *Finance: Theory and Practice, 23*(2), 105–116. https://doi.org/10.26794/2587-5671-2019-23-2-105-116

Mikhaylov, A., Bhatti, I. M., Dinçer, H., & Yüksel, S. (2022a). Integrated decision recommendation system using iteration-enhanced collaborative filtering, golden cut bipolar for analyzing the risk-based oil market spillovers. *Computational Economics*. https://doi.org/10.1007/s10614-022-10341-8

Mikhaylov, A., Yumashev, A., & Kolpak, E. (2022b). Quality of life, anxiety and depressive disorders in patients with extrasystolic arrhythmia. *Archives of Medical Science, 18*(2). https://doi.org/10.5114/aoms.2020.101359

Mikhaylov, A., Dinçer, H., & Yüksel, S. (2023a). Analysis of financial development and open innovation oriented fintech potential for emerging economies using an integrated decision-making approach of MF-X-DMA and golden cut bipolar q-ROFSs. *Financial Innovation, 9*(1), 10. https://doi.org/10.1186/s40854-022-00234-w

Mikhaylov, A., Dinçer, H., Yüksel, S., Pinter, G., & Shaikh, Z. A. (2023b). Bitcoin mempool growth and trading volumes: Integrated approach based on QROF multi-SWARA and aggregation operators. *Journal of Innovation & Knowledge, 8*(3), 100378.

Moiseev, N., Mikhaylov, A., Varyash, I., & Saqib, A. (2020). Investigating the relation of GDP per capita and corruption index. *Entrepreneurship and Sustainability Issues, 8*(1), 780–794.

Moiseev, N., Sorokin, A., Zvezdina, N., Mikhaylov, A., Khomyakova, L., & Danish, M. S. S. (2021). Credit risk theoretical model on the base of DCC-GARCH in time-varying parameters framework. *Mathematics, 9*, 2423. https://doi.org/10.3390/math9192423

Moiseev, N., Mikhaylov, A., Dinçer, H., & Yüksel, S. (2023a). Effects of market capitalization shocks for open innovation models in e-commerce industry: Golden cut QROF MCDM analysis. *Financial Innovation, 9*(1), 32.

Moiseev, N., Mikhaylov, A., Dinçer, H., & Yüksel, S. (2023b). Market capitalization shock effects on open innovation models in e-commerce: Golden cut q-rung orthopair fuzzy multicriteria decision-making analysis. *Financial Innovation, 9*(1), 55.

Mukhametov, A., Bekhorashvili, A., Avdeenko, A., & Mikhaylov, A. (2021). The impact of growing legume plants under conditions of biologization and soil cultivation on Chernozem fertility and productivity of rotation crops. *Legume Research, 44*, 12.

Mutalimov, V., Kovaleva, I., Mikhaylov, A., & Stepanova, D. (2021). Assessing regional growth of small business in Russia. *Entrepreneurial Business and Economics Review, 9*(3), 119–133. https://doi.org/10.15678/EBER.2021.090308

Nie, D., Panfilova, E., Samusenkov, V., & Mikhaylov, A. (2020). E-Learning financing models in Russia for sustainable development. *Sustainability, 12*(11), 4412.

Niu, X., Yüksel, S., & Dinçer, H. (2023). Emission strategy selection for the circular economy-based production investments with the enhanced decision support system. *Energy, 274*, 127446.

Nyangarika, A., Mikhaylov, A., & Tang, B.-J. (2018). Correlation of oil prices and gross domestic product in oil producing countries. *International Journal of Energy Economics and Policy, 8*(5), 42–48.

Nyangarika, A., Mikhaylov, A., & Richter, U. (2019a). Influence oil price towards economic indicators in Russia. *International Journal of Energy Economics and Policy, 9*(1), 123–129.

Nyangarika, A., Mikhaylov, A., & Richter, U. (2019b). Oil price factors: Forecasting on the base of modified auto-regressive integrated moving average model. *International Journal of Energy Economics and Policy, 9*(1), 149–159.

Nyangarika, A., Mikhaylov, A., Muyeen, S. M., Yadykin, V., Mottaeva, A., Pryadko, I., Barykin, S., Fomenko, N., Rykov, G., & Shvandar, K. (2022). Energy stability and decarbonization in developing countries: Random Forest approach for forecasting of crude oil trade flows and macro indicators. *Frontiers in Environmental Science, 10*, 1031343. https://doi.org/10.3389/fenvs.2022.1031343

Saqib, A., Chan, T.-H., Mikhaylov, A., & Lean, H. H. (2021). Are the responses of sectoral energy imports asymmetric to exchange rate volatilities in Pakistan? Evidence from recent foreign exchange regime. *Frontiers in Energy Research, 9*, 614463. https://doi.org/10.3389/fenrg.2021.614463

Sediqi, M. M., Nakadomari, A., Mikhaylov, A., Krishnan, N., Lotfy, M. E., Yona, A., & Senjyu, T. (2022). Impact of time-of-use demand response program on optimal operation of Afghanistan real power system. *Energies, 15*, 296.

Shaikh, Z. A., Moiseev, N., Mikhaylov, A., & Yüksel, S. (2021). Facile synthesis of copper oxide-cobalt oxide/nitrogen-doped carbon (Cu2O-Co3O4/CN) composite for efficient water splitting. *Applied Sciences, 11*, 9974. https://doi.org/10.3390/app11219974

Shaikh, Z. A., Kraikin, A., Mikhaylov, A., & Pinter, G. (2022). Forecasting stock prices of companies producing solar panels using machine learning methods. *Complexity, 2022*, 9186265. https://doi.org/10.1155/2022/9186265

Tamashiro, K., Alharbi, T., Mikhaylov, A., Hemeida, A. M., Krishnan, N., Lotfy, M. E., & Senjyu, T. (2021). Investigation of home energy management with advanced direct load control and optimal scheduling of controllable loads. *Energies, 14*, 7314. https://doi.org/10.3390/en14217314

Tamashiro, K., Omine, E., Krishnan, N., Mikhaylov, A., Hemeida, A. M., & Senjyu, T. (2023). Optimal components capacity based multi-objective optimization and optimal scheduling based MPC-optimization algorithm in smart apartment buildings. *Energy & Buildings, 278*, 112616. https://doi.org/10.1016/j.enbuild.2022.112616

Uyeh, D. D., Asem-Hiablie, S., Park, T., Kim, K.-M., Mikhaylov, A., Woo, S., & Ha, Y. (2021). Could Japonica rice be an alternative variety for increased global food security and climate change mitigation? *Food, 10*, 1869. https://doi.org/10.3390/foods10081869

Varyash, I., Mikhaylov, A., Moiseev, N., & Aleshin, K. (2020). Triple bottom line and corporate social responsibility performance indicators for Russian companies. *Entrepreneurship and Sustainability Issues, 8*(1), 313–331.

Wan, Q., Miao, X., Wang, C., Dinçer, H., & Yüksel, S. (2023). A hybrid decision support system with golden cut and bipolar q-ROFSs for evaluating the risk-based strategic priorities of fintech lending for clean energy projects. *Financial Innovation, 9*(1), 1–25.

Wang, Q., Guo, J., Li, R., Mikhaylov, A., & Moiseev, N. (2023). Does technical assistance alleviate energy poverty in sub-Saharan African countries? A new perspective on spatial spillover effects of technical assistance. *Energy Strategy Reviews, 45*, 101047. https://doi.org/10.1016/j.esr.2022.101047

Xu, X., Yüksel, S., & Dinçer, H. (2023). An integrated decision-making approach with golden cut and bipolar q-ROFSs to renewable energy storage investments. *International Journal of Fuzzy Systems, 25*(1), 168–181.

Yüksel, S., & Dinçer, H. (2023). Sustainability analysis of digital transformation and circular industrialization with quantum spherical fuzzy modeling and golden cuts. *Applied Soft Computing, 138*, 110192.

Yüksel, S., Dinçer, H., & Uluer, G. S. (2023). Socio-economic indicators of nuclear energy investments: A study on countries. In *Inclusive developments through socio-economic indicators: New theoretical and empirical insights* (pp. 33–44). Emerald Publishing.

Yumashev, A., & Mikhaylov, A. (2020). Development of polymer film coatings with high adhesion to steel alloys and high wear resistance. *Polymer Composites, 41*(7), 2875–2880.

Yumashev, A., Ślusarczyk, B., Kondrashev, S., & Mikhaylov, A. (2020). Global indicators of sustainable development: Evaluation of the influence of the human development index on consumption and quality of energy. *Energies, 13*, 2768.

Zhao, Q., Cherkasov, S., Avdeenko, A., Kondratenko, L., & Mikhaylov, A. (2021). Integral estimate of the added value of agricultural crops in the synergetic agriculture on the example of Vicia faba (Vicia faba L.). *Polish Journal of Environmental Studies, 30*(5), 1–9. https://doi.org/10.15244/pjoes/131949

Chapter 18
Evaluation of Investment Projects by Analyzing a System of Financial Coefficients

Md. Mominur Rahman and Rizwan Ullah Khan

18.1 Introduction

The mechanisms for investing in electricity require that investments can be provided for to increase the tariff by issuing securities. In their opinion, this practice encourages energy companies to access the means of regulating activities at the expense of economic benefits received in addition to tariff and state regulation (Yüksel et al., 2023). Therefore, attracting and stimulating investments in the form of investment projects is considered an important task of state policy in the field of the country's energy complex (Xu et al., 2023). In order to overcome financial crises in the energy sector, it is extremely necessary to develop and improve methods for evaluating projects that are taken into account when taking into account the use of additional sources of investment aimed at their implementation. When producing electricity, reducing energy costs allows you to get additional funds to ensure the nominal level of depreciation of fixed and production assets. For these projects under development, due to the specialization of technological processes at different stages of electricity production, they bring an economic effect.

Conditions for the development of production infrastructure are an important condition for attracting investment in the energy sector. This potential is described and explained by the economic and geographical position of the republic and the development of infrastructures (Han et al., 2023). Thus, over the past decade, 23 state investment projects totaling 32 billion somoni have been implemented in order to free the republic from communication isolation and access to world seaports. As a result of these projects, 1650 kms of highways have been reconstructed, and 109

M. M. Rahman (✉)
Research Wing (Accounting), BIGM, University of Dhaka (Affiliated), Dhaka, Bangladesh

R. U. Khan
Graduate School of Business, Universiti Sains Malaysia, Gelugor, Penang, Malaysia

H. Dinçer et al. (eds.), *Decision Making in Interdisciplinary Renewable Energy Projects*, Contributions to Management Science,
https://doi.org/10.1007/978-3-031-51532-3_18

bridges and 27 kms of tunnels have been built and put into operation. Currently, 11 state investment projects worth 3.5 billion somoni are being implemented in the transport sector. In order to attract investors, free economic zones have been created in four regions of the republic: Sogdian, Panj, Dangara, Ishkoshim. All this makes up the overall investment attractiveness in the energy sector. Demographic power of the republic remains another condition for attracting investments. For investors, not only quantitative characteristics are important but also qualitative characteristics of the labor market, which may include the level of professional training, pretraining of personnel in various areas, education, age, and health status (Kou et al., 2023a, b).

When assessing them, the following indicators are important: availability of labor resources, skilled labor, demand and supply of labor in industries, etc. In modern conditions, the number of skilled labor is considered an important component of attracting investment. It should be noted that 67.5% of the country's population is young people, who have become the main force determining the sustainability of socioeconomic development. In this regard, the implementation of investment projects requires an unresolved and unplanned state policy of attracting investment resources and the level of investment incentives in industries, the main directions of which are improving the training and raising the standard of living of the population, creating or building a training system for the needs of the labor market through the investment activities of the energy complex of the republic.

18.2 Literature Review

The transition to innovative development creates conditions for the beneficial use of innovative energy and allows for the sustainable development of the economy and its important industries (Eti et al., 2022). Along with organizational and economic resources, innovative factors are also important in energy management. In relation to this level of investment, the nature of the use of innovative energy is determined (Bhuiyan et al., 2021, 2022b; Dong et al., 2021; Mikhaylov, 2021b; Barykin et al., 2022a; Liu et al., 2022a, b; Danish et al., 2022a, b; Saqib et al., 2021; Mukhametov et al., 2021; Candila et al., 2021; Mikhaylov & Grilli, 2022; Li et al., 2022).

Therefore, innovative energy, as the main indicator of the level of development of science and technology and the introduction of new technologies in the production of electricity, in most cases becomes one of the important elements of the reproduction of the natural resource potential of industries (Moiseev et al., 2023a, b). It is important to note that attracting investments in the energy sector depends on the influence of factors, analysis and evaluation, and forecasting of investment projects (Nie et al. 2020; Mikhaylov et al., 2023a; Mikhaylov, 2021a; Varyash et al., 2020; Zhao et al., 2021; An & Mikhaylov, 2020, 2021; Alwaelya et al., 2021; Yumashev & Mikhaylov, 2020; Yumashev et al., 2020; Mutalimov et al., 2021).

This makes it possible to determine the main directions and ways to activate investment resources and stimulate investment activity (Khan et al., 2022; Dinçer et al., 2022a, b; Badr et al., 2022; Barykin et al., 2022b; Mehta et al., 2022; Kalinina et al. 2022; Shaikh et al., 2022; Mikhaylov et al., 2022a, b; Nyangarika et al., 2022).

Another condition for attracting investment is the availability of information about the political and economic situation of the investment force, access to Internet resources, and the image of the industry as a whole; the creation of favorable conditions for attracting investment increases the level of return on investment activity (Denisova et al., 2019; Nyangarika et al., 2019a, b; Huang et al., 2021a, b; Mikhaylov, 2018, 2022; Mikhaylov et al., 2019; Conteh et al., 2021; Sediqi et al., 2022, Khan et al., 2021; Bhuiyan et al., 2022a; Liu et al., 2021a; b; Daniali et al., 2021; Moiseev et al., 2023a, b).

Another condition for investors is a stable international experience of the regulatory framework governing the movement of investment resources, especially foreign investments (Yüksel & Dinçer, 2023). Constant changes in the legislation regulating the attraction of investments, regulatory legal acts, with the legal development of foreign investment activities in the energy sector. Before investing, a willing investor weighs the legal risks and the state's attitude to investment activities (Mikhaylov et al., 2023b; Wang et al., 2023; He et al., 2023; Irshad et al., 2023; Iqbal et al., 2023; Aksoy et al., 2023).

That is why the investment power of the industries is completely under strict control and a scientifically sound policy in the field of investment promotion and investment risk management (Dayong et al., 2020; Mikhaylov et al., 2018; Nyangarika et al., 2018; Danish et al., 2020, 2021; An et al., 2021; Uyeh et al., 2021; Tamashiro et al., 2021, 2023; Shaikh et al., 2021).

Thus, it is necessary to consider many conditions affecting the attraction of investments and note that with a detailed study of each of these conditions, it is possible to achieve stable investment activity of the economy for the preparation of investment projects (Eti et al., 2023). The economic development of energy is a process that further aggravates the uneven development of the national economy and depends on many positive and negative conditions of the economy and its branches (An et al., 2019a, b, 2020a, b, c, 2022; Mikhaylov & Tarakanov, 2020; Moiseev et al., 2020, 2021; Gura et al., 2022; Dooyum et al., 2020; Mikhaylov et al., 2020; Mikhaylov, 2020a, b, c, 2023; Dınçer et al., 2023a, b; Ivleva et al., 2023; Aissou et al., 2023).

When justifying the evaluation of investment projects, it is necessary to determine what are the quantitative and qualitative indicators of the sources of efficiency improvement (Wan et al., 2023). In such conditions, the source of attracting investments can be free investment, if there is a greater opportunity to generate income and higher profitability than investments in other sectors of the economy and their diversification (Ai et al., 2023). At the same time, the condition for ensuring the profitability of the industry is the use of cheap energy resources, raw materials and labor, as well as the possibility of purchasing power in the market, rapid production growth, and preservation of the initial cost (Niu et al., 2023).

18.3 Results

Below is a detailed methodology for measuring the effectiveness of investments in the energy sector. The quantitative assessment should take into account the main criteria that determine the composition of investment income and expenses, as well as their effectiveness. In this regard, the payback period of the investment project, from the date of financing of the investment project through direct investment until the moment when the difference between the accumulated amount of net profit minus the depreciation fund and the amount of production costs should be positive (Martínez et al., 2023). This pattern is determined by the formula for calculating the net discounted value of NPV. Here $t = 1,2\ldots n$ is the period of implementation of the investment project in years; Dt is the annual cash flow, including net profit and depreciation; $\sum$ is the discount coefficient, which in this case is assumed to be constant throughout the entire period of the investment program, taking into account the added value; t is the amount of investments in the implementation of the investment project. It should be noted that when evaluating the effectiveness of investment projects in the energy sector, similar to the calculation formula, there are features of the formation of external costs, which are presented in this formula. First of all, the payback period (t) cannot be arbitrary and is set to attract investments for at least seven years. As for the annual cash flow (Dt), when analyzing the evaluation of the effectiveness of investments, the characteristics of the methods for determining net profit depend on the result of the implementation of the project of investment resources in a new industry by attracting investment or electricity generation by attracting investment. Both quantitative and cost indicators can be used to assess the effectiveness of investment projects (Aksoy et al., 2023).

In October 2021, European gas prices set a historical record (almost $ 2000 per 1000 cubic meters). They have grown more than 20 times (compared to July last year). This rapid growth can be explained by several reasons. Among the main reasons are the sharp increase in demand for LNG in Asia, the weather, and the actions of speculators. Then, the price dropped slightly but still remains above $ 1000. The following conclusions follow from the above: (1) the global economy and gas demand are recovering much faster than expected. This leads to a general increase in prices for all commodities, including natural gas. Asian economies are also growing rapidly, which therefore buy more LNG. The Asian market is traditionally premium, so in summer in 2021, "free" LNG supplies were reoriented to Asia. (2) Quotas for greenhouse gas emissions in Europe have almost tripled in price over the past year, which also contributed to an increase in gas prices. The more electricity producers have to pay for emissions, the higher the competitiveness of gas generation compared to coal. (3) Speculators also played a role. Volatility in the gas futures market has increased sharply, which, apparently, attracted additional speculative capital. This was reflected in the growth of trading volumes and open positions on gas futures on the Dutch index TTF is the main European gas distribution center and a benchmark for the European market. (4) The reason can be called a change in Gazprom's long-term strategy (Russia). During this time, the share of gas supplies under contracts linked to oil prices from Gazprom has decreased dramatically—primarily under pressure from European regulators. Under the conditions of the "old" system, when the contract price depended on the oil market and was fairly stable, when prices jumped, it was advantageous for Gazprom to increase exports—the only way it could get additional profit.

18.4 Conclusion

One of the important indicators in this case is the increase in labor productivity, the increase of which at all levels is calculated by ΔQ and the increase in production volume by Q: Here, Qt, Qt-1—the volume of production, respectively, in the t-th period after the implementation of foreign investments and in the t-1 first period—before their implementation; Pt-1 is the average annual number of employees, respectively, during these periods. At the same time, in this case, the factor of price inflation should be taken into account. Because as a result of the implementation of the investment project, there will be a transition to the production of products that meet consumer requirements and environmentally friendly products that provide the electricity market with more favorable prices (Dinçer et al., 2023a, b, c, d, e, f). Thus, the assessment of the effectiveness of attracting investments in the energy sector is a complex process that requires an analysis of the features of attracting investments and taking into account numerous factors and conditions for the effectiveness of investment projects (Carayannis et al., 2023).

References

Ai, R., Zheng, Y., Yüksel, S., & Dinçer, H. (2023). Investigating the components of fintech ecosystem for distributed energy investments with an integrated quantum spherical decision support system. *Financial Innovation, 9*(1), 27.

Aissou, A., Ahmed, H. S., Hussein, M. E., Hassan, M. R., Mikhaylov, A., & Mahmoud, E. (2023). Optimizing reliability of flow networks using RWGA subject to Total capacity constraint. *Applied Mathematics and Information Sciences, 7*(4), 559–565. https://doi.org/10.18576/amis/170404

Aksoy, T., Yüksel, S., Dınçer, H., Hacioglu, U., Mikhaylov, A., & Muyeen, S. M. (2023). Assessing the electricity production capacities of emerging markets for the sustainable investment. *IEEE Access, 11*, 24574–24590.

Alwaelya, S. A., Yousif, N. B. A., & Mikhaylov, A. (2021). Emotional development in preschoolers and socialization. *Early Child Development and Care, 191*, 16.

An, J., & Mikhaylov, A. (2020). Russian energy projects in South Africa. *Journal of Energy in Southern Africa, 31*(3), 58–64.

An, J., & Mikhaylov, A. (2021). Analysis of energy projects financial efficiency and renewable energy generation in Russia. *Finance: Theory and Practice, 25*(5), 79–91.

An, J., Mikhaylov, A., & Moiseev, N. (2019a). Oil price predictors: Machine learning approach. *International Journal of Energy Economics and Policy, 9*(5), 1–6.

An, J., Mikhaylov, A., & Sokolinskaya, N. (2019b). Machine learning in economic planning: Ensembles of algorithms. *Journal of Physics: Conference Series, 1353*, 012126.

An, J., Mikhaylov, A., & Jung, S.-U. (2020a). The strategy of South Korea in the global oil market. *Energies, 13*(10), 2491.

An, J., Mikhaylov, A., & Kim, K. (2020b). Machine learning approach in heterogeneous group of algorithms for transport safety-critical system. *Applied Sciences, 10*(8), 2670.

An, J., Mikhaylov, A., & Richter, U. H. (2020c). Trade war effects: Evidence from sectors of energy and resources in Africa. *Heliyon, 6*, e05693. https://doi.org/10.1016/j.heliyon.2020.e05693

An, J., Mikhaylov, A., & Jung, S.-U. (2021). A linear programming approach for robust network revenue management in the airline industry. *Journal of Air Transport Management, 91*, 101979.

An, J., Mikhaylov, A., Dinçer, H., & Yüksel, S. (2022). Economic modelling of electricity generation: Long short-term memory and Q-rung orthopair fuzzy sets. *Heliyon, 8*, e12345.

Badr, A. M. N. A. A., Mohamed, T. I., Osman, N. A. Q., & Mikhaylov, A. (2022). A review of social media website users' interaction paths with governmental accounts during the COVID-19 pandemic. *Informatics, 9*, 50. https://doi.org/10.3390/informatics9030050

Barykin, S., Mikheev, A., Kiseleva, E., Putikhin, Y., Alekseeva, N., & Mikhaylov, A. (2022a). An empirical analysis of Russian regions' debt sustainability. *Economies, 10*(4), 296.

Barykin, S. E., Kapustina, I. V., Sergeev, S. M., Daniali, S. M., Kopteva, L. A., Semenova, G. N., Pryadko, I. P., Mikhaylov, A., Baboshkin, P., Datsyuk, P., & Senjyu, T. (2022b). Financial logistics models based on systematic approach improving management solutions. *F1000Research, 11*, 570. https://doi.org/10.12688/f1000research.111252.1

Bhuiyan, M. A., An, J., Mikhaylov, A., Moiseev, N., & Danish, M. S. S. (2021). Renewable energy deployment and COVID-19 measures for sustainable development. *Sustainability, 13*, 4418.

Bhuiyan, M. A., Dinçer, H., Yüksel, S., Mikhaylov, A., Danish, M. S. S., Pinter, G., Uyeh, D. D., & Stepanova, D. (2022a). Economic indicators and bioenergy supply in developed economies: QROF-DEMATEL and random forest models. *Energy Reports, 8*(2022), 561–570.

Bhuiyan, M. A., Zhang, Q., Khare, V., Mikhaylov, A., Pinter, G., & Huang, X. (2022b). Renewable energy consumption and economic growth nexus—A systematic literature review. *Frontiers in Environmental Science., 10*, 878394. https://doi.org/10.3389/fenvs.2022.878394

Candila, V., Maximov, D., Mikhaylov, A., Moiseev, N., Senjyu, T., & Tryndina, N. (2021). On the relationship between oil and exchange rates of oil-exporting and oil-importing countries: From the great recession period to the Covid-19 era. *Energies, 14*, 8065.

Carayannis, E., Kostis, P., Dinçer, H., & Yüksel, S. (2023). Quality function deployment-oriented strategic outlook to sustainable energy policies based on quintuple innovation helix. *Journal of the Knowledge Economy*, 1–19.

Conteh, F., Takahashi, H., Hemeida, A. M., Krishnan, N., Mikhaylov, A., & Senjyu, T. (2021). Hybrid grid-connected renewable power generation for sustainable electricity supply in Sierra Leone: Case study Lungi Town. *Sustainability, 13*(18).

Daniali, S. M., Barykin, S. E., Kapustina, I. V., Khortabi, F. M., Sergeev, S. M., Kalinina, O. V., Mikhaylov, A., Veynberg, R., Zasova, L., & Senjyu, T. (2021). Predicting volatility index according to technical index and economic indicators on the basis of deep learning algorithm. *Sustainability, 13*, 14011.

Danish, M. S. S., Bhattacharya, A., Stepanova, D., Mikhaylov, A., Grilli, M. L., Khosravy, M., & Senjyu, T. (2020). A systematic review of metal oxide applications for energy and environmental sustainability. *Metals, 10*(12), 1604.

Danish, M. S. S., Senjyu, T., Sabory, N. R., Khosravy, M., Grilli, M. L., Mikhaylov, A., & Majidi, H. (2021). A forefront framework for sustainable aquaponics modeling and design. *Sustainability, 13*, 9313.

Danish, M. S., Senjyu, T. S., Sabory, N. R., & Mikhaylov, A. (2022a). Lessons from COVID-19 conferring environmental re-engineering opportunity. In M. Danish & T. Senjyu (Eds.), *Eco-friendly and agile energy strategies and policy development* (pp. 99–118). IGI Global. https://doi.org/10.4018/978-1-7998-9502-2.ch005

Danish, M. S., Senjyu, T. S., Sabory, N. R., & Mikhaylov, A. (2022b). Interplay consequences of COVID-19 on global environmental sustainability. In M. Danish & T. Senjyu (Eds.), *Eco-friendly and agile energy strategies and policy development*. IGI Global. https://doi.org/10.4018/978-1-7998-9502-2.ch003

Dayong, N., Mikhaylov, A., Bratanovsky, S., Shaikh, Z. A., & Stepanova, D. (2020). Mathematical modeling of the technological processes of catering products production. *Journal of Food Process Engineering, 43*(2), 1–9.

Denisova, V., Mikhaylov, A., & Lopatin, E. (2019). Blockchain infrastructure and growth of global power consumption. *International Journal of Energy Economics and Policy, 9*(4), 22–29.

Dinçer, H., Yüksel, S., Mikhaylov, A., Pinter, G., & Shaikh, Z. A. (2022a). Analysis of renewable-friendly smart grid technologies for the distributed energy investment projects using a hybrid picture fuzzy rough decision-making approach. *Energy Reports, 8*, 11466–11477. https://doi.org/10.1016/j.egyr.2022.08.275

Dinçer, H., Yüksel, S., Mikhaylov, A., Barykin, S. E., Aksoy, T., & Hacioğlu, U. (2022b). Analysis of environmental priorities for green project investments using an integrated q-rung orthopair fuzzy modelling. *IEEE Access, 10*, 50996–51007. https://doi.org/10.1109/ACCESS.2022.3174058

Dinçer, H., Yüksel, S., Hacioglu, U., Yilmaz, M. K., & Delen, D. (2023a). Development of a sustainable corporate social responsibility index for performance evaluation of the energy industry: A hybrid decision-making methodology. *Resources Policy, 85*, 103940.

Dinçer, H., Eti, S., Gökalp, Y., Yüksel, S., & Çelebi, B. (2023b). The role of small-scale wind turbines in increasing the competitiveness of hospitals. In *Handbook of research on quality and competitiveness in the healthcare services sector* (pp. 336–352). IGI Global.

Dinçer, H., Yüksel, S., Mikhaylov, A., Muyeen, S. M., Chang, T., Barykin, S., & Kalinina, O. (2023c). CO2 emissions integrated fuzzy model: A case of seven emerging economies. *Energy Reports, 9*, 5741–5751.

Dinçer, H., Yüksel, S., Aksoy, T., Hacıoğlu, Ü., Mikhaylov, A., & Pinter, G. (2023d). Analysis of solar module alternatives for efficiency-based energy investments with hybrid 2-tuple IVIF modeling. *Energy Reports, 10*, 61–71.

Dinçer, H., Eti, S., Yüksel, S., Özdemir, S., Yílmaz, A. E., & Ergün, E. (2023e). Integrating data mining and fuzzy decision-making techniques for analyzing the key minimizing factors of carbon emissions. *Journal of Intelligent & Fuzzy Systems, 45*, 1–17.

Dinçer, H., Eti, S., Aksoy, T., Yüksel, S., Hacioglu, U., Mikhaylov, A., & Muyeen, S. M. (2023f). Analysis of environmental impact for material production investments using a novel soft computing methodology. *IEEE Access, 11*, 37987–38001. https://doi.org/10.1109/ACCESS.2023.3266524

Dong, B., Ikonnikova, I., Rogulin, R., Sakulyeva, T., & Mikhaylov, A. (2021). Environmental economic approach to optimization of transport communication in megacities. *Journal of Environmental Science and Health, Part A, 56*(6), 660–666.

Dooyum, U. D., Mikhaylov, A., & Varyash, I. (2020). Energy security concept in Russia and South Korea. *International Journal of Energy Economics and Policy, 10*(4), 102–107.

Eti, S., Dinçer, H., Yüksel, S., & Gökalp, Y. (2022). Analysis of the suitability of the solar panels for hospitals: A new fuzzy decision-making model proposal with the T-spherical TOP-DEMATEL method. *Journal of Intelligent & Fuzzy Systems, 44*, 1–13.

Eti, S., Dinçer, H., Gökalp, Y., Yüksel, S., & Kararoğlu, D. (2023). Identifying key issues to handle the inflation problem in the healthcare industry caused by energy prices: An evaluation with decision-making models. In *Managing inflation and supply chain disruptions in the global economy* (pp. 162–178). Springer.

Gura, D., Mikhaylov, A., Glushkov, S., Zaikov, M., & Shaikh, Z. A. (2022). Model for estimating power dissipation along the interconnect length in single on-chip topology. *Evolutionary Intelligence, 15*, 2369–2373. https://doi.org/10.1007/s12065-020-00407-7

Han, Z., Gong, L., Chen, H., & Yüksel, S. (2023). The role of mineral and oil resources in achieving a clean environment in emerging economies: An asymmetric analysis. *Resources Policy, 86*, 104099.

He, H., Huang, Y., Nakadomari, A., Masrur, H., Krishnan, N., Hemeida, A. M., Mikhaylov, A., Omine, E., & Senjyu, T. (2023). Potential and economic viability of green hydrogen production from seawater electrolysis using renewable energy in remote Japanese islands. *Renewable Energy, 202*, 1436–1447. https://doi.org/10.1016/j.renene.2022.12.046

Huang, Y., Masrur, H., Shigenobu, R., Hemeida, A. M., Mikhaylov, A., & Senjyu, T. (2021a). A comparative design of a campus microgrid considering a multi-scenario and multi-objective approach. *Energies, 14*, 2853.

Huang, Y., Yona, A., Takahashi, H., Hemeida, A. M., Mandal, P., Mikhaylov, A., Senjyu, T., & Lotfy, M. E. (2021b). Energy management system optimization of drug store operate electric vehicles charging station. *Sustainability, 14*, 1217836.

Iqbal, K. M. J., Khan, M. I., Mikhaylov, A., Shah, A. A., Yadykin, V., Filho, W. L., Tariq, M. A. U. R., & Ullah, W. (2023). Modeling principles, criteria and indicators to assess water sector governance for climate compatibility and sustainability. *Frontiers in Environmental Science, 11*, 989930.

Irshad, A. S., Ludin, G. A., Masrur, H., Ahmadi, M., Yona, A., Mikhaylov, A., Krishnan, N., & Senjyu, T. (2023). Optimization of grid-photovoltaic and battery hybrid system with most technically efficient PV technology after the performance analysis. *Renewable Energy, 207*, 714–730.

Ivleva, S., Hajimajeed, M., Kartashova, O., & Mikhaylov, A. (2023). A comparative study of regional variations in common visual impairments among school-age children. Electronic journal of. *General Medicine, 20*(5), em504. https://doi.org/10.29333/ejgm/13264

Kalinina, O. V., Barykin, S. E., Sergeev, S. M., Semenova, G. N., Fatkullina, A., Mikhaylov, A., & De La Poza, E. (2022). Smart city perspectives in post-pandemic governance: Externalities reduction policy. *F1000Research, 11*, 1032. https://doi.org/10.12688/f1000research.123195.1

Khan, A. A., Shaikh, Z. A., Baitenova, L., Mutaliyeva, L., Moiseev, N., Mikhaylov, A., Laghari, A. A., Idris, S. A., & Alshazly, H. (2021). QoS-ledger: Smart contracts and metaheuristic for secure quality-of-service and cost-efficient scheduling of medical-data processing. *Electronics, 10*, 3083.

Khan, A. A., Laghari, A. A., Gadekallu, T. R., Shaikh, Z. A., Javed, A. R., Rashid, M., Estrela, V. V., & Mikhaylov, A. (2022). A drone-based data management and optimization using metaheuristic algorithms and blockchain smart contracts in a secure fog environment. *Computers and Electrical Engineering, 102*, 108234. https://doi.org/10.1016/j.compeleceng.2022.108234

Kou, G., Pamucar, D., Dinçer, H., & Yüksel, S. (2023a). From risks to rewards: A comprehensive guide to sustainable investment decisions in renewable energy using a hybrid facial expression-based fuzzy decision-making approach. *Applied Soft Computing, 142*, 110365.

Kou, G., Yuksel, S., & Dincer, H. (2023b). A facial expression and expert recommendation fuzzy decision-making approach for sustainable business investments within the metaverse world. *Applied Soft Computing, 148*, 110849.

Li, J., Dinçer, H., Yüksel, S., Mikhaylov, A., & Barykin, S. E. (2022). Bipolar q-ROF hybrid decision making model with golden cut for analyzing the levelized cost of renewable energy alternatives. *IEEE Access, 10*, 35900–35900. https://doi.org/10.1109/ACCESS.2022.3168315

Liu, L., Kato, T, Mandal, P., Mikhaylov, A, Hemeida, A. M., & Senjyu T. (2021a). Multi-area power system load frequency control by using model predictive control by introducing renewable energy sources. In *2021 International Conference on Science & Contemporary Technologies (ICSCT)*, pp. 1–6. https://doi.org/10.1109/ICSCT53883.2021.9642521

Liu, L., Kato, T., Mandal, P., Mikhaylov, A., Hemeida, A. M., Takahashi, H., & Senjyu, T. (2021b). Two cases studies of model predictive control approach for hybrid renewable energy systems. *AIMS Energy, 9*(6), 1241–1259.

Liu, Z.-J., Panfilova, E., Mikhaylov, A., & Kurilova, A. (2022a). COVID-19 crisis impact on the stability between parties in crowdfunding and crowdsourcing. *Wireless Personal Communications, 122*, 915–930.

Liu, Z., Panfilova, E., Mikhaylov, A., & Kurilova, A. (2022b). Assessing stability in the relationship between parties in crowdfunding and crowdsourcing projects during the COVID-19 crisis. *Journal of Global Information Management (JGIM), 30*(4), 1–18. https://doi.org/10.4018/JGIM.297905

Martínez, L., Dinçer, H., & Yüksel, S. (2023). A hybrid decision making approach for new service development process of renewable energy investment. *Applied Soft Computing, 133*, 109897.

Mehta, R., Verghese, J., Mahajan, S., Barykin, S., Bozhuk, S., Kozlova, N., Kapustina, I., Mikhaylov, A., Naumova, E., & Dedyukhina, N. (2022). Consumers' behavior in conversational commerce marketing based on messenger chatbots. *F1000Research, 11*(647). https://doi.org/10.12688/f1000research.122037.1

Mikhaylov, A. (2018). The yield of the carry trade strategy. *Finance: Theory and Practice, 22*(3), 52–63. https://doi.org/10.26794/2587-5671-2018-22-3-52-63

Mikhaylov, A. (2020a). Cryptocurrency market analysis from the open innovation perspective. *Journal of Open Innovation: Technology, Market, and Complexity, 6*(4), 197.

Mikhaylov, A. (2020b). Geothermal energy development in Iceland. *International Journal of Energy Economics and Policy, 10*(4), 31–35.

Mikhaylov, A. (2020c). Cryptocurrency market development: Hurst method. *Finance: Theory and Practice, 24*(3), 81–91.

Mikhaylov, A. (2021a). Development of Friedrich von Hayek's theory of private money and economic implications for digital currencies. *Terra Economicus, 19*(1), 53–62.

Mikhaylov, A. (2021b). Lichens as indicators of atmospheric pollution in urban ecosystems. *Israel Journal of Ecology & Evolution, 67*(1–2), 60–68. https://doi.org/10.1163/22244662-bja10016

Mikhaylov, A. (2022). Sustainable development and renewable energy: A new view to a global problem. *Energies, 15*, 1397. https://doi.org/10.3390/en15041397

Mikhaylov, A. (2023). Understanding the risks associated with wallets, depository services, trading, lending, and borrowing in the crypto space. *Journal of Infrastructure, Policy and Development, 6*, 2.

Mikhaylov, A., & Grilli, M. L. (2022). Machine learning methods and sustainable development: Metal oxides and multilayer metal-oxides. *Metals, 12*(4).

Mikhaylov, A., & Tarakanov, S. (2020). Development of Levenberg-Marquardt theoretical approach for electric network. *Journal of Physics: Conference Series, 1515*, 052006.

Mikhaylov, A., Sokolinskaya, N., & Nyangarika, A. (2018). Optimal carry trade strategy based on currencies of energy and developed economies. *Journal of Reviews on Global Economics, 7*, 582–592.

Mikhaylov, A., Burakov, D., & Didenko, V. (2019). Relationship between oil price and macroeconomic indicators in Russia. *Finance: Theory and Practice, 23*(2), 105–116. https://doi.org/10.26794/2587-5671-2019-23-2-105-116

Mikhaylov, A., Moiseev, N., Aleshin, K., & Burkhardt, T. (2020). Global climate change and greenhouse effect. *Entrepreneurship and Sustainability Issues, 7*(4), 2897–2913. https://doi.org/10.9770/jesi.2020.7.4(21)

Mikhaylov, A., Bhatti, I. M., Dinçer, H., & Yüksel, S. (2022a). Integrated decision recommendation system using iteration-enhanced collaborative filtering, golden cut bipolar for analyzing the risk-based oil market spillovers. *Computational Economics*, 1. https://doi.org/10.1007/s10614-022-10341-8

Mikhaylov, A., Yumashev, A., & Kolpak, E. (2022b). Quality of life, anxiety and depressive disorders in patients with extrasystolic arrhythmia. *Archives of Medical Science, 18*(2). https://doi.org/10.5114/aoms.2020.101359

Mikhaylov, A., Dinçer, H., & Yüksel, S. (2023a). Analysis of financial development and open innovation oriented fintech potential for emerging economies using an integrated decision-making approach of MF-X-DMA and golden cut bipolar q-ROFSs. *Financial Innovation, 9*(1), 10. https://doi.org/10.1186/s40854-022-00234-w

Mikhaylov, A., Dinçer, H., Yüksel, S., Pinter, G., & Shaikh, Z. A. (2023b). Bitcoin mempool growth and trading volumes: Integrated approach based on QROF multi-SWARA and aggregation operators. *Journal of Innovation & Knowledge, 8*(3), 100378.

Moiseev, N., Mikhaylov, A., Varyash, I., & Saqib, A. (2020). Investigating the relation of GDP per capita and corruption index. *Entrepreneurship and Sustainability Issues, 8*(1), 780–794.

Moiseev, N., Sorokin, A., Zvezdina, N., Mikhaylov, A., Khomyakova, L., & Danish, M. S. S. (2021). Credit risk theoretical model on the base of DCC-GARCH in time-varying parameters framework. *Mathematics, 9*, 2423. https://doi.org/10.3390/math9192423

Moiseev, N., Mikhaylov, A., Dinçer, H., & Yüksel, S. (2023a). Effects of market capitalization shocks for open innovation models in e-commerce industry: Golden cut QROF MCDM analysis. *Financial Innovation, 9*(1), 32.

Moiseev, N., Mikhaylov, A., Dinçer, H., & Yüksel, S. (2023b). Market capitalization shock effects on open innovation models in e-commerce: Golden cut q-rung orthopair fuzzy multicriteria decision-making analysis. *Financial Innovation, 9*(1), 55.

Mukhametov, A., Bekhorashvili, A., Avdeenko, A., & Mikhaylov, A. (2021). The impact of growing legume plants under conditions of biologization and soil cultivation on Chernozem fertility and productivity of rotation crops. *Legume Research, 44*, 12.

Mutalimov, V., Kovaleva, I., Mikhaylov, A., & Stepanova, D. (2021). Assessing regional growth of small business in Russia. *Entrepreneurial Business and Economics Review, 9*(3), 119–133. https://doi.org/10.15678/EBER.2021.090308

Nie, D., Panfilova, E., Samusenkov, V., & Mikhaylov, A. (2020). E-learning financing models in Russia for sustainable development. *Sustainability, 12*(11), 4412.

Niu, X., Yüksel, S., & Dinçer, H. (2023). Emission strategy selection for the circular economy-based production investments with the enhanced decision support system. *Energy, 274*, 127446.

Nyangarika, A., Mikhaylov, A., & Tang, B.-J. (2018). Correlation of oil prices and gross domestic product in oil producing countries. *International Journal of Energy Economics and Policy, 8*(5), 42–48.

Nyangarika, A., Mikhaylov, A., & Richter, U. (2019a). Influence oil price towards economic indicators in Russia. *International Journal of Energy Economics and Policy, 9*(1), 123–129.

Nyangarika, A., Mikhaylov, A., & Richter, U. (2019b). Oil price factors: Forecasting on the base of modified auto-regressive integrated moving average model. *International Journal of Energy Economics and Policy, 9*(1), 149–159.

Nyangarika, A., Mikhaylov, A., Muyeen, S. M., Yadykin, V., Mottaeva, A., Pryadko, I., Barykin, S., Fomenko, N., Rykov, G., & Shvandar, K. (2022). Energy stability and decarbonization in developing countries: Random Forest approach for forecasting of crude oil trade flows and macro indicators. *Frontiers in Environmental Science, 10*, 1031343. https://doi.org/10.3389/fenvs.2022.1031343

Saqib, A., Chan, T.-H., Mikhaylov, A., & Lean, H. H. (2021). Are the responses of sectoral energy imports asymmetric to exchange rate volatilities in Pakistan? Evidence from recent foreign exchange regime. *Frontiers in Energy Research, 9*, 614463. https://doi.org/10.3389/fenrg.2021.614463

Sediqi, M. M., Nakadomari, A., Mikhaylov, A., Krishnan, N., Lotfy, M. E., Yona, A., & Senjyu, T. (2022). Impact of time-of-use demand response program on optimal operation of Afghanistan real power system. *Energies, 15*, 296.

Shaikh, Z. A., Moiseev, N., Mikhaylov, A., & Yüksel, S. (2021). Facile synthesis of copper oxide-cobalt oxide/nitrogen-doped carbon (Cu2O-Co3O4/CN) composite for efficient water splitting. *Applied Sciences, 11*, 9974. https://doi.org/10.3390/app11219974

Shaikh, Z. A., Kraikin, A., Mikhaylov, A., & Pinter, G. (2022). Forecasting stock prices of companies producing solar panels using machine learning methods. *Complexity, 2022*, 9186265. https://doi.org/10.1155/2022/9186265

Tamashiro, K., Alharbi, T., Mikhaylov, A., Hemeida, A. M., Krishnan, N., Lotfy, M. E., & Senjyu, T. (2021). Investigation of home energy management with advanced direct load control and optimal scheduling of controllable loads. *Energies, 14*, 7314. https://doi.org/10.3390/en14217314

Tamashiro, K., Omine, E., Krishnan, N., Mikhaylov, A., Hemeida, A. M., & Senjyu, T. (2023). Optimal components capacity based multi-objective optimization and optimal scheduling based MPC-optimization algorithm in smart apartment buildings. *Energy & Buildings, 278*, 112616. https://doi.org/10.1016/j.enbuild.2022.112616

Uyeh, D. D., Asem-Hiablie, S., Park, T., Kim, K.-M., Mikhaylov, A., Woo, S., & Ha, Y. (2021). Could Japonica rice be an alternative variety for increased global food security and climate change mitigation? *Food, 10*, 1869. https://doi.org/10.3390/foods10081869

Varyash, I., Mikhaylov, A., Moiseev, N., & Aleshin, K. (2020). Triple bottom line and corporate social responsibility performance indicators for Russian companies. *Entrepreneurship and Sustainability Issues, 8*(1), 313–331.

Wan, Q., Miao, X., Wang, C., Dinçer, H., & Yüksel, S. (2023). A hybrid decision support system with golden cut and bipolar q-ROFSs for evaluating the risk-based strategic priorities of fintech lending for clean energy projects. *Financial Innovation, 9*(1), 1–25.

Wang, Q., Guo, J., Li, R., Mikhaylov, A., & Moiseev, N. (2023). Does technical assistance alleviate energy poverty in sub-Saharan African countries? A new perspective on spatial spillover effects of technical assistance. *Energy Strategy Reviews, 45*, 101047. https://doi.org/10.1016/j.esr.2022.101047

Xu, X., Yüksel, S., & Dinçer, H. (2023). An integrated decision-making approach with golden cut and bipolar q-ROFSs to renewable energy storage investments. *International Journal of Fuzzy Systems, 25*(1), 168–181.

Yüksel, S., & Dinçer, H. (2023). Sustainability analysis of digital transformation and circular industrialization with quantum spherical fuzzy modeling and golden cuts. *Applied Soft Computing, 138*, 110192.

Yüksel, S., Dinçer, H., & Uluer, G. S. (2023). Socio-economic indicators of nuclear energy investments: A study on countries. In *Inclusive developments through socio-economic indicators: New theoretical and empirical insights* (pp. 33–44). Emerald Publishing.

Yumashev, A., & Mikhaylov, A. (2020). Development of polymer film coatings with high adhesion to steel alloys and high wear resistance. *Polymer Composites, 41*(7), 2875–2880.

Yumashev, A., Ślusarczyk, B., Kondrashev, S., & Mikhaylov, A. (2020). Global indicators of sustainable development: Evaluation of the influence of the human development index on consumption and quality of energy. *Energies, 13*, 2768.

Zhao, Q., Cherkasov, S., Avdeenko, A., Kondratenko, L., & Mikhaylov, A. (2021). Integral estimate of the added value of agricultural crops in the synergetic agriculture on the example of Vicia faba (Vicia faba L.). *Polish Journal of Environmental Studies, 30*(5), 1–9. https://doi.org/10.15244/pjoes/131949

Chapter 19
Financial Regulation of Household Poverty and Income Inequality: ESG Sustainability and Effectiveness of Social Security Systems

Mikhail Dorofeev and Vikas Khare

19.1 Introduction

Conventional approaches to tackle these issues often rely on static economic models that fail to capture the complex dynamics of real-world systems. AB-SFC modeling provides a promising framework to better understand and address these challenges. By considering the heterogeneity and interactions of agents within an economy, AB-SFC modeling enables a more comprehensive analysis of poverty and income inequality dynamics (Kou et al., 2023a, b).

By following these procedures, the AB-SCF modeling approach aims to capture the complex dynamics of an economy, taking into account the interactions between heterogeneous economic agents and the macroeconomic environment. The integration of agent-based and stock-flow consistent methodologies offers a promising avenue for understanding and analyzing real-world economic systems in a more comprehensive and realistic manner (Moiseev et al., 2023a, b).

In the second study, an attempt was made to demonstrate how different scenarios of technological progress and investment levels can impact the level of household income inequality (Caiani et al., 2017). Overall, most parameters from the previous concept remained unchanged; however, the initial model underwent several modifications. Tax rates for different taxable entities became diverse, introducing the factor of progressive tax rates (Eti et al., 2022). Parameters describing innovation processes and investments were added. The household collection also underwent

M. Dorofeev (✉)
Financial University Under the Government of the Russian Federation, Moscow, Russia
e-mail: mldorofeev@fa.ru

V. Khare
School of Technology, Management and Engineering NMIMS, Indore, Indore, India
e-mail: vikas.khare@nmims.edu

H. Dinçer et al. (eds.), *Decision Making in Interdisciplinary Renewable Energy Projects*, Contributions to Management Science,
https://doi.org/10.1007/978-3-031-51532-3_19

changes, including various types of hired workers (the working class, office workers, researchers, managers) selling their labor to two types of corporations (companies producing means of production and companies manufacturing goods for household consumption). These differences are clearly visible when comparing the flow and stock diagrams between the original and subsequent models of the economy (Yüksel & Dinçer, 2023).

In the fourth study, the authors present a more complex AB-SCF model with a multicountry economic structure developed for the EU (Caiani et al., 2019). This work introduces a model of an open economy, which was previously demonstrated in the foundational work by Godley and Lavoie (2007). In this article, the authors did not address the issue of economic inequality, as it had already been addressed in previous studies. Nevertheless, we consider it an important logical step in the development of the AB-SCF modeling methodology.

19.2 Literature Review

Based on the conducted review analysis, the main shortcomings of SCF models can be grouped in three groups of problems:

1. Excessive theorization, insufficient realism, and an excessive number of simplifications and assumptions in the model (Denisova et al., 2019; Nyangarika et al., 2019a, b; Huang et al., 2021a, b; Mikhaylov, 2018, 2022; Mikhaylov et al., 2019; Conteh et al., 2021; Sediqi et al., 2022, Khan et al., 2021; Bhuiyan et al., 2022a; Liu et al., 2021a, b; Daniali et al., 2021; Moiseev et al., 2023a, b)
2. Utilization of nonuniversal/controversial economic hypotheses and assumptions in model development, along with logical inconsistencies in the models (Nie et al. 2020; Mikhaylov et al., 2023a; Mikhaylov, 2021a; Varyash et al., 2020; Zhao et al., 2021; An & Mikhaylov, 2020, 2021; Alwaelya et al., 2021; Yumashev & Mikhaylov, 2020; Yumashev et al., 2020; Mutalimov et al., 2021)
3. Poorly developed conceptual framework in the economic section of the model and vague formulations of economic categories (Bhuiyan et al., 2021; Dong et al., 2021, Mikhaylov, 2021b; Barykin et al., 2022a; Liu et al., 2022a, b; Bhuiyan et al., 2022b; Danish et al., 2022a, b; Saqib et al., 2021; Mukhametov et al., 2021, Candila et al., 2021; Mikhaylov & Grilli, 2022; Li et al., 2022)

The first group of problems remains the most challenging to overcome among all, given the complexity of income processes in the global economy (Wan et al., 2023). It is still difficult to fully capture the entirety and specificity of flows and changes in stocks on a global scale (Ai et al., 2023). Despite the increasing level of detail in each subsequent AB-SCF model, even the most authoritative authors often acknowledge its excessive theorization and simplified functional features of economic agents and regulators, which limits the potential use of this scientific research method (Eti et al., 2023).

However, AB-SCF modeling still deserves attention and has the potential to evolve into more sophisticated and comprehensive theoretical ecosystems, enabling realistic economic simulations and justifications of policy decisions (Niu et al., 2023). Currently, this methodology is suitable for conducting theoretical laboratory research and modeling household income inequality (Khan et al., 2022; Dinçer et al., 2022a, b; Badr et al., 2022; Barykin et al., 2022b; Mehta et al., 2022; Kalinina et al. 2022; Shaikh et al., 2022; Mikhaylov et al., 2022a, b; Nyangarika et al., 2022).

The second group of problems also requires attention from users of the resulting information derived from the use of AB-SCF models (Aksoy et al., 2023). Caution should be exercised in applying the obtained results in practice, as both the data itself and the unrealistic nature of the model can lead to incorrect reforms (Mikhaylov et al., 2018, 2023b; Wang et al. 2023; He et al., 2023; Irshad et al., 2023; Iqbal et al., 2023; Aksoy et al., 2023; Dayong et al., 2020; Nyangarika et al., 2018; Danish et al., 2020, 2021; An et al., 2021; Uyeh et al., 2021; Tamashiro et al., 2021, 2023; Shaikh et al., 2021).

The third group of concerns has already been addressed in Yakovenko (2009), where the author discussed the problem of using fuzzy terminology for various economic categories such as money, income, wealth, capital, etc. This problem is the least significant among all. It has persisted even after 2009. Practically in all the articles we analyzed, a formal approach to defining inequality as an economic category is observed. The term "inequality" was mainly used by the authors, apparently referring to the collective concept of household economic inequality in terms of income and wealth, although its meaning is much broader. The concept of inequality has too many dimensions and types to be used without a concretization (Dorofeev, 2021, 2022, 2023). We also noticed inaccuracies in defining households and labor force in their SCF models. It remains unclear whether they referred to an individual worker selling their labor to firms in the labor market or a household consisting of a group of individuals living in the same territory and managing their common economy.

Most misunderstandings between economists and physicists do not arise because they have different answers to the same question, but rather because their areas of scientific interest do not align very well, and scientists pose different tasks and ask different questions. To avoid confusion in building SFC models, it is advisable to provide more detailed and clear definitions of key terms, making it easier for specialists from other fields of scientific interest to integrate their ideas with the developments of econophysicists (An et al., 2019a, b, 2020a, b, 2020c, 2022; Mikhaylov & Tarakanov, 2020; Moiseev et al., 2020, 2021; Gura et al., 2022; Dooyum et al., 2020; Mikhaylov et al., 2020; Mikhaylov, 2020a, b, c).

19.3 ESG Sustainable Modeling Method for Poverty and Income Inequality Modeling

To effectively apply the AB-SFC modeling method for poverty and income inequality modeling, it is essential to follow a systematic approach and carefully plan the modeling process. Creating a comprehensive AB-SFC model should be based on accurate picture and realistic specifics of poverty and income inequality problems of a certain country. The model should incorporate a diverse range of agents that represent different population groups, considering their heterogeneity and interactions (Dinçer et al., 2023a, b, c, d, e, f). Furthermore, it should account for the key factors influencing poverty and income inequality in a country, including standard of living, income levels, wealth distribution disparities, social mobility, and other relevant socioeconomic variables. It should consider actual country's social security system and fiscal parameters, such as the scale of a progressive tax system (Martínez et al., 2023).

Such models can be useful if there is a certain degree of confidence that key macroeconomic variables and model parameters of the country will remain relatively stable within the modeling horizon, regardless of its duration, for example, in a situation where the central bank communicates its intention to maintain interest rates unchanged to the market or the government decides to adhere to legislatively approved policies limiting the government debt as a percentage of GDP at a certain level.

To ensure the reliability and accuracy of the AB-SFC model, it is necessary to calibrate and validate it using relevant data. The availability and quality of data is the primary challenge in applying AB-SFC modeling to poverty and income inequality analysis. Calibration of data involves adjusting the model parameters to match the characteristics of the specific country or region under study. Validation, on the other hand, involves comparing the model's outputs with empirical data to assess its ability to replicate real-world phenomena. This iterative process of calibration and validation enhances the model's credibility and its capacity to produce reliable insights (Yüksel et al., 2023).

The author's main thesis is that deregulation of the financial market, relaxed credit requirements for borrowers, the erosion of union bargaining power, and the increasing profits of investors and shareholders contribute to the growth of income inequality. The problem with the rising income inequality lies in the fact that the growing debt in the economy allows for an increase in consumer demand and accelerates economic growth in the short term. However, simultaneously, debt levels increase, and the net personal wealth of the working class decreases.

Calculations conducted using the example of populations in developed countries (mainly Great Britain and the USA) revealed that the income and wealth distribution among households in the lower segment of the sample follows an exponential Boltzmann-Gibbs law, while the upper segment follows a power-law distribution function known as Pareto distribution. Furthermore, the author examines the obtained results based on astrophysics theory and classifies income distribution into a "thermal" group and a "superthermal" group. Thus, the proposal suggests defining the boundaries between the lower class (97% of the population) and the upper class (3% of the population) based on the transition point of the distribution function from exponential to power law.

The reality that scientists attempted to model in the aforementioned works is so multifaceted and complex that the assumptions and simplifications made appear to be a reasonable compromise necessary for the initial development of the theory. Continuing research in this direction has led to the creation of more intricate SFC models with an open economy, gradually addressing the aforementioned limitations. Let us now examine a typical SFC model of income distribution in an open economy.

19.4 Conclusion

If it is possible to solve the problem of data, which significantly limits the potential of the use of the AB-SFC model in real world, this methodology can be very useful to explore the impacts of different policy measures, socioeconomic factors, and external shocks on poverty and income inequality dynamics. Simulations may include examining the effects of changes in taxation policies, social security systems, investment in human capital, and other relevant variables that have the potential to influence poverty and income inequality outcomes (Xu et al., 2023).

Despite the mentioned limitations, we consider the methodology of SFC modeling to be interesting and suitable for building acceptable theoretical financial-investment model of social security and finding optimal parameters for better regulation of poverty and income inequality (Han et al., 2023).FundingThe publication is based on the results of budgetary-supported research according to the state task carried out by the Financial University under the Government of the Russian Federation, Moscow, Russia.

References

Ai, R., Zheng, Y., Yüksel, S., & Dinçer, H. (2023). Investigating the components of fintech ecosystem for distributed energy investments with an integrated quantum spherical decision support system. *Financial Innovation, 9*(1), 27.

Aksoy, T., Yüksel, S., Dınçer, H., Hacioglu, U., Mikhaylov, A., & Muyeen, S. M. (2023). Assessing the electricity production capacities of emerging markets for the sustainable investment. *IEEE Access, 11*, 24574–24590.

Alwaelya, S. A., Yousif, N. B. A., & Mikhaylov, A. (2021). Emotional development in preschoolers and socialization. *Early Child Development and Care, 191*, 16.

An, J., & Mikhaylov, A. (2020). Russian energy projects in South Africa. *Journal of Energy in Southern Africa, 31*(3), 58–64.

An, J., & Mikhaylov, A. (2021). Analysis of energy projects financial efficiency and renewable energy generation in Russia. *Finance: Theory and Practice, 25*(5), 79–91.

An, J., Mikhaylov, A., & Moiseev, N. (2019a). Oil price predictors: Machine learning approach. *International Journal of Energy Economics and Policy, 9*(5), 1–6.

An, J., Mikhaylov, A., & Sokolinskaya, N. (2019b). Machine learning in economic planning: Ensembles of algorithms. *Journal of Physics: Conference Series, 1353*, 012126.

An, J., Mikhaylov, A., & Jung, S.-U. (2020a). The strategy of South Korea in the global oil market. *Energies, 13*(10), 2491.

An, J., Mikhaylov, A., & Kim, K. (2020b). Machine learning approach in heterogeneous group of algorithms for transport safety-critical system. *Applied Sciences, 10*(8), 2670.

An, J., Mikhaylov, A., & Richter, U. H. (2020c). Trade war effects: Evidence from sectors of energy and resources in Africa. *Heliyon, 6*, e05693. https://doi.org/10.1016/j.heliyon.2020.e05693

An, J., Mikhaylov, A., & Jung, S.-U. (2021). A linear programming approach for robust network revenue management in the airline industry. *Journal of Air Transport Management, 91*, 101979.

An, J., Mikhaylov, A., Dinçer, H., & Yüksel, S. (2022). Economic modelling of electricity generation: Long short-term memory and Q-rung Orthopair fuzzy sets. *Heliyon, 8*, e12345.

Badr, A. M. N. A. A., Mohamed, T. I., Osman, N. A. Q., & Mikhaylov, A. (2022). A review of social media website users' interaction paths with governmental accounts during the COVID-19 pandemic. *Informatics, 9*, 50. https://doi.org/10.3390/informatics9030050

Barykin, S., Mikheev, A., Kiseleva, E., Putikhin, Y., Alekseeva, N., & Mikhaylov, A. (2022a). An empirical analysis of Russian regions' debt sustainability. *Economies, 10*(4), 296.

Barykin, S. E., Kapustina, I. V., Sergeev, S. M., Daniali, S. M., Kopteva, L. A., Semenova, G. N., Pryadko, I. P., Mikhaylov, A., Baboshkin, P., Datsyuk, P., & Senjyu, T. (2022b). Financial logistics models based on systematic approach improving management solutions. *F1000Research, 11*, 570. https://doi.org/10.12688/f1000research.111252.1

Bhuiyan, M. A., An, J., Mikhaylov, A., Moiseev, N., & Danish, M. S. S. (2021). Renewable energy deployment and COVID-19 measures for sustainable development. *Sustainability, 13*, 4418.

Bhuiyan, M. A., Dinçer, H., Yüksel, S., Mikhaylov, A., Danish, M. S. S., Pinter, G., Uyeh, D. D., & Stepanova, D. (2022a). Economic indicators and bioenergy supply in developed economies: QROF-DEMATEL and random forest models. *Energy Reports, 8*(2022), 561–570.

Bhuiyan, M. A., Zhang, Q., Khare, V., Mikhaylov, A., Pinter, G., & Huang, X. (2022b). Renewable energy consumption and economic growth nexus—A systematic literature review. *Frontiers in Environmental Science., 10*, 878394. https://doi.org/10.3389/fenvs.2022.878394

Caiani, A., Russo, A., & Gallegati, M. (2017). *Are higher wages good for business? An assessment under alternative innovation and investment scenarios*. Retrieved from SSRN https://ssrn.com/abstract=3009593 or https://doi.org/10.2139/ssrn.3009593

Caiani, A., Catullo, E., & Gallegati, M. (2019). The effects of alternative wage regimes in a monetary union: A multi-country agent based-stock flow consistent model. *Journal of Economic Behavior & Organization, 162*, 389–416.

Candila, V., Maximov, D., Mikhaylov, A., Moiseev, N., Senjyu, T., & Tryndina, N. (2021). On the relationship between oil and exchange rates of oil-exporting and oil-importing countries: From the great recession period to the Covid-19 era. *Energies, 14*, 8065.

Conteh, F., Takahashi, H., Hemeida, A. M., Krishnan, N., Mikhaylov, A., & Senjyu, T. (2021). Hybrid grid-connected renewable power generation for sustainable electricity supply in Sierra Leone: Case study Lungi Town. *Sustainability, 13*(18).

Daniali, S. M., Barykin, S. E., Kapustina, I. V., Khortabi, F. M., Sergeev, S. M., Kalinina, O. V., Mikhaylov, A., Veynberg, R., Zasova, L., & Senjyu, T. (2021). Predicting volatility index according to technical index and economic indicators on the basis of deep learning algorithm. *Sustainability, 13*, 14011.

Danish, M. S. S., Bhattacharya, A., Stepanova, D., Mikhaylov, A., Grilli, M. L., Khosravy, M., & Senjyu, T. (2020). A systematic review of metal oxide applications for energy and environmental sustainability. *Metals, 10*(12), 1604.

Danish, M. S. S., Senjyu, T., Sabory, N. R., Khosravy, M., Grilli, M. L., Mikhaylov, A., & Majidi, H. (2021). A forefront framework for sustainable aquaponics modeling and design. *Sustainability, 13*, 9313.

Danish, M. S., Senjyu, T. S., Sabory, N. R., & Mikhaylov, A. (2022a). Lessons from COVID-19 conferring environmental re-engineering opportunity. In M. Danish & T. Senjyu (Eds.), *Eco-friendly and agile energy strategies and policy development* (pp. 99–118). IGI Global. https://doi.org/10.4018/978-1-7998-9502-2.ch005

Danish, M. S., Senjyu, T. S., Sabory, N. R., & Mikhaylov, A. (2022b). Interplay consequences of COVID-19 on global environmental sustainability. In M. Danish & T. Senjyu (Eds.), *Eco-friendly and agile energy strategies and policy development*. IGI Global. https://doi.org/10.4018/978-1-7998-9502-2.ch003

Dayong, N., Mikhaylov, A., Bratanovsky, S., Shaikh, Z. A., & Stepanova, D. (2020). Mathematical modeling of the technological processes of catering products production. *Journal of Food Process Engineering, 43*(2), 1–9.

Denisova, V., Mikhaylov, A., & Lopatin, E. (2019). Blockchain infrastructure and growth of global power consumption. *International Journal of Energy Economics and Policy, 9*(4), 22–29.

Dinçer, H., Yüksel, S., Mikhaylov, A., Pinter, G., & Shaikh, Z. A. (2022a). Analysis of renewable-friendly smart grid technologies for the distributed energy investment projects using a hybrid picture fuzzy rough decision-making approach. *Energy Reports, 8*, 11466–11477. https://doi.org/10.1016/j.egyr.2022.08.275

Dinçer, H., Yüksel, S., Mikhaylov, A., Barykin, S. E., Aksoy, T., & Hacioğlu, U. (2022b). Analysis of environmental priorities for green project investments using an integrated q-rung orthopair fuzzy modelling. *IEEE Access, 10*, 50996–51007. https://doi.org/10.1109/ACCESS.2022.3174058

Dinçer, H., Yüksel, S., Hacioglu, U., Yilmaz, M. K., & Delen, D. (2023a). Development of a sustainable corporate social responsibility index for performance evaluation of the energy industry: A hybrid decision-making methodology. *Resources Policy, 85*, 103940.

Dinçer, H., Eti, S., Gökalp, Y., Yüksel, S., & Çelebi, B. (2023b). The role of small-scale wind turbines in increasing the competitiveness of hospitals. In *Handbook of research on quality and competitiveness in the healthcare services sector* (pp. 336–352). IGI Global.

Dinçer, H., Yüksel, S., Mikhaylov, A., Muyeen, S. M., Chang, T., Barykin, S., & Kalinina, O. (2023c). CO2 emissions integrated fuzzy model: A case of seven emerging economies. *Energy Reports, 9*, 5741–5751.

Dinçer, H., Yüksel, S., Aksoy, T., Hacıoğlu, Ü., Mikhaylov, A., & Pinter, G. (2023d). Analysis of solar module alternatives for efficiency-based energy investments with hybrid 2-tuple IVIF modeling. *Energy Reports, 10*, 61–71.

Dinçer, H., Eti, S., Yüksel, S., Özdemir, S., Yïlmaz, A. E., & Ergün, E. (2023e). Integrating data mining and fuzzy decision-making techniques for analyzing the key minimizing factors of carbon emissions. *Journal of Intelligent & Fuzzy Systems, 45*, 1–17.

Dinçer, H., Etı, S., Aksoy, T., Yüksel, S., Hacioglu, U., Mikhaylov, A., & Muyeen, S. M. (2023f). *Analysis of environmental impact for material production investments using a novel soft computing methodology* (Vol. 11, p. 37987). IEEE Access.

Dong, B., Ikonnikova, I., Rogulin, R., Sakulyeva, T., & Mikhaylov, A. (2021). Environmental economic approach to optimization of transport communication in megacities. *Journal of Environmental Science and Health, Part A, 56*(6), 660–666.

Dooyum, U. D., Mikhaylov, A., & Varyash, I. (2020). Energy security concept in Russia and South Korea. *International Journal of Energy Economics and Policy, 10*(4), 102–107.

Dorofeev, M. L. (2021). Does income inequality create excessive threats to the sustainable development of Russia? Evidence from intercountry comparisons via analysis of inequality heatmaps. *Economies, 9*(4), 166.

Dorofeev, M. L. (2022). Interrelations between income inequality and sustainable economic growth: Contradictions of empirical research and new results. *Economies, 10*(2), 44.

Dorofeev, M. L. (2023). Comprehensive analysis of the effectiveness of social security regional financial models in Russia based on the DEA method. *Voprosy Ekonomiki, 6*, 117–137. (In Russian. https://doi.org/10.32609/0042-8736-2023-6-117-137

Eti, S., Dinçer, H., Yüksel, S., & Gökalp, Y. (2022). Analysis of the suitability of the solar panels for hospitals: A new fuzzy decision-making model proposal with the T-spherical TOP-DEMATEL method. *Journal of Intelligent & Fuzzy Systems, 44*, 1–13.
Eti, S., Dinçer, H., Gökalp, Y., Yüksel, S., & Kararoğlu, D. (2023). Identifying key issues to handle the inflation problem in the healthcare industry caused by energy prices: An evaluation with decision-making models. In *Managing inflation and supply chain disruptions in the global economy* (pp. 162–178). Springer.
Godley, W., & Lavoie, M. (2007). *Monetary economics: An integrated approach to credit, money, income, production and wealth*. Palgrave Macmillan.
Gura, D., Mikhaylov, A., Glushkov, S., Zaikov, M., & Shaikh, Z. A. (2022). Model for estimating power dissipation along the interconnect length in single on-chip topology. *Evolutionary Intelligence, 15*, 2369–2373. https://doi.org/10.1007/s12065-020-00407-7
Han, Z., Gong, L., Chen, H., & Yüksel, S. (2023). The role of mineral and oil resources in achieving a clean environment in emerging economies: An asymmetric analysis. *Resources Policy, 86*, 104099.
He, H., Huang, Y., Nakadomari, A., Masrur, H., Krishnan, N., Hemeida, A. M., Mikhaylov, A., Omine, E., & Senjyu, T. (2023). Potential and economic viability of green hydrogen production from seawater electrolysis using renewable energy in remote Japanese islands. *Renewable Energy, 202*, 1436–1447. https://doi.org/10.1016/j.renene.2022.12.046
Huang, Y., Masrur, H., Shigenobu, R., Hemeida, A. M., Mikhaylov, A., & Senjyu, T. (2021a). A comparative design of a campus microgrid considering a multi-scenario and multi-objective approach. *Energies, 14*, 2853.
Huang, Y., Yona, A., Takahashi, H., Hemeida, A. M., Mandal, P., Mikhaylov, A., Senjyu, T., & Lotfy, M. E. (2021b). Energy management system optimization of drug store operate electric vehicles charging station. *Sustainability, 14*, 1217836.
Iqbal, K. M. J., Khan, M. I., Mikhaylov, A., Shah, A. A., Yadykin, V., Filho, W. L., Tariq, M. A. U. R., & Ullah, W. (2023). Modeling principles, criteria and indicators to assess water sector governance for climate compatibility and sustainability. *Frontiers in Environmental Science, 11*, 989930.
Irshad, A. S., Ludin, G. A., Masrur, H., Ahmadi, M., Yona, A., Mikhaylov, A., Krishnan, N., & Senjyu, T. (2023). Optimization of grid-photovoltaic and battery hybrid system with most technically efficient PV technology after the performance analysis. *Renewable Energy, 207*, 714–730.
Kalinina, O. V., Barykin, S. E., Sergeev, S. M., Semenova, G. N., Fatkullina, A., Mikhaylov, A., & De La Poza, E. (2022). Smart city perspectives in post-pandemic governance: Externalities reduction policy. *F1000Research, 11*, 1032. https://doi.org/10.12688/f1000research.123195.1
Khan, A. A., Shaikh, Z. A., Baitenova, L., Mutaliyeva, L., Moiseev, N., Mikhaylov, A., Laghari, A. A., Idris, S. A., & Alshazly, H. (2021). QoS-ledger: Smart contracts and metaheuristic for secure quality-of-service and cost-efficient scheduling of medical-data processing. *Electronics, 10*, 3083.
Khan, A. A., Laghari, A. A., Gadekallu, T. R., Shaikh, Z. A., Javed, A. R., Rashid, M., Estrela, V. V., & Mikhaylov, A. (2022). A drone-based data management and optimization using metaheuristic algorithms and blockchain smart contracts in a secure fog environment. *Computers and Electrical Engineering, 102*, 108234. https://doi.org/10.1016/j.compeleceng.2022.108234
Kou, G., Pamucar, D., Dinçer, H., & Yüksel, S. (2023a). From risks to rewards: A comprehensive guide to sustainable investment decisions in renewable energy using a hybrid facial expression-based fuzzy decision-making approach. *Applied Soft Computing, 142*, 110365.
Kou, G., Yuksel, S., & Dincer, H. (2023b). A facial expression and expert recommendation fuzzy decision-making approach for sustainable business investments within the metaverse world. *Applied Soft Computing, 148*, 110849.
Li, J., Dinçer, H., Yüksel, S., Mikhaylov, A., & Barykin, S. E. (2022). Bipolar q-ROF hybrid decision making model with golden cut for analyzing the levelized cost of renewable energy alternatives. *IEEE Access, 10*, 35900. https://doi.org/10.1109/ACCESS.2022.3168315

Liu, L., Kato, T., Mandal, P., Mikhaylov, A, Hemeida, A. M., & Senjyu T. (2021a). Multi-area power system load frequency control by using model predictive control by introducing renewable energy sources. In *2021 International Conference on Science & Contemporary Technologies (ICSCT)*, pp. 1–6. https://doi.org/10.1109/ICSCT53883.2021.9642521

Liu, L., Kato, T., Mandal, P., Mikhaylov, A., Hemeida, A. M., Takahashi, H., & Senjyu, T. (2021b). Two cases studies of model predictive control approach for hybrid renewable energy systems. *AIMS Energy, 9*(6), 1241–1259.

Liu, Z.-J., Panfilova, E., Mikhaylov, A., & Kurilova, A. (2022a). COVID-19 crisis impact on the stability between parties in crowdfunding and crowdsourcing. *Wireless Personal Communications, 122*, 915–930.

Liu, Z., Panfilova, E., Mikhaylov, A., & Kurilova, A. (2022b). Assessing stability in the relationship between parties in crowdfunding and crowdsourcing projects during the COVID-19 crisis. *Journal of Global Information Management (JGIM), 30*(4), 1–18. https://doi.org/10.4018/JGIM.297905

Martínez, L., Dinçer, H., & Yüksel, S. (2023). A hybrid decision making approach for new service development process of renewable energy investment. *Applied Soft Computing, 133*, 109897.

Mehta, R., Verghese, J., Mahajan, S., Barykin, S., Bozhuk, S., Kozlova, N., Kapustina, I., Mikhaylov, A., Naumova, E., & Dedyukhina, N. (2022). Consumers' behavior in conversational commerce marketing based on messenger chatbots. *F1000Research, 11*(647), 10.12688/f1000research.122037.1.

Mikhaylov, A. (2018). The yield of the carry trade strategy. *Finance: Theory and Practice, 22*(3), 52–63. https://doi.org/10.26794/2587-5671-2018-22-3-52-63

Mikhaylov, A. (2020a). Cryptocurrency market analysis from the open innovation perspective. *Journal of Open Innovation: Technology, Market, and Complexity, 6*(4), 197.

Mikhaylov, A. (2020b). Geothermal energy development in Iceland. *International Journal of Energy Economics and Policy, 10*(4), 31–35.

Mikhaylov, A. (2020c). Cryptocurrency market development: Hurst method. *Finance: Theory and Practice, 24*(3), 81–91.

Mikhaylov, A. (2021a). Development of Friedrich von Hayek's theory of private money and economic implications for digital currencies. *Terra Economicus, 19*(1), 53–62.

Mikhaylov, A. (2021b). Lichens as indicators of atmospheric pollution in urban ecosystems. *Israel Journal of Ecology & Evolution, 67*(1–2), 60–68. https://doi.org/10.1163/22244662-bja10016

Mikhaylov, A. (2022). Sustainable development and renewable energy: A new view to a global problem. *Energies, 15*, 1397. https://doi.org/10.3390/en15041397

Mikhaylov, A., & Grilli, M. L. (2022). Machine learning methods and sustainable development: Metal oxides and multilayer metal-oxides. *Metals, 12*(4).

Mikhaylov, A., & Tarakanov, S. (2020). Development of Levenberg-Marquardt theoretical approach for electric network. *Journal of Physics: Conference Series, 1515*, 052006.

Mikhaylov, A., Sokolinskaya, N., & Nyangarika, A. (2018). Optimal carry trade strategy based on currencies of energy and developed economies. *Journal of Reviews on Global Economics, 7*, 582–592.

Mikhaylov, A., Burakov, D., & Didenko, V. (2019). Relationship between oil price and macroeconomic indicators in Russia. *Finance: Theory and Practice, 23*(2), 105–116. https://doi.org/10.26794/2587-5671-2019-23-2-105-116

Mikhaylov, A., Moiseev, N., Aleshin, K., & Burkhardt, T. (2020). Global climate change and greenhouse effect. *Entrepreneurship and Sustainability Issues, 7*(4), 2897–2913. https://doi.org/10.9770/jesi.2020.7.4(21)

Mikhaylov, A., Bhatti, I. M., Dinçer, H., & Yüksel, S. (2022a). Integrated decision recommendation system using iteration-enhanced collaborative filtering, golden cut bipolar for analyzing the risk-based oil market spillovers. *Computational Economics*. https://doi.org/10.1007/s10614-022-10341-8

Mikhaylov, A., Yumashev, A., & Kolpak, E. (2022b). Quality of life, anxiety and depressive disorders in patients with extrasystolic arrhythmia. *Archives of Medical Science, 18*(2), 328. https://doi.org/10.5114/aoms.2020.101359

Mikhaylov, A., Dinçer, H., & Yüksel, S. (2023a). Analysis of financial development and open innovation oriented fintech potential for emerging economies using an integrated decision-making approach of MF-X-DMA and golden cut bipolar q-ROFSs. *Financial Innovation, 9*(1), 10. https://doi.org/10.1186/s40854-022-00234-w

Mikhaylov, A., Dinçer, H., Yüksel, S., Pinter, G., & Shaikh, Z. A. (2023b). Bitcoin mempool growth and trading volumes: Integrated approach based on QROF multi-SWARA and aggregation operators. *Journal of Innovation & Knowledge, 8*(3), 100378.

Moiseev, N., Mikhaylov, A., Varyash, I., & Saqib, A. (2020). Investigating the relation of GDP per capita and corruption index. *Entrepreneurship and Sustainability Issues, 8*(1), 780–794.

Moiseev, N., Sorokin, A., Zvezdina, N., Mikhaylov, A., Khomyakova, L., & Danish, M. S. S. (2021). Credit risk theoretical model on the base of DCC-GARCH in time-varying parameters framework. *Mathematics, 9*, 2423. https://doi.org/10.3390/math9192423

Moiseev, N., Mikhaylov, A., Dinçer, H., & Yüksel, S. (2023a). Effects of market capitalization shocks for open innovation models in e-commerce industry: Golden cut QROF MCDM analysis. *Financial Innovation, 9*(1), 32.

Moiseev, N., Mikhaylov, A., Dinçer, H., & Yüksel, S. (2023b). Market capitalization shock effects on open innovation models in e-commerce: Golden cut q-rung orthopair fuzzy multicriteria decision-making analysis. *Financial Innovation, 9*(1), 55.

Mukhametov, A., Bekhorashvili, A., Avdeenko, A., & Mikhaylov, A. (2021). The impact of growing legume plants under conditions of biologization and soil cultivation on Chernozem fertility and productivity of rotation crops. *Legume Research, 44*, 12.

Mutalimov, V., Kovaleva, I., Mikhaylov, A., & Stepanova, D. (2021). Assessing regional growth of small business in Russia. *Entrepreneurial Business and Economics Review, 9*(3), 119–133. https://doi.org/10.15678/EBER.2021.090308

Nie, D., Panfilova, E., Samusenkov, V., & Mikhaylov, A. (2020). E-Learning financing models in Russia for sustainable development. *Sustainability, 12*(11), 4412.

Niu, X., Yüksel, S., & Dinçer, H. (2023). Emission strategy selection for the circular economy-based production investments with the enhanced decision support system. *Energy, 274*, 127446.

Nyangarika, A., Mikhaylov, A., & Tang, B.-J. (2018). Correlation of oil prices and gross domestic product in oil producing countries. *International Journal of Energy Economics and Policy, 8*(5), 42–48.

Nyangarika, A., Mikhaylov, A., & Richter, U. (2019a). Influence oil price towards economic indicators in Russia. *International Journal of Energy Economics and Policy, 9*(1), 123–129.

Nyangarika, A., Mikhaylov, A., & Richter, U. (2019b). Oil price factors: Forecasting on the base of modified auto-regressive integrated moving average model. *International Journal of Energy Economics and Policy, 9*(1), 149–159.

Nyangarika, A., Mikhaylov, A., Muyeen, S. M., Yadykin, V., Mottaeva, A., Pryadko, I., Barykin, S., Fomenko, N., Rykov, G., & Shvandar, K. (2022). Energy stability and decarbonization in developing countries: Random Forest approach for forecasting of crude oil trade flows and macro indicators. *Frontiers in Environmental Science, 10*, 1031343. https://doi.org/10.3389/fenvs.2022.1031343

Saqib, A., Chan, T.-H., Mikhaylov, A., & Lean, H. H. (2021). Are the responses of sectoral energy imports asymmetric to exchange rate volatilities in Pakistan? Evidence from recent foreign exchange regime. *Frontiers in Energy Research, 9*, 614463. https://doi.org/10.3389/fenrg.2021.614463

Sediqi, M. M., Nakadomari, A., Mikhaylov, A., Krishnan, N., Lotfy, M. E., Yona, A., & Senjyu, T. (2022). Impact of time-of-use demand response program on optimal operation of Afghanistan real power system. *Energies, 15*, 296.

Shaikh, Z. A., Moiseev, N., Mikhaylov, A., & Yüksel, S. (2021). Facile synthesis of copper oxide-cobalt oxide/nitrogen-doped carbon (Cu2O-Co3O4/CN) composite for efficient water splitting. *Applied Sciences, 11*, 9974. https://doi.org/10.3390/app11219974

Shaikh, Z. A., Kraikin, A., Mikhaylov, A., & Pinter, G. (2022). Forecasting stock prices of companies producing solar panels using machine learning methods. *Complexity, 2022*, 9186265. https://doi.org/10.1155/2022/9186265

Tamashiro, K., Alharbi, T., Mikhaylov, A., Hemeida, A. M., Krishnan, N., Lotfy, M. E., & Senjyu, T. (2021). Investigation of home energy management with advanced direct load control and optimal scheduling of controllable loads. *Energies, 14*, 7314. https://doi.org/10.3390/en14217314

Tamashiro, K., Omine, E., Krishnan, N., Mikhaylov, A., Hemeida, A. M., & Senjyu, T. (2023). Optimal components capacity based multi-objective optimization and optimal scheduling based MPC-optimization algorithm in smart apartment buildings. *Energy & Buildings, 278*, 112616. https://doi.org/10.1016/j.enbuild.2022.112616

Uyeh, D. D., Asem-Hiablie, S., Park, T., Kim, K.-M., Mikhaylov, A., Woo, S., & Ha, Y. (2021). Could Japonica rice be an alternative variety for increased global food security and climate change mitigation? *Food, 10*, 1869. https://doi.org/10.3390/foods10081869

Varyash, I., Mikhaylov, A., Moiseev, N., & Aleshin, K. (2020). Triple bottom line and corporate social responsibility performance indicators for Russian companies. *Entrepreneurship and Sustainability Issues, 8*(1), 313–331.

Wan, Q., Miao, X., Wang, C., Dinçer, H., & Yüksel, S. (2023). A hybrid decision support system with golden cut and bipolar q-ROFSs for evaluating the risk-based strategic priorities of fintech lending for clean energy projects. *Financial Innovation, 9*(1), 1–25.

Wang, Q., Guo, J., Li, R., Mikhaylov, A., & Moiseev, N. (2023). Does technical assistance alleviate energy poverty in sub-Saharan African countries? A new perspective on spatial spillover effects of technical assistance. *Energy Strategy Reviews, 45*, 101047. https://doi.org/10.1016/j.esr.2022.101047

Xu, X., Yüksel, S., & Dinçer, H. (2023). An integrated decision-making approach with golden cut and bipolar q-ROFSs to renewable energy storage investments. *International Journal of Fuzzy Systems, 25*(1), 168–181.

Yakovenko, V. (2009). Econophysics, statistical mechanics approach to. In *Encyclopedia of complexity and system science* (pp. 2800–2826). Springer. https://doi.org/10.1007/978-0-387-30440-3_169

Yüksel, S., & Dinçer, H. (2023). Sustainability analysis of digital transformation and circular industrialization with quantum spherical fuzzy modeling and golden cuts. *Applied Soft Computing, 138*, 110192.

Yüksel, S., Dinçer, H., & Uluer, G. S. (2023). Socio-economic indicators of nuclear energy investments: A study on countries. In *Inclusive developments through socio-economic indicators: New theoretical and empirical insights* (pp. 33–44). Emerald Publishing.

Yumashev, A., & Mikhaylov, A. (2020). Development of polymer film coatings with high adhesion to steel alloys and high wear resistance. *Polymer Composites, 41*(7), 2875–2880.

Yumashev, A., Ślusarczyk, B., Kondrashev, S., & Mikhaylov, A. (2020). Global indicators of sustainable development: Evaluation of the influence of the human development index on consumption and quality of energy. *Energies, 13*, 2768.

Zhao, Q., Cherkasov, S., Avdeenko, A., Kondratenko, L., & Mikhaylov, A. (2021). Integral estimate of the added value of agricultural crops in the synergetic agriculture on the example of Vicia faba (Vicia faba L.). *Polish Journal of Environmental Studies, 30*(5), 1–9. https://doi.org/10.15244/pjoes/131949

Chapter 20
Challenges and Solutions of AB-SFC Methodology for ESG Sustainable Social Security Systems

Mikhail Dorofeev and Hooi Hooi Lean

20.1 Introduction

The AB-SCF modeling methodology has emerged at the intersection of economics and physics and is a developing direction of scientific research. It involves attempts to examine well-known phenomena in sociology and economics from a different perspective, through the prism of methods and models that have proven themselves in the natural sciences.

The theory and methodology of AB-SFC modeling aim to formalize economics as a system of economic agents and the interconnections between them, manifested in transactions involving the exchange of stocks of goods, services, and financial assets, as well as the reciprocal movement of monetary flows (incomes), described based on econometric models. The founder of this scientific approach is Copeland (1949), whose works served as the basis for the development of the first macroeconomic models for the US Federal Reserve System. Currently, this research method is gaining popularity and attracting increasing attention from scholars and practitioners.

AB-SFC modeling can simulate the effects of various policies aimed at reducing poverty (Aksoy et al., 2023). By incorporating multiple agent types representing different income groups, the model can capture the dynamics of wealth accumulation, income distribution, and poverty rates (Niu et al., 2023). Simulations can explore the potential impacts of policies such as progressive taxation, targeted social

M. Dorofeev (✉)
Financial University Under the Government of the Russian Federation, Moscow, Russia
e-mail: mldorofeev@fa.ru

H. H. Lean
School of Social Sciences, Universiti Sains Malaysia, Gelugor, Penang, Malaysia

H. Dinçer et al. (eds.), *Decision Making in Interdisciplinary Renewable Energy Projects*, Contributions to Management Science,
https://doi.org/10.1007/978-3-031-51532-3_20

welfare programs, and investments in education and skills development (Ponkratov et al., 2018; Kakaulina, 2021; Popkova et al., 2022; Tyurina et al., 2023).

The work of Caverzasi and Godin (2014) provides a systematic overview of scientific research in the field of AB-SFC modeling of income processes. Currently, this field is represented by two major scientific schools: (1) the North American school (Wynne Godley, Marc Lavoie, Gennaro Zezza, Claudio Dos Santos) and (2) the European school (Jacques Mazier, Stephen Kinsella, and Edwin Le Heron). It is worth noting that in more recent studies, the works of Russian scientists have also been found, indicating the existence of a Russian scientific school (Karayev, 2018).

According to the classification by Caverzasi and Godin (2014), the existing AB-SFC models in the scientific literature can be grouped into three main categories: (1) theoretical models with discursive solutions, (2) theoretical models with simulation of results, and (3) empirical models. The third group of models holds the highest practical significance among all the others, as these models rely on parameters and assumptions that are most closely aligned with reality (Ai et al., 2023). The most extensive representation of the third group of models can be found in the works of authors from the North American school of AB-SFC modeling at the Levy Institute, including prominent names such as Godley, Zezza, Papadimitriou, Kinsella, Tiou-Tagba Aliti, Clévenot, Dos Santos, and others.

The quantity of scientific works and the progress in the development of the AB-SFC modeling method indicate its demand and relevance in modern conditions. In more recent studies dedicated to the advancement of this method, progress can be observed in the detailed elaboration of models in terms of (1) expanding the range of financial instruments through which transactions between economic agents take place, (2) the composition of economic agents, and (3) the openness and connection of the model to the external world. New models attempt to explore more complex forms of interactions among economic agents in the economy (Wan et al., 2023).

In early models, there were four types of economic agents: households, firms, government, and banks. Contemporary models have also started to incorporate central banks, the external world, and even multiple regions within a single country (Yüksel & Dinçer, 2023). This group of authors proposed a model for an open economy with multiple countries, introducing corresponding export-import operations and exchange rates for different currencies, using the example of the USA and the EU (Godley & Lavoie, 2012). This advancement has propelled the theory of AB-SFC modeling to a fundamentally new level of theoretical and practical significance.

20.2 Standard SFC Model of Income Distribution in a Closed Economy

Standard SFC model of a closed economy was proposed by the so-called European school in the article by Kinsella et al. (2011). In this model, the authors conducted a simulation of the income distribution process among households and described the factors that influence the dynamics of income inequality. The institutional units in this model include households, firms, government, and banks. Transactions between economic agents are described by a matrix of flows and stocks of goods and services, monetary funds, and financial instruments.

The total sum of all operations for a given period is zero, meaning that the actual business activity in this described closed economic system is a game among its agents with a zero-sum outcome (An et al., 2019a, b, 2020a, b; c, 2022; Mikhaylov & Tarakanov, 2020; Moiseev et al., 2020, 2021; Gura et al., 2022; Dooyum et al., 2020; Mikhaylov et al., 2020; Mikhaylov, 2020a, b, c).

The simulation results demonstrated that the dynamics of income inequality exhibit a countercyclical pattern relative to the dynamics of unemployment (Moiseev et al., 2023a, b). In the initial period, all institutional units have the same level of material equality (Khan et al., 2022; Dinçer et al., 2022a, b; Badr et al., 2022; Barykin et al., 2022b; Mehta et al., 2022; Kalinina et al. 2022; Shaikh et al., 2022; Mikhaylov et al., 2022a, b; Nyangarika et al., 2022).

However, after a series of simulations, the economic system starts to lose this balance, and household incomes become more differentiated (Mikhaylov et al., 2018, 2023b; Wang et al. 2023; He et al., 2023; Irshad et al., 2023; Iqbal et al., 2023; Aksoy et al., 2023; Dayong et al., 2020; Nyangarika et al., 2018; Danish et al., 2020, 2021; An et al., 2021; Uyeh et al., 2021; Tamashiro et al., 2021, 2023; Shaikh et al., 2021).

At the same time, government intervention, which is assigned a passive role in the model, allows for mitigating these distortions to some extent through the redistribution of a portion of national income (Denisova et al., 2019; Nyangarika et al., 2019a, b; Huang et al., 2021a, b; Mikhaylov, 2018, 2022b; Mikhaylov et al., 2019; Conteh et al., 2021; Sediqi et al., 2022, Khan et al., 2021; Bhuiyan et al., 2022a; Liu et al., 2021a, b; Daniali et al., 2021; Moiseev et al., 2023a, b).

The government acts as a passive (automatic) stabilizer by increasing taxes to finance government expenditures (Nie et al. 2020; Mikhaylov et al., 2023a; Mikhaylov, 2021a; Varyash et al., 2020; Zhao et al., 2021; An & Mikhaylov, 2020, 2021; Alwaelya et al., 2021; Yumashev & Mikhaylov, 2020; Yumashev et al., 2020; Mutalimov et al., 2021).

The increase in tax rates corresponds to the required amounts for supporting households during periods of rising unemployment in the economy (Kou et al., 2023a, b). The authors acknowledge that the developed model can be used to describe the dynamics of income inequality in a long-term retrospective and can demonstrate how its level changes over successive generations (Han et al., 2023). In other words, this model can serve as an artificial laboratory within which experiments on the modeling of income inequality can be conducted (Bhuiyan et al., 2021, 2022b; Dong et al., 2021, Mikhaylov, 2021b; Barykin et al., 2022a; Liu et al., 2022a, b; Danish et al., 2022a, b; Saqib et al., 2021; Mukhametov et al., 2021, Candila et al., 2021; Mikhaylov & Grilli, 2022, Li et al., 2022).

There are indeed several limitations to the typical SFC model. The most significant one is that the authors make numerous simplifications and assumptions in their models, intentionally reducing their realism (Xu et al., 2023). Naturally, this raises skepticism regarding their practical significance (Yüksel et al., 2023). Some of the main simplifications include (1) passive role of the government with limited range of capabilities and actions comparing with the real economy; (2) fixed parameters of social security model; (3) simplified financial sector, which does not reflect all the complexities and risks associated with financial operations and instruments; (4) flat tax system, disregarding the variety of marginal tax rates and exemptions present in different countries; and (5) insufficient disclosure of variable determination principles (Dinçer et al., 2023a, b, c, d, e, f). All these limitations should be considered when interpreting and applying the typical SFC model, and further developments of the method could aim to incorporate more realistic aspects and account for additional variables to enhance its accuracy and applicability (Carayannis et al., 2023).

20.3 Typical AB-SFC Model of Income Distribution in an Open Economy

Another representative of the European school, Detzer (2016), presented an AB-SFC model of an open economy with the external world in his article. The research focuses on the interrelationships between debt, inequality, and economic growth (Palley, 1994; Dutt, 2006; Zezza, 2008; Hein 2012; Kim et al. 2014).

In the literature review, Detzer (2016) examined nine scientific works that investigated the interplay between consumption levels, household debt, economic inequality, and economic growth from various perspectives. In most cases, the authors concluded that the growth of debt burden has a negative impact on income inequality and economic growth. Similar results are also obtained through AB-SFC modeling by Karayev (2018) and Cardaci (2018).

A detailed analysis of the author's SCF model also reveals that it contains simplifications and assumptions regarding the parameters included in the model, which prevent it from being considered sufficiently realistic for addressing the issue of income inequality. However, dynamic modeling based on the SFC method is not necessarily a futile endeavor (Dos Santos & Macedoe Silva, 2009).

20.4 Conclusion

The authors introduced a new level of modeling by proposing a method to combine agent-based (AB) and SCF modeling approaches (Caiani et al., 2016a, b). The article presents a universal model consisting of 16 equations and 18 unknowns, which underwent a series of calibration and testing procedures. According to the authors, this model can serve as a suitable foundation for the development of future theoretical and practice-oriented AB-SCF models. The authors also formulated universal guidelines for developing and testing the realism of new models in this field. The strategy for developing an AB-SCF model involves the following sequential procedures:

Development of a basic aggregated version of the AB-SCF model: This involves creating a graphical representation of flows and stocks between economic agents, a matrix of intersectoral balances, and textual and formulaic descriptions of participant behaviors and transactions (Martínez et al., 2023; Chang et al., 2023; Eti et al., 2023b).

1. Introduction of constraints for the model's growth rates and stationarity: Nominal growth rates of the economy, production volumes, prices, wages, etc. should be stationary, and constants should remain unchanged.
2. Search for quantitative solutions in the model with exogenously determined adequate values for parameters that have empirical quantitative information (interest rates, unemployment, inflation, tax rates, etc.). This provides initial input data for stocks and flows.
3. In the initial period, income and wealth are evenly distributed among all economic agents within sectors. Historical data is used as the expectations of economic agents.

To maintain the initial sizes of noncirculating assets, productive capacity, and durable stocks, it is assumed that firms reinvest profits and/or loans to maintain their initial productive capacity. It is also assumed that the real cost of equity and debt financial resources directed toward maintaining productive capacities is constant and does not change throughout the simulation process (Eti et al., 2022; Danish et al., 2022c; Lavoie, 2008; Mikhaylov, 2022a).

4. To configure the initial interaction between economic agents, suppliers and consumers are randomly assigned for each market. Each supplier has an equal number of buyers (customers). Household deposits and firm loans are randomly distributed among commercial banks, so that each bank has an equal amount of deposits and loans.

After model calibration is complete, 100 Monte Carlo simulations are conducted over 400 periods. The results are then compared with real-world data. The final stage involves sensitivity analysis of the model and testing the stability of the results by considering 25 scenarios (Eti et al., 2023a; An et al., 2024; An and Mikhaylov, 2024; Jargalsaikhan et al., 2024; Eti et al., 2024; Yuksel et al., 2024;Rahadian et al., 2024).**FundingThe publication is based on the results of budgetary-supported research according to the state task carried out by the Financial University under the Government of the Russian Federation, Moscow, Russia.**

References

Ai, R., Zheng, Y., Yüksel, S., & Dinçer, H. (2023). Investigating the components of fintech ecosystem for distributed energy investments with an integrated quantum spherical decision support system. *Financial Innovation, 9*(1), 27.

Aksoy, T., Yüksel, S., Dınçer, H., Hacioglu, U., Mikhaylov, A., & Muyeen, S. M. (2023). Assessing the electricity production capacities of emerging markets for the sustainable investment. *IEEE Access, 11*, 24574–24590.

Alwaelya, S. A., Yousif, N. B. A., & Mikhaylov, A. (2021). Emotional development in preschoolers and socialization. *Early Child Development and Care, 191*, 16.

An, J., & Mikhaylov, A. (2020). Russian energy projects in South Africa. *Journal of Energy in Southern Africa, 31*(3), 58–64.

An, J., & Mikhaylov, A. (2021). Analysis of energy projects financial efficiency and renewable energy generation in Russia. *Finance: Theory and Practice, 25*(5), 79–91.

An, J., & Mikhaylov, A. (2024). Technology-based forecasting approach for recognizing trade-off between time-to-market reduction and devising a scheduling process in open innovation management. *Journal of Open Innovation: Technology, Market, and Complexity, 10*(1), 100207. https://doi.org/10.1016/j.joitmc.2024.100207

An, J., Mikhaylov, A., & Moiseev, N. (2019a). Oil price predictors: Machine learning approach. *International Journal of Energy Economics and Policy, 9*(5), 1–6.

An, J., Mikhaylov, A., & Sokolinskaya, N. (2019b). Machine learning in economic planning: Ensembles of algorithms. *Journal of Physics: Conference Series, 1353*, 012126.

An, J., Mikhaylov, A., & Jung, S.-U. (2020a). The strategy of South Korea in the global oil market. *Energies, 13*(10), 2491.

An, J., Mikhaylov, A., & Kim, K. (2020b). Machine learning approach in heterogeneous group of algorithms for transport safety-critical system. *Applied Sciences, 10*(8), 2670.

An, J., Mikhaylov, A., & Richter, U. H. (2020c). Trade war effects: Evidence from sectors of energy and resources in Africa. *Heliyon, 6*, e05693. https://doi.org/10.1016/j.heliyon.2020.e05693

An, J., Mikhaylov, A., & Jung, S.-U. (2021). A linear programming approach for robust network revenue management in the airline industry. *Journal of Air Transport Management, 91*, 101979.

An, J., Mikhaylov, A., Dinçer, H., & Yüksel, S. (2022). Economic modelling of electricity generation: Long short-term memory and Q-rung orthopair fuzzy sets. *Heliyon, 8*, e12345.

An, J., Mikhaylov, A., & Chang, T. (2024). Relationship between the popularity of a platform and the price of NFT assets. *Finance Research Letters, 61*(3), 105057. https://doi.org/10.1016/j.frl.2024.105057

Badr, A. M. N. A. A., Mohamed, T. I., Osman, N. A. Q., & Mikhaylov, A. (2022). A review of social media website users' interaction paths with governmental accounts during the COVID-19 pandemic. *Informatics, 9*, 50. https://doi.org/10.3390/informatics9030050

Barykin, S., Mikheev, A., Kiseleva, E., Putikhin, Y., Alekseeva, N., & Mikhaylov, A. (2022a). An empirical analysis of Russian regions' debt sustainability. *Economies, 10*(4), 296.

Barykin, S. E., Kapustina, I. V., Sergeev, S. M., Daniali, S. M., Kopteva, L. A., Semenova, G. N., Pryadko, I. P., Mikhaylov, A., Baboshkin, P., Datsyuk, P., & Senjyu, T. (2022b). Financial logistics models based on systematic approach improving management solutions. *F1000Research, 11*, 570. https://doi.org/10.12688/f1000research.111252.1

Bhuiyan, M. A., An, J., Mikhaylov, A., Moiseev, N., & Danish, M. S. S. (2021). Renewable energy deployment and COVID-19 measures for sustainable development. *Sustainability, 13*, 4418.

Bhuiyan, M. A., Dinçer, H., Yüksel, S., Mikhaylov, A., Danish, M. S. S., Pinter, G., Uyeh, D. D., & Stepanova, D. (2022a). Economic indicators and bioenergy supply in developed economies: QROF-DEMATEL and random forest models. *Energy Reports, 8*(2022), 561–570.

Bhuiyan, M. A., Zhang, Q., Khare, V., Mikhaylov, A., Pinter, G., & Huang, X. (2022b). Renewable energy consumption and economic growth nexus—A systematic literature review. *Frontiers in Environmental Science., 10*, 878394. https://doi.org/10.3389/fenvs.2022.878394

Caiani, A., Godin, A., Caverzasi, E., Gallegati, M., Kinsella, S., & Stiglitz, J. E. (2016a). Agent based-stock flow consistent macroeconomics: Towards a benchmark model. *Journal of Economic Dynamics and Control, 69*, 375–408. Retrieved from SSRN https://ssrn.com/abstract=2664125 or https://doi.org/10.2139/ssrn.2664125

Caiani, A., Russo, A., & Gallegati, M. (2016b). *Does inequality hamper innovation and growth?* Retrieved from SSRN 2790911.

Candila, V., Maximov, D., Mikhaylov, A., Moiseev, N., Senjyu, T., & Tryndina, N. (2021). On the relationship between oil and exchange rates of oil-exporting and oil-importing countries: From the great recession period to the Covid-19 era. *Energies, 14*, 8065.

Carayannis, E., Kostis, P., Dinçer, H., & Yüksel, S. (2023). Quality function deployment-oriented strategic outlook to sustainable energy policies based on quintuple innovation helix. *Journal of the Knowledge Economy*, 1–19.

Cardaci, A. (2018). Inequality, household debt and financial instability: An agent-based perspective. *Journal of Economic Behavior and Organization, 149*, 434. https://doi.org/10.1016/j.jebo.2018.01.010

Caverzasi, E., & Godin, A. (2014). Post-Keynesian stock-flow-consistent modelling: A survey. *Cambridge Journal of Economics., 39*, 157–187. https://doi.org/10.1093/cje/beu021

Chang, H.-W., Chang, T., Xiang, F., Mikhaylov, A., & Grigorescu, A. (2023). Revisiting R&D intensity and CO2 emissions link in the USA using time varying granger causality test: 1870~2020. *Heliyon, 9*(10), e20319. https://doi.org/10.1016/j.heliyon.2023.e20319

Conteh, F., Takahashi, H., Hemeida, A. M., Krishnan, N., Mikhaylov, A., & Senjyu, T. (2021). Hybrid grid-connected renewable power generation for sustainable electricity supply in Sierra Leone: Case study Lungi Town. *Sustainability, 13*(18).

Copeland, M. A. (1949). Social accounting for money flows. *Accounting Review, 24*(3), 254–264.

Daniali, S. M., Barykin, S. E., Kapustina, I. V., Khortabi, F. M., Sergeev, S. M., Kalinina, O. V., Mikhaylov, A., Veynberg, R., Zasova, L., & Senjyu, T. (2021). Predicting volatility index according to technical index and economic indicators on the basis of deep learning algorithm. *Sustainability, 13*, 14011.

Danish, M. S. S., Bhattacharya, A., Stepanova, D., Mikhaylov, A., Grilli, M. L., Khosravy, M., & Senjyu, T. (2020). A systematic review of metal oxide applications for energy and environmental sustainability. *Metals, 10*(12), 1604.

Danish, M. S. S., Senjyu, T., Sabory, N. R., Khosravy, M., Grilli, M. L., Mikhaylov, A., & Majidi, H. (2021). A forefront framework for sustainable aquaponics modeling and design. *Sustainability, 13*, 9313.

Danish, M. S., Senjyu, T. S., Sabory, N. R., & Mikhaylov, A. (2022a). Lessons from COVID-19 conferring environmental re-engineering opportunity. In M. Danish & T. Senjyu (Eds.), *Eco-friendly and agile energy strategies and policy development* (pp. 99–118). IGI Global. https://doi.org/10.4018/978-1-7998-9502-2.ch005

Danish, M. S., Senjyu, T. S., Sabory, N. R., & Mikhaylov, A. (2022b). Interplay consequences of COVID-19 on global environmental sustainability. In M. Danish & T. Senjyu (Eds.), *Eco-friendly and agile energy strategies and policy development*. IGI Global. https://doi.org/10.4018/978-1-7998-9502-2.ch003

Danish, M. S. S., Estrella-Pajulas, L. L., Alemaida, I. M. A., Grilli, M. L., Mikhaylov, A., & Senjyu, T. (2022c). Green synthesis of silver oxide nanoparticles for photocatalytic environmental remediation and biomedical applications. *Metals, 12*(4).

Datsyuk, P., Mikhaylov, A., & Pinter, G. (2023b). Group correlation analysis of socio-economic indicators and the volume of collected E-waste in European countries. *Environmental Research Communications, 5*(11), 115004. https://doi.org/10.1088/2515-7620/ad0850

Dayong, N., Mikhaylov, A., Bratanovsky, S., Shaikh, Z. A., & Stepanova, D. (2020). Mathematical modeling of the technological processes of catering products production. *Journal of Food Process Engineering, 43*(2), 1–9.

Denisova, V., Mikhaylov, A., & Lopatin, E. (2019). Blockchain infrastructure and growth of global power consumption. *International Journal of Energy Economics and Policy, 9*(4), 22–29.

Detzer, D. (2016). *Financialisation, debt and inequality: Scenarios based on a stock flow consistent model, Working Paper, No. 64/2016*. Hochschule für Wirtschaft und Recht Berlin, Institute for International Political Economy (IPE).

Dinçer, H., Yüksel, S., Mikhaylov, A., Pinter, G., & Shaikh, Z. A. (2022a). Analysis of renewable-friendly smart grid technologies for the distributed energy investment projects using a hybrid picture fuzzy rough decision-making approach. *Energy Reports, 8*, 11466–11477. https://doi.org/10.1016/j.egyr.2022.08.275

Dinçer, H., Yüksel, S., Mikhaylov, A., Barykin, S. E., Aksoy, T., & Hacioğlu, U. (2022b). Analysis of environmental priorities for green project investments using an integrated q-rung orthopair fuzzy modelling. *IEEE Access, 10*, 50996–51007. https://doi.org/10.1109/ACCESS.2022.3174058

Dinçer, H., Yüksel, S., Hacioglu, U., Yilmaz, M. K., & Delen, D. (2023a). Development of a sustainable corporate social responsibility index for performance evaluation of the energy industry: A hybrid decision-making methodology. *Resources Policy, 85*, 103940.

Dinçer, H., Eti, S., Gökalp, Y., Yüksel, S., & Çelebi, B. (2023b). The role of small-scale wind turbines in increasing the competitiveness of hospitals. In *Handbook of research on quality and competitiveness in the healthcare services sector* (pp. 336–352). IGI Global.

Dinçer, H., Yüksel, S., Mikhaylov, A., Muyeen, S. M., Chang, T., Barykin, S., & Kalinina, O. (2023c). CO2 emissions integrated fuzzy model: A case of seven emerging economies. *Energy Reports, 9*, 5741–5751.

Dinçer, H., Yüksel, S., Aksoy, T., Hacıoğlu, Ü., Mikhaylov, A., & Pinter, G. (2023d). Analysis of solar module alternatives for efficiency-based energy investments with hybrid 2-tuple IVIF modeling. *Energy Reports, 10*, 61–71.

Dinçer, H., Eti, S., Yüksel, S., Özdemir, S., Yîlmaz, A. E., & Ergün, E. (2023e). Integrating data mining and fuzzy decision-making techniques for analyzing the key minimizing factors of carbon emissions. *Journal of Intelligent & Fuzzy Systems, 45*, 1–17.

Dinçer, H., Etı, S., Aksoy, T., Yüksel, S., Hacioglu, U., Mikhaylov, A., & Muyeen, S. M. (2023f). *Analysis of environmental impact for material production investments using a novel soft computing methodology* (Vol. 11, p. 37987). IEEE Access.

Dong, B., Ikonnikova, I., Rogulin, R., Sakulyeva, T., & Mikhaylov, A. (2021). Environmental economic approach to optimization of transport communication in megacities. *Journal of Environmental Science and Health, Part A, 56*(6), 660–666.

Dooyum, U. D., Mikhaylov, A., & Varyash, I. (2020). Energy security concept in Russia and South Korea. *International Journal of Energy Economics and Policy, 10*(4), 102–107.

Dos Santos, C., & Macedoe Silva, A. C. (2009). *Revisiting (and connecting) Marglin–Bhaduri and Minsky: An SFC look at Financialization and profit-led growth, working paper no. 567*. Levy Economics Institute of Bard College.

Dutt, A. K. (2006). Maturity, stagnation and consumer debt: A Steindlian approach. *Metroeconomica, 57*, 339–364.

Eti, S., Dinçer, H., Yüksel, S., & Gökalp, Y. (2022). Analysis of the suitability of the solar panels for hospitals: A new fuzzy decision-making model proposal with the T-spherical TOP-DEMATEL method. *Journal of Intelligent & Fuzzy Systems, 44*, 1–13.

Eti, S., Dinçer, H., Gökalp, Y., Yüksel, S., & Kararoğlu, D. (2023a). Identifying key issues to handle the inflation problem in the healthcare industry caused by energy prices: An evaluation with decision-making models. In *Managing inflation and supply chain disruptions in the global economy* (pp. 162–178). Springer.

Eti, S., Dincer, H., Yuksel, S., Uslu, Y. D., Gokalp, Y., Kalkavan, H., Mikhaylov, A., & Pinter, G. (2023b). Determination of priority criteria in green building transformation: An analysis on the service industry. *Research in Globalization, 7*(12), 100164. https://doi.org/10.1016/j.resglo.2023.100164

Eti, S., Yüksel, S., Dinçer, H., Kalkavan, H., Hacioglu, U., Mikhaylov, A., Danish, M. S. S., & Pinter, G. (2024). Assessment of technical and financial challenges for renewable energy project alternatives. *Cleaner Engineering and Technology, 18*, 100719. https://doi.org/10.1016/j.clet.2023.100719

Godley, W., & Lavoie, M. (2012). A simple model of three economies with two currencies: The Eurozone and the USA. In *The stock-flow consistent approach* (pp. 159–188). Springer. https://doi.org/10.1057/9780230353848

Gura, D., Mikhaylov, A., Glushkov, S., Zaikov, M., & Shaikh, Z. A. (2022). Model for estimating power dissipation along the interconnect length in single on-chip topology. *Evolutionary Intelligence, 15*, 2369–2373. https://doi.org/10.1007/s12065-020-00407-7

Han, Z., Gong, L., Chen, H., & Yüksel, S. (2023). The role of mineral and oil resources in achieving a clean environment in emerging economies: An asymmetric analysis. *Resources Policy, 86*, 104099.

He, H., Huang, Y., Nakadomari, A., Masrur, H., Krishnan, N., Hemeida, A. M., Mikhaylov, A., Omine, E., & Senjyu, T. (2023). Potential and economic viability of green hydrogen production from seawater electrolysis using renewable energy in remote Japanese islands. *Renewable Energy, 202*, 1436–1447. https://doi.org/10.1016/j.renene.2022.12.046

Hein, E. (2012). Finance-dominated capitalism, re-distribution, household debt and financial fragility in a Kaleckian distribution and growth model. *PSL Quarterly Review, 65*(260), 11–51.

Huang, Y., Masrur, H., Shigenobu, R., Hemeida, A. M., Mikhaylov, A., & Senjyu, T. (2021a). A comparative design of a campus microgrid considering a multi-scenario and multi-objective approach. *Energies, 14*, 2853.

Huang, Y., Yona, A., Takahashi, H., Hemeida, A. M., Mandal, P., Mikhaylov, A., Senjyu, T., & Lotfy, M. E. (2021b). Energy management system optimization of drug store operate electric vehicles charging station. *Sustainability, 14*, 1217836.

Iqbal, K. M. J., Khan, M. I., Mikhaylov, A., Shah, A. A., Yadykin, V., Filho, W. L., Tariq, M. A. U. R., & Ullah, W. (2023). Modeling principles, criteria and indicators to assess water sector governance for climate compatibility and sustainability. *Frontiers in Environmental Science, 11*, 989930.

Irshad, A. S., Ludin, G. A., Masrur, H., Ahmadi, M., Yona, A., Mikhaylov, A., Krishnan, N., & Senjyu, T. (2023). Optimization of grid-photovoltaic and battery hybrid system with most technically efficient PV technology after the performance analysis. *Renewable Energy, 207*, 714–730.

Jargalsaikhan, N., Ueda, S., Masahiro, F., Matayoshi, H., Mikhaylov, A., Byambaa, S., & Senjyu, T. (2024). Exploring influence of air density deviation on power production of wind energy conversion system: Study on correction method. *Renewable Energy, 2120*, 119636. https://doi.org/10.1016/j.renene.2023.119636

Kakaulina, M. O. (2021). Projected shortfall in personal income tax revenues of regional governments in Russia due to the COVID-19 pandemic. *Journal of Tax Reform, 7*(1), 39–54.

Kalinina, O. V., Barykin, S. E., Sergeev, S. M., Semenova, G. N., Fatkullina, A., Mikhaylov, A., & De La Poza, E. (2022). Smart city perspectives in post-pandemic governance: Externalities reduction policy. *F1000Research, 11*, 1032. https://doi.org/10.12688/f1000research.123195.1

Karayev, A. K. (2018). The impact of financialization and income inequality on economic growth, investment processes and innovation. *Economics, Taxes & Law, 11*(6), 57–68. https://doi.org/10.26794/1999-849X-2018-11-6-57-68

Khan, A. A., Shaikh, Z. A., Baitenova, L., Mutaliyeva, L., Moiseev, N., Mikhaylov, A., Laghari, A. A., Idris, S. A., & Alshazly, H. (2021). QoS-ledger: Smart contracts and metaheuristic for secure quality-of-service and cost-efficient scheduling of medical-data processing. *Electronics, 10*, 3083.

Khan, A. A., Laghari, A. A., Gadekallu, T. R., Shaikh, Z. A., Javed, A. R., Rashid, M., Estrela, V. V., & Mikhaylov, A. (2022). A drone-based data management and optimization using metaheuristic algorithms and blockchain smart contracts in a secure fog environment. *Computers and Electrical Engineering, 102*, 108234. https://doi.org/10.1016/j.compeleceng.2022.108234

Khan, R. U., Saqib, A., Abbasi, M. A., Mikhaylov, A., & Pinter, G. (2023). Green leadership, environmental knowledge sharing, and sustainable performance in manufacturing industry: Application from upper echelon theory. *Sustainable Energy Technologies and Assessments, 60*, 103540. https://doi.org/10.1016/j.seta.2023.103540

Kim, Y. K., Setterfield, M., & Mei, Y. (2014). A theory of aggregate consumption. *European Journal of Economics and Economic Policies: Intervention, 11*(1), 31–49.

Kinsella, S., Greiff, M., & Nell, E. (2011). Income distribution in a stock-flow consistent model with education and technological change. *Eastern Economic Journal., 37*, 134. https://doi.org/10.1057/eej.2010.31

Kou, G., Pamucar, D., Dinçer, H., & Yüksel, S. (2023a). From risks to rewards: A comprehensive guide to sustainable investment decisions in renewable energy using a hybrid facial expression-based fuzzy decision-making approach. *Applied Soft Computing, 142*, 110365.

Kou, G., Yuksel, S., & Dincer, H. (2023b). A facial expression and expert recommendation fuzzy decision-making approach for sustainable business investments within the metaverse world. *Applied Soft Computing, 148*, 110849.

Lavoie, M. (2008). Financialisation issues in a post-Keynesian stock-flow consistent model. *Intervention, 5*(2), 331–356.

Li, J., Dinçer, H., Yüksel, S., Mikhaylov, A., & Barykin, S. E. (2022). Bipolar q-ROF hybrid decision making model with golden cut for analyzing the levelized cost of renewable energy alternatives. *IEEE Access, 10*, –35900. https://doi.org/10.1109/ACCESS.2022.3168315

Liu, L., Kato, T., Mandal, P., Mikhaylov, A., Hemeida, A. M., Takahashi, H., & Senjyu, T. (2021a). Two cases studies of model predictive control approach for hybrid renewable energy systems. *AIMS Energy, 9*(6), 1241–1259.

Liu, L., Kato, T, Mandal, P., Mikhaylov, A, Hemeida, A. M., & Senjyu T. (2021b). Multi-area power system load frequency control by using model predictive control by introducing renewable energy sources. In *2021 International Conference on Science & Contemporary Technologies (ICSCT)*, pp. 1–6. https://doi.org/10.1109/ICSCT53883.2021.9642521

Liu, Z.-J., Panfilova, E., Mikhaylov, A., & Kurilova, A. (2022a). COVID-19 crisis impact on the stability between parties in crowdfunding and crowdsourcing. *Wireless Personal Communications, 122*, 915–930.

Liu, Z., Panfilova, E., Mikhaylov, A., & Kurilova, A. (2022b). Assessing stability in the relationship between parties in crowdfunding and crowdsourcing projects during the COVID-19 crisis. *Journal of Global Information Management (JGIM), 30*(4), 1–18. https://doi.org/10.4018/JGIM.297905

Martínez, L., Dinçer, H., & Yüksel, S. (2023). A hybrid decision making approach for new service development process of renewable energy investment. *Applied Soft Computing, 133*, 109897.

Mehta, R., Verghese, J., Mahajan, S., Barykin, S., Bozhuk, S., Kozlova, N., Kapustina, I., Mikhaylov, A., Naumova, E., & Dedyukhina, N. (2022). Consumers' behavior in conversational commerce marketing based on messenger chatbots. *F1000Research, 11*(647). https://doi.org/10.12688/f1000research.122037.1

Mikhaylov, A. (2018). The yield of the carry trade strategy. *Finance: Theory and Practice, 22*(3), 52–63. https://doi.org/10.26794/2587-5671-2018-22-3-52-63

Mikhaylov, A. (2020a). Cryptocurrency market analysis from the open innovation perspective. *Journal of Open Innovation: Technology, Market, and Complexity, 6*(4), 197.

Mikhaylov, A. (2020b). Geothermal energy development in Iceland. *International Journal of Energy Economics and Policy, 10*(4), 31–35.

Mikhaylov, A. (2020c). Cryptocurrency market development: Hurst method. *Finance: Theory and Practice, 24*(3), 81–91.

Mikhaylov, A. (2021a). Development of Friedrich von Hayek's theory of private money and economic implications for digital currencies. *Terra Economicus, 19*(1), 53–62.

Mikhaylov, A. (2021b). Lichens as indicators of atmospheric pollution in urban ecosystems. *Israel Journal of Ecology & Evolution, 67*(1–2), 60–68. https://doi.org/10.1163/22244662-bja10016

Mikhaylov, A. (2022a). Efficiency of renewable energy plants in Russia. *Anais da Academia Brasileira de Ciências, 94*(4), e20191226. https://doi.org/10.1590/0001-3765202220191226

Mikhaylov, A. (2022b). Sustainable development and renewable energy: A new view to a global problem. *Energies, 15*, 1397. https://doi.org/10.3390/en15041397

Mikhaylov, A. (2023). Understanding the risks associated with wallets, depository services, trading, lending, and borrowing in the crypto space. *Journal of Infrastructure, Policy and Development, 7*(3), 2223. https://doi.org/10.24294/jipd.v7i3.2223

Mikhaylov, A., & Grilli, M. L. (2022). Machine learning methods and sustainable development: Metal oxides and multilayer metal-oxides. *Metals, 12*(4).

Mikhaylov, A., & Tarakanov, S. (2020). Development of Levenberg-Marquardt theoretical approach for electric network. *Journal of Physics: Conference Series, 1515*, 052006.

Mikhaylov, A., Sokolinskaya, N., & Nyangarika, A. (2018). Optimal carry trade strategy based on currencies of energy and developed economies. *Journal of Reviews on Global Economics, 7*, 582–592.

Mikhaylov, A., Burakov, D., & Didenko, V. (2019). Relationship between oil price and macroeconomic indicators in Russia. *Finance: Theory and Practice, 23*(2), 105–116. https://doi.org/10.26794/2587-5671-2019-23-2-105-116

Mikhaylov, A., Moiseev, N., Aleshin, K., & Burkhardt, T. (2020). Global climate change and greenhouse effect. *Entrepreneurship and Sustainability Issues, 7*(4), 2897–2913. https://doi.org/10.9770/jesi.2020.7.4(21)

Mikhaylov, A., Bhatti, I. M., Dinçer, H., & Yüksel, S. (2022a). Integrated decision recommendation system using iteration-enhanced collaborative filtering, golden cut bipolar for analyzing the risk-based oil market spillovers. *Computational Economics*. https://doi.org/10.1007/s10614-022-10341-8

Mikhaylov, A., Yumashev, A., & Kolpak, E. (2022b). Quality of life, anxiety and depressive disorders in patients with extrasystolic arrhythmia. *Archives of Medical Science, 18*(2), 328. https://doi.org/10.5114/aoms.2020.101359

Mikhaylov, A., Dinçer, H., & Yüksel, S. (2023a). Analysis of financial development and open innovation oriented fintech potential for emerging economies using an integrated decision-making approach of MF-X-DMA and golden cut bipolar q-ROFSs. *Financial Innovation, 9*(1), 10. https://doi.org/10.1186/s40854-022-00234-w

Mikhaylov, A., Dinçer, H., Yüksel, S., Pinter, G., & Shaikh, Z. A. (2023b). Bitcoin mempool growth and trading volumes: Integrated approach based on QROF multi-SWARA and aggregation operators. *Journal of Innovation & Knowledge, 8*(3), 100378.

Moiseev, N., Mikhaylov, A., Varyash, I., & Saqib, A. (2020). Investigating the relation of GDP per capita and corruption index. *Entrepreneurship and Sustainability Issues, 8*(1), 780–794.

Moiseev, N., Sorokin, A., Zvezdina, N., Mikhaylov, A., Khomyakova, L., & Danish, M. S. S. (2021). Credit risk theoretical model on the base of DCC-GARCH in time-varying parameters framework. *Mathematics, 9*, 2423. https://doi.org/10.3390/math9192423

Moiseev, N., Mikhaylov, A., Dinçer, H., & Yüksel, S. (2023a). Effects of market capitalization shocks for open innovation models in e-commerce industry: Golden cut QROF MCDM analysis. *Financial Innovation, 9*(1), 32.

Moiseev, N., Mikhaylov, A., Dinçer, H., & Yüksel, S. (2023b). Market capitalization shock effects on open innovation models in e-commerce: Golden cut q-rung orthopair fuzzy multicriteria decision-making analysis. *Financial Innovation, 9*(1), 55.

Mukhametov, A., Bekhorashvili, A., Avdeenko, A., & Mikhaylov, A. (2021). The impact of growing legume plants under conditions of biologization and soil cultivation on Chernozem fertility and productivity of rotation crops. *Legume Research, 44*, 12.

Mutalimov, V., Kovaleva, I., Mikhaylov, A., & Stepanova, D. (2021). Assessing regional growth of small business in Russia. *Entrepreneurial Business and Economics Review, 9*(3), 119–133. https://doi.org/10.15678/EBER.2021.090308

Nie, D., Panfilova, E., Samusenkov, V., & Mikhaylov, A. (2020). E-learning financing models in Russia for sustainable development. *Sustainability, 12*(11), 4412.

Niu, X., Yüksel, S., & Dinçer, H. (2023). Emission strategy selection for the circular economy-based production investments with the enhanced decision support system. *Energy, 274*, 127446.

Nyangarika, A., Mikhaylov, A., & Tang, B.-J. (2018). Correlation of oil prices and gross domestic product in oil producing countries. *International Journal of Energy Economics and Policy, 8*(5), 42–48.

Nyangarika, A., Mikhaylov, A., & Richter, U. (2019a). Influence oil price towards economic indicators in Russia. *International Journal of Energy Economics and Policy, 9*(1), 123–129.

Nyangarika, A., Mikhaylov, A., & Richter, U. (2019b). Oil price factors: Forecasting on the base of modified auto-regressive integrated moving average model. *International Journal of Energy Economics and Policy, 9*(1), 149–159.

Nyangarika, A., Mikhaylov, A., Muyeen, S. M., Yadykin, V., Mottaeva, A., Pryadko, I., Barykin, S., Fomenko, N., Rykov, G., & Shvandar, K. (2022). Energy stability and decarbonization in developing countries: Random forest approach for forecasting of crude oil trade flows and macro indicators. *Frontiers in Environmental Science, 10*, 1031343. https://doi.org/10.3389/fenvs.2022.1031343

Palley, T. I. (1994). Debt, aggregate demand, and the business cycle: An analysis in the Spirit of Kaldor and Minsky. *Journal of Post Keynesian Economics, 16*(3), 371–390.

Ponkratov, V., Karaev, A., Solyannikova, S., Gorlova, O., Smirnov, D., & Makashina, O. (2018). Impact of fiscal, monetary and structural economic policies on the growth rates of leading economies: 2000-2016. *European Research Studies Journal, 21*(Special Issue 3), 333–343.

Popkova, E. G., De Bernardi, P., Tyurina, Y. G., & Sergi, B. S. (2022). A theory of digital technology advancement to address the grand challenges of sustainable development. *Technology in Society, 68*, 101831.

Rahadian, D., Firli, A., Dinçer, H., Yüksel, S., Mikhaylov, A., & Ecer, F. (2024). A hybrid neuro fuzzy decision-making approach to the participants of derivatives market for fintech investors in emerging economies. *Financial Innovation, 10*(1), 37. https://doi.org/10.1186/s40854-023-00563-6

Saqib, A., Chan, T.-H., Mikhaylov, A., & Lean, H. H. (2021). Are the responses of sectoral energy imports asymmetric to exchange rate volatilities in Pakistan? Evidence from recent foreign exchange regime. *Frontiers in Energy Research, 9*, 614463. https://doi.org/10.3389/fenrg.2021.614463

Sediqi, M. M., Nakadomari, A., Mikhaylov, A., Krishnan, N., Lotfy, M. E., Yona, A., & Senjyu, T. (2022). Impact of time-of-use demand response program on optimal operation of Afghanistan real power system. *Energies, 15*, 296.

Shaikh, Z. A., Moiseev, N., Mikhaylov, A., & Yüksel, S. (2021). Facile synthesis of copper oxide-cobalt oxide/nitrogen-doped carbon (Cu2O-Co3O4/CN) composite for efficient water splitting. *Applied Sciences, 11*, 9974. https://doi.org/10.3390/app11219974

Shaikh, Z. A., Kraikin, A., Mikhaylov, A., & Pinter, G. (2022). Forecasting stock prices of companies producing solar panels using machine learning methods. *Complexity, 2022*, 9186265. https://doi.org/10.1155/2022/9186265

Tamashiro, K., Alharbi, T., Mikhaylov, A., Hemeida, A. M., Krishnan, N., Lotfy, M. E., & Senjyu, T. (2021). Investigation of home energy management with advanced direct load control and optimal scheduling of controllable loads. *Energies, 14*, 7314. https://doi.org/10.3390/en14217314

Tamashiro, K., Omine, E., Krishnan, N., Mikhaylov, A., Hemeida, A. M., & Senjyu, T. (2023). Optimal components capacity based multi-objective optimization and optimal scheduling based MPC-optimization algorithm in smart apartment buildings. *Energy & Buildings, 278*, 112616. https://doi.org/10.1016/j.enbuild.2022.112616

Tyurina, Y., Frumina, S., Demidova, S., Kairbekuly, A., & Kakaulina, M. (2023). Estimation of tax expenditures stimulating the energy sector development and the use of alternative energy sources in OECD countries. *Energies, 16*, 2652.

Uyeh, D. D., Asem-Hiablie, S., Park, T., Kim, K.-M., Mikhaylov, A., Woo, S., & Ha, Y. (2021). Could Japonica rice be an alternative variety for increased global food security and climate change mitigation? *Food, 10*, 1869. https://doi.org/10.3390/foods10081869

Varyash, I., Mikhaylov, A., Moiseev, N., & Aleshin, K. (2020). Triple bottom line and corporate social responsibility performance indicators for Russian companies. *Entrepreneurship and Sustainability Issues, 8*(1), 313–331.

Wan, Q., Miao, X., Wang, C., Dinçer, H., & Yüksel, S. (2023). A hybrid decision support system with golden cut and bipolar q-ROFSs for evaluating the risk-based strategic priorities of fintech lending for clean energy projects. *Financial Innovation, 9*(1), 1–25.

Wang, Q., Guo, J., Li, R., Mikhaylov, A., & Moiseev, N. (2023). Does technical assistance alleviate energy poverty in sub-Saharan African countries? A new perspective on spatial spillover effects of technical assistance. *Energy Strategy Reviews, 45*, 101047. https://doi.org/10.1016/j.esr.2022.101047

Xu, X., Yüksel, S., & Dinçer, H. (2023). An integrated decision-making approach with golden cut and bipolar q-ROFSs to renewable energy storage investments. *International Journal of Fuzzy Systems, 25*(1), 168–181.

Yüksel, S., & Dinçer, H. (2023). Sustainability analysis of digital transformation and circular industrialization with quantum spherical fuzzy modeling and golden cuts. *Applied Soft Computing, 138*, 110192.

Yüksel, S., Dinçer, H., & Uluer, G. S. (2023). Socio-economic indicators of nuclear energy investments: A study on countries. In *Inclusive developments through socio-economic indicators: New theoretical and empirical insights* (pp. 33–44). Emerald Publishing.

Yüksel, S., Eti, S., Dinçer, H., Gokalp, Y., Yavuz, D., Mikhaylov, A., & Pinter, G. (2024). Prioritizing the indicators of energy performance management: A novel fuzzy decision-making approach for G7 service industries. *Environmental Research Communications, 6*(1), 015003. https://doi.org/10.1088/2515-7620/ad1c07

Yumashev, A., & Mikhaylov, A. (2020). Development of polymer film coatings with high adhesion to steel alloys and high wear resistance. *Polymer Composites, 41*(7), 2875–2880.

Yumashev, A., Ślusarczyk, B., Kondrashev, S., & Mikhaylov, A. (2020). Global indicators of sustainable development: Evaluation of the influence of the human development index on consumption and quality of energy. *Energies, 13*, 2768.

Zezza, G. (2008). U.S. growth, the housing market, and the distribution of income. *Journal of Post Keynesian Economics, 30*(3), 375–401.

Zhao, Q., Cherkasov, S., Avdeenko, A., Kondratenko, L., & Mikhaylov, A. (2021). Integral estimate of the added value of agricultural crops in the synergetic agriculture on the example of Vicia faba (Vicia faba L.). *Polish Journal of Environmental Studies, 30*(5), 1–9. https://doi.org/10.15244/pjoes/131949

Chapter 21
Investigating Energy Consumption in Hospitals by Text Mining Method

Yaşar Gökalp and Serkan Eti

21.1 Introduction

Hospitals are institutions where individuals apply to regain their health (Hiwar et al., 2021). Healthy individuals directly affect the level of development of a country (Dinçer et al., 2023a, b, c, d). Therefore, it is important for individuals to be healthy. Accordingly, it becomes even more important for hospitals to provide sustainable services (Carino et al., 2020). However, many issues affect the sustainability of hospitals (Khahro et al., 2021). Factors such as proper waste management policies, ensuring resource efficiency, establishing an efficient logistics network, effective cost management, water efficiency, and community engagement affect the sustainability of hospitals (Dinçer et al., 2023a, b, c, d). In addition to these aspects, the energy consumption dimension and energy efficiency are important aspects that will contribute to the sustainability of hospitals (Montiel-Santiago et al., 2020).

Hospitals are places with high energy consumption (Zor et al., 2020). Electricity, natural gas, and renewable energy sources are the energy types that meet the energy needs of hospitals (Dinçer et al., 2023a, b, c, d). The energy consumption of hospitals may vary according to factors such as size, infrastructure, and the efficiency of the technological devices used (Coskun et al., 2022). In addition, technological devices used for diagnosis and treatment increase the energy consumption dimension. Especially X-ray equipment, magnetic resonance imaging equipment, and laboratory equipment cause high energy consumption. In addition, ventilation systems consume high amounts of energy to provide the heating and cooling processes

Y. Gökalp
The School of Health Science, İstanbul Medipol University, İstanbul, Turkey

S. Eti (✉)
Vocational School, İstanbul Medipol University, İstanbul, Turkey
e-mail: seti@medipol.edu.tr

H. Dinçer et al. (eds.), *Decision Making in Interdisciplinary Renewable Energy Projects*, Contributions to Management Science,
https://doi.org/10.1007/978-3-031-51532-3_21

required to provide healthcare services (William et al., 2020). Since hospitals are 24/7 institutions, lighting also causes significant energy consumption. Such high energy consumption endangers the sustainability of hospitals (Fan et al., 2021). Developing measures and strategies for such situations is important for hospitals.

It is important to achieve an income-expense balance to ensure a sustainable service without compromising quality healthcare (Liu et al., 2021). There are many cost factors that can be mentioned for hospitals (Eti et al., 2023). Personnel expenses, equipment purchases, and consumables are among the leading cost items for hospitals (Addison et al., 2022). However, it is not possible to intervene in these items or to purchase the cheaper device. This is because human resources such as doctors, nurses, and other health personnel or medical devices are needed to provide health services (Gökalp, 2021). Moreover, while ensuring efficiency in health services, service quality should not be compromised (Dinçer et al., 2020). Therefore, it is not easy to intervene in these cost items. Energy costs are among the most manageable cost items (Gökalp et al., 2022).

Accordingly, ensuring efficiency in energy consumption in hospitals is important for sustainability (Eti et al., 2022). In this context, the aim of this study is to reveal the dimensions of energy consumption of hospitals and to develop strategies to ensure efficiency. For this purpose, 740 articles scanned with the keywords "hospital" and "energy consumption" in the Web of Science database are analyzed. These studies are text mined, and predictions are made to provide future projections.

The contribution of this study to the literature can be expressed as follows:

(i) With this study, studies on energy consumption in hospitals are analyzed by text mining method. In this way, it is aimed to provide future projections on energy consumption in hospitals.
(ii) Apart from this, the methodological contribution of this study is the use of the KNIME method. This method provides many pre-built nodes for data processing and analysis. Thanks to these nodes, data cleaning, transformation, filtering, scaling, merging, and visualization operations can be performed.

21.2 Literature Review

Energy consumption is of great importance for hospitals (Dinçer & Yüksel, 2023; Dinçer et al., 2022). Energy is needed to provide health services with a sustainability approach (Wan et al., 2023). Therefore, hospitals are energy dependent to provide health services. The devices and laboratories used to provide diagnosis, treatment, and care services need electrical energy. In addition, backup power sources must be available to ensure that health services are not disrupted (Dinçer et al., 2023a, b, c, d). Intercalarily, the necessary hygiene conditions must be met to provide healthcare services in appropriate conditions (Gökalp et al., 2021). These conditions are provided by energy-consuming devices such as water heating systems and cleaning devices (Dinçer et al., 2023a, b, c, d). Accordingly, energy-demanding actions such as lighting, heating, and cooling need to be implemented to ensure patient satisfaction and provide quality healthcare services.

Ji and Qu (2019) conducted a study to investigate and evaluate the energy consumption performance of hospital buildings in China. It is stated that energy use in hospitals is higher than in other public buildings. Therefore, it is mentioned in the article that it is important to investigate and evaluate energy consumption performance for energy saving. The study indicates that improving air conditioning systems is the most direct and effective way to save energy in a hospital building. Shi et al. (2021) carried out a study investigating the energy consumption of 30 public hospitals in the cold region of China. In the study, which is done by using the data of 30 hospitals for the years 2015–2017, it is underlined that energy consumption is quite high in terms of hospital usage.

Shen et al. (2019) analyzed building energy consumption in a hospital to investigate whether energy consumption varies climatologically. The study states that the energy consumption per unit area of hospital buildings is twice that of general public buildings. Sukarno et al. (2021) carried out a study to reduce the energy consumption of ventilation systems. It is reported that energy efficiency in the ventilation systems of hospitals can be increased by using a heat pipe heat exchanger. Panagiotou and Dounis (2022) performed a study on the comparison of energy consumption estimates of hospitals using artificial neural networks. It is noted that the energy efficiency of buildings should be ensured as expressed in the Green Deal of the European Union.

Sun et al. (2023) conducted a study involving the analysis and evaluation of the indoor environment, occupant satisfaction, and energy consumption in hospitals in China. The study utilized an online satisfaction survey, indoor environment data for each typical season, and energy consumption data. The study underlines that the amount of energy consumed in hospitals is very high. Cygańska and Kludacz-Alessandri (2021) investigated the determinants of energy consumption in hospitals by climate zones in Poland. It is emphasized that parameters such as hospital size and number of beds affect the size of energy consumed and the size of energy consumption in hospitals is high.

García-Sanz-Calcedo et al. (2019) aimed to find energy consumption indicators focused on healthcare activities in a hospital. The study is carried out by analyzing data on annual thermal and electrical energy consumption in 13 Spanish private hospitals between 2008 and 2017. It is emphasized that energy consumption is high, and it is important to ensure efficiency. Ma et al. (2019) conducted a study covering 50 public buildings in Tianjin, China. It is underlined that building size directly affects energy consumption and hospitals are the most energy-consuming public buildings. de Oliveira et al. (2021) conducted a study investigating efficient and sustainable energy management in hospital buildings. It is stated that high energy consumption in hospitals leads to increased operating costs, reduced competitiveness, and negative environmental impacts.

Ma et al. (2021) conducted a study to examine the thermal comfort and energy savings of healthcare buildings. While it is stated that ventilation systems are the most energy-consuming tool in hospitals, it is also stated that the energy consumption of hospitals is high. Fatema et al. (2020) performed energy analysis in multistory hospitals using big data analytics. Ventilation, service costs, energy end-uses, monthly peak energy demand, energy sources, water consumption, and gas emissions are the criteria taken into account. It is stated that energy consumption is very important for hospitals. Bawaneh et al. (2019) aimed to analyze the energy consumption of healthcare facilities in the United States of America (USA). It is noted in the study that US healthcare facilities have higher energy intensity than those of many other countries, especially European countries.

Besbas et al. (2022) and Yuan et al. (2022) state that energy consumption is very important for hospitals. Liu et al. (2022) and Uddin et al. (2022) indicate that energy has an important place in the provision of health services. Santamato et al. (2023) and Jemmad et al. (2020) report that energy consumption is higher in hospitals than in other buildings. Montiel-Santiago (2020) and Squire et al. (2021) note that energy is important for hospitals and energy efficiency should be ensured.

As a result of the literature review, the following conclusions are reached:

(i) Energy consumption is a very important issue for hospitals.
(ii) Energy is especially needed to ensure the sustainability of health services.
(iii) However, energy efficiency is also needed to ensure sustainability.
(iv) In addition, energy consumption in hospitals is high.
(v) Therefore, ensuring energy efficiency is a very important issue for hospitals.

21.3 Methodology

Data mining can be defined as data analysis that uses statistical and computer science techniques to discover patterns, information, and meaningful relationships within a big dataset. These methods help businesses improve their decision-making processes and make predictive analyses (Eti et al., 2023). Data mining, unlike statistics, can also analyze with unstructured data. Structured data refers to the type of data that is represented in a regular structure and can be easily processed, while unstructured data represents data that is irregular or in a free format. In other words, data such as text documents, visual contents, or audio recordings are unstructured data types (Eti, 2019, 2021).

Text mining is a data mining method that aims to extract meaningful information and patterns by analyzing big text data. This method uses language processing and machine learning techniques on documents. Some of the methods include applications such as sentiment analysis, topic modeling, and information extraction (Adalı et al., 2022). The preparation process for text mining consists of data collection, text cleaning, and transformation steps. In these stages of the method, text data is drawn from data sources; unnecessary characters are cleaned. As a result of this process, unstructured data is converted into structured data and made ready for analysis. One of the text mining methods is frequency analysis (Dinçer et al., 2023a, b, c, d).

Frequency analysis is a method used to determine how often words or terms occur in a document. This analysis helps identify important words or topics in texts. A different version of this method is Ngram analysis. Ngram analysis is defined as a method that examines consecutive word groups or character sequences within a text (Eti & Mızrak, 2020). It is often preferred especially for language modeling and understanding word relationships. Here, the N value is the number of words. When analyzing with binary word groups, N value is taken as 2. When this value is taken as 3, the frequencies of consecutive three-word groups are obtained.

21.4 Findings

Studies on energy consumption in hospitals have been analyzed with the text mining method. Thanks to the analysis, it is aimed to make a future projection for energy use in hospitals. In this context, the keywords "hospital" and "energy consumption" were searched in the Web of Science database. As a result, 740 articles are obtained. The abstracts of the articles are analyzed, and the most repetitive single, double, and triple words are scanned. Accordingly, the most recurring single words in the articles in the Web of Science database are given in Table 21.1.

Accordingly, apart from the scanned words "energy," "hospital," and "consumption," words such as "reduce," "performance," and "efficiency" are the most repeated words. Accordingly, it can be concluded that energy consumption in hospitals is high and should be reduced. Apart from this, it can also be said that the majority of the studies conducted are in the direction of ensuring energy consumption dimension and efficiency. Table 21.2 shows the most recurring binary words.

Table 21.1 Most repeated single words

Term	Sum (TF abs)
Energy	5874
Hospital	2736
Consumption	2678
System	2300
Build	1306
Building	1260
Patient	970
Propose	790
Network	752
Reduce	716
Performance	708
Efficiency	682
Design	678
Health	660
Increase	656
Healthcare	612

Table 21.2 Most repetitive double words

Ngram	Corpus frequency
Energy consumption	2218
Energy efficiency	284
Reduce energy	181
Build energy	168
Energy performance	160
Energy demand	148
Energy saving	145
Thermal comfort	130
Healthcare facility	126
Renewable energy	116
Hospital build	115
Electrical energy	114
Hospital building	110
Energy system	108
Consumption hospital	105
Environmental impact	104
Primary energy	104
Heate coole	104
Total energy	102
Paper present	100
Energy conservation	100

When the most recurring binary word groups are analyzed, it can be mentioned word groups such as "energy efficiency," "reduce energy," "energy performance," and "energy saving." Accordingly, it can be said that the studies are carried out in the direction of ensuring energy efficiency and reducing energy consumption. Apart from this, "renewable energy" is among the most repeated words. Therefore, it can be stated that the use of renewable energy is important for hospitals. The most repeated triple words are also given in Table 21.3.

When the most recurring triplet words are analyzed, apart from the scanned words such as "energy consumption hospital" and "build energy consumption," binaries such as "reduce energy consumption," "total energy consumption," and "annual energy consumption" can be mentioned. Accordingly, it is possible to conclude that energy consumption should be reduced. In addition, most of the studies focus on total energy consumption and annual energy consumption. Table 21.4 shows the most recurring quadruple words.

When the most recurring quadruple-word groups are examined, word groups such as "average annual energy consumption," "heat ventilation conditioning system," and "energy consumption hospital building" can be mentioned. Apart from this, it can be stated that studies involving the reduction of energy consumption such as "carbon dioxide emission reduction" and "reduce primary energy consumption" have been carried out.

Table 21.3 Most repeated triple words

Ngram	Corpus frequency
Reduce energy consumption	146
Energy consumption hospital	93
Build energy consumption	66
Total energy consumption	62
Primary energy consumption	60
Annual energy consumption	54
Electrical energy consumption	50
Energy consumption building	42
Energy consumption build	38
Carbon dioxide emission	36
Increase energy consumption	34
Lower energy consumption	34
Heate ventilation conditione	32
Energy consumption system	26
Energy consumption reduction	26
Reduction energy consumption	26
Hospital energy consumption	26
Average energy consumption	26

Table 21.4 Most repeated quadruple words

Ngram	Corpus frequency
Average annual energy consumption	22
Heate ventilation conditione system	20
Energy consumption hospital building	14
Carbon dioxide emission reduction	12
Health force density index	12
Energy consumption public building	12
Reduce primary energy consumption	10
Combine coole heate power	10
Heate ventilation air-conditione system	10
Energy consumption hospital build	10

21.5 Conclusion

Hospitals need many things to deliver health services. Some of these needs are qualified personnel, medical devices, and consumables. In addition to these, one of the most important factors needed to provide these services is energy. In order to ensure the sustainability of health services, they need uninterrupted energy. Therefore, hospitals consume very high amounts of energy. Since this energy consumption is high, it is important to ensure efficiency in order to control costs. Accordingly, in this study, 740 articles are analyzed by searching the Web of Science database with the keywords "hospital" and "energy consumption." The 740 scanned articles are text mined using the KNIME program. In this way, a summary of the studies that have been conducted so far in the Web of Science database with the keywords "hospital" and "energy consumption" has been summarized.

Accordingly, the most recurring single words are "energy," "hospital," and "consumption." Apart from the scanned words, words such as "reduce," "performance," and "efficiency" are the most recurring words. Therefore, it can be concluded that energy consumption in hospitals is high and should be reduced. It can also be said that the studies have developed within this framework. When we look at the most recurring binary words, we can mention word groups such as "energy efficiency," "reduce energy," "energy performance," and "energy saving." Thus, it can be said that the studies are carried out in the direction of ensuring energy efficiency and reducing energy consumption. Apart from this, "renewable energy" is among the most repeated words. Consequently, it can be stated that the use of renewable energy is also important for hospitals.

When the most recurring triadic phrases are analyzed, apart from the scanned words such as "energy consumption hospital" and "build energy consumption," binaries such as "reduce energy consumption," "total energy consumption," and "annual energy consumption" can be mentioned. Hence, it can be said that the studies focus on total energy consumption and average energy consumption. When the quadruple word groups are analyzed, it is possible to mention the word group "carbon dioxide emission reduction" unlike other repeating word groups. Considering all repeating word groups, it is possible to say that the studies focus on reducing energy consumption and calculating total energy consumption.

The studies so far have focused on general issues such as calculating total energy consumption or ensuring energy efficiency. Future studies may focus on unit-specific energy consumption calculation. Apart from this, there are few studies on how energy efficiency should be achieved in hospitals. The authors can focus on these issues for future processes. Another understudied area for hospitals is the use of renewable energy in hospitals. Studies on the use of renewable energy for hospitals and the reduction of carbon emissions will become more important in the future. In addition, there are few studies on energy storage. Studies can be conducted on energy storage for periods of interruption on behalf of hospitals.

References

Adalı, Z., Dinçer, H., Eti, S., Mikhaylov, A., & Yüksel, S. (2022). Identifying new perspectives on geothermal energy investments. In H. Dinçer & S. Yüksel (Eds.), *Multidimensional strategic outlook on global competitive energy economics and finance* (pp. 1–11). Emerald Publishing.

Addison, D., Cheng, E., Forrest, P., Livingstone, A., Morton, R. L., & Dennis, M. (2022). Cost-effectiveness of extracorporeal cardiopulmonary resuscitation for adult out-of-hospital cardiac arrest: A systematic review. *Resuscitation, 178*, 19.

Bawaneh, K., Ghazi Nezami, F., Rasheduzzaman, M., & Deken, B. (2019). Energy consumption analysis and characterization of healthcare facilities in the United States. *Energies, 12*(19), 3775.

Besbas, S., Nocera, F., Zemmouri, N., Khadraoui, M. A., & Besbas, A. (2022). Parametric-based multi-objective optimization workflow: Daylight and energy performance study of hospital building in Algeria. *Sustainability, 14*(19), 12652.

Carino, S., Porter, J., Malekpour, S., & Collins, J. (2020). Environmental sustainability of hospital foodservices across the food supply chain: A systematic review. *Journal of the Academy of Nutrition and Dietetics, 120*(5), 825–873.

Coskun, C., Oktay, Z., Özbek, H., & Sogut, M. Z. (2022, April). Hospital energy analysis in Turkey. In *International symposium on energy management and sustainability* (pp. 643–654). Springer.

Cygańska, M., & Kludacz-Alessandri, M. (2021). Determinants of electrical and thermal energy consumption in hospitals according to climate zones in Poland. *Energies, 14*(22), 7585.

de Oliveira, K. B., dos Santos, E. F., Neto, A. F., de Mello Santos, V. H., & de Oliveira, O. J. (2021). Guidelines for efficient and sustainable energy management in hospital buildings. *Journal of Cleaner Production, 329*, 129644.

Dinçer, H., & Yüksel, S. (Eds.). (2023). *Economic development and the environmental ecosystem: The role of energy policy in economic growth*. Springer Nature.

Dinçer, H., Yüksel, S., Gökalp, Y., & Eti, S. (2020). SERVQUAL-based evaluation of service quality in Turkish health industry with fuzzy logic. In *Interdisciplinary perspectives on operations management and service evaluation* (Vol. 32, pp. 213–233). Springer.

Dinçer, H., Eti, S., Yüksel, S., Gokalp, Y., & Celebi, B. (2022). Strategy generation for risk minimization of renewable energy technology investments in hospitals with SF TOP-DEMATEL methodology. *Journal of Computational and Cognitive Engineering*.

Dinçer, H., Eti, S., Yüksel, S., Özdemir, S., Yîlmaz, A. E., & Ergün, E. (2023a). Integrating data mining and fuzzy decision-making techniques for analyzing the key minimizing factors of carbon emissions. *Journal of Intelligent & Fuzzy Systems, 45*, 1–17.

Dinçer, H., Eti, S., Gökalp, Y., Yüksel, S., & Çelebi, B. (2023b). The role of small-scale wind turbines in increasing the competitiveness of hospitals. In *Handbook of research on quality and competitiveness in the healthcare services sector* (pp. 336–352). IGI Global.

Dinçer, H., Gökalp, Y., Eti, S., & Yüksel, S. (2023c). Strategy generation for minimizing health risks to obtain economic development with pedroni panel cointegration analysis and AHP model. In *Social sector development and governance* (pp. 161–176). Routledge India.

Dinçer, H., Yüksel, S., & Eti, S. (2023d). Identifying the right policies for increasing the efficiency of the renewable energy transition with a novel fuzzy decision-making model. *Journal of Soft Computing and Decision Analytics, 1*(1), 50–62.

Eti, S. (2019). ULAKBİM İndeksinde Taranan Sosyal Bilimler Alanındaki Dergilerde Öne Çıkan Konu Ve Yöntemlerin Metin Madenciliği Yaklaşımı ile Belirlenmesi. *Uluslararası Hukuk ve Sosyal Bilim Araştırmaları Dergisi, 1*(1), 61–66.

Eti, S. (2021). *Veri madenciliği ile markaların dijital pazarlama kanallarının performansının değerlendirilmesi üzerine bir uygulama (Order No. 29442857)*. Retrieved from ProQuest Dissertations & Theses Global (2724234885).

Eti, S., & Mızrak, F. (2020). Analysing customer satisfaction of civil aviation companies of Turkey with text mining. In H. Dincer & S. Yüksel (Eds.), *Strategic outlook for innovative work behaviours. Contributions to management science*. Springer.

Eti, S., Dinçer, H., Yüksel, S., & Gökalp, Y. (2022). Analysis of the suitability of the solar panels for hospitals: A new fuzzy decision-making model proposal with the T-spherical TOP-DEMATEL method. *Journal of Intelligent & Fuzzy Systems, 44*, 1–13.

Eti, S., Dinçer, H., Gökalp, Y., Yüksel, S., & Kararoğlu, D. (2023). Identifying key issues to handle the inflation problem in the healthcare industry caused by energy prices: An evaluation with decision-making models. In *Managing inflation and supply chain disruptions in the global economy* (pp. 162–178). Springer.

Fan, M., Cao, G., Pedersen, C., Lu, S., Stenstad, L. I., & Skogås, J. G. (2021). Suitability evaluation on laminar airflow and mixing airflow distribution strategies in operating rooms: A case study at St. Olavs Hospital. *Building and Environment, 194*, 107677.

Fatema, N., Malik, H., & Iqbal, A. (2020). Big-data analytics based energy analysis and monitoring for multi-storey hospital buildings: Case study. In *Soft computing in condition monitoring and diagnostics of electrical and mechanical systems: Novel methods for condition monitoring and diagnostics* (pp. 325–343). Springer.

García-Sanz-Calcedo, J., Gómez-Chaparro, M., & Sanchez-Barroso, G. (2019). Electrical and thermal energy in private hospitals: Consumption indicators focused on healthcare activity. *Sustainable Cities and Society, 47*, 101482.

Gökalp, Y. (2021). Identifying innovative financial health management strategies for Turkey. In *Financial strategies in competitive markets: multidimensional approaches to financial policies for local companies* (pp. 207–222). Springer.

Gökalp, Y., Dinçer, H., & Eti, S. (2021). The role of health policies for the strategic investment decisions of the international companies. In *Management strategies to survive in a competitive environment: How to improve company performance* (pp. 321–333). Springer.

Gökalp, Y., Yüksel, S., & Dinçer, H. (2022). Balanced scorecard-based cost analysis of service industry using a novel hybrid decision making approach based on golden cut-oriented bipolar and q-ROF sets. *Journal of Intelligent & Fuzzy Systems, 43*(4), 4709–4722.

Hiwar, W., King, M. F., Shuweihdi, F., Fletcher, L. A., Dancer, S. J., & Noakes, C. J. (2021). What is the relationship between indoor air quality parameters and airborne microorganisms in hospital environments? A systematic review and meta-analysis. *Indoor Air, 31*(5), 1308–1322.

Jemmad, K., Hmidat, A., & Saad, A. (2020). A novel systematic and comprehensive method for identification of energy-saving measures. *EAI Endorsed Transactions on Energy Web, 7*(30).

Ji, R., & Qu, S. (2019). Investigation and evaluation of energy consumption performance for hospital buildings in China. *Sustainability, 11*(6), 1724.

Khahro, S. H., Kumar, D., Siddiqui, F. H., Ali, T. H., Raza, M. S., & Khoso, A. R. (2021). Optimizing energy use, cost and carbon emission through building information modelling and a sustainability approach: A case-study of a hospital building. *Sustainability, 13*(7), 3675.

Liu, J., Capurro, D., Nguyen, A., & Verspoor, K. (2021). Early prediction of diagnostic-related groups and estimation of hospital cost by processing clinical notes. *npj Digital Medicine, 4*(1), 103.

Liu, X., Ding, Y., Tang, H., Fan, L., & Lv, J. (2022). Investigating the effects of key drivers on energy consumption of nonresidential buildings: A data-driven approach integrating regularization and quantile regression. *Energy, 244*, 122720.

Ma, H., Li, C., Lai, J., Yang, F., & Li, Z. (2019). Investigation on energy consumption of public buildings in Tianjin. *Energy Procedia, 158*, 3427–3432.

Ma, S., Ma, Y., Zhang, Q., Deng, W., Lu, J., & Zhou, T. (2021, October). Thermal comfort and energy consumption in healthcare buildings – a review. In *International conference on architecture, materials and construction* (pp. 381–390). Springer.

Montiel-Santiago, F. J., Hermoso-Orzáez, M. J., & Terrados-Cepeda, J. (2020). Sustainability and energy efficiency: BIM 6D. Study of the BIM methodology applied to hospital buildings. Value of interior lighting and daylight in energy simulation. *Sustainability, 12*(14), 5731.

Panagiotou, D. K., & Dounis, A. I. (2022). Comparison of hospital building's energy consumption prediction using artificial neural networks, ANFIS, and LSTM network. *Energies, 15*(17), 6453.

Santamato, V., Esposito, D., Tricase, C., Faccilongo, N., Marengo, A., & Pange, J. (2023, June). Assessment of public health performance in relation to hospital energy demand, socio-economic efficiency and quality of services: An Italian case study. In *International conference on computational science and its applications* (pp. 505–522). Springer Nature Switzerland.

Shen, C., Zhao, K., Ge, J., & Zhou, Q. (2019). Analysis of building energy consumption in a hospital in the hot summer and cold winter area. *Energy Procedia, 158*, 3735–3740.

Shi, Y., Yan, Z., Li, C., & Li, C. (2021). Energy consumption and building layouts of public hospital buildings: A survey of 30 buildings in the cold region of China. *Sustainable Cities and Society, 74*, 103247.
Squire, M. M., Munsamy, M., Lin, G., Telukdarie, A., & Igusa, T. (2021). Modeling hospital energy and economic costs for COVID-19 infection control interventions. *Energy and Buildings, 242*, 110948.
Sukarno, R., Putra, N., Hakim, I. I., Rachman, F. F., & Mahlia, T. M. I. (2021). Utilizing heat pipe heat exchanger to reduce the energy consumption of airborne infection isolation hospital room HVAC system. *Journal of Building Engineering, 35*, 102116.
Sun, Y., Kojima, S., Nakaohkubo, K., Zhao, J., & Ni, S. (2023). Analysis and evaluation of indoor environment, occupant satisfaction, and energy consumption in General Hospital in China. *Buildings, 13*(7), 1675.
Uddin, M. N., Ruva, I. J., Syed, M. A., Hossain, D., Akter, R., Tamanna, N., Rahman, A., & Saka, A. (2022). Occupant centric energy renovation strategy for hospital and restaurant building envelop using distinct modellingtools: A case study from low-income cultural context. *Energy and Buildings, 272*, 112338.
Wan, Q., Miao, X., Wang, C., Dinçer, H., & Yüksel, S. (2023). A hybrid decision support system with golden cut and bipolar q-ROFSs for evaluating the risk-based strategic priorities of fintech lending for clean energy projects. *Financial Innovation, 9*(1), 1–25.
William, M. A., Elharidi, A. M., Hanafy, A. A., Attia, A., & Elhelw, M. (2020). Energy-efficient retrofitting strategies for healthcare facilities in hot-humid climate: Parametric and economical analysis. *Alexandria Engineering Journal, 59*(6), 4549–4562.
Yuan, F., Yao, R., Sadrizadeh, S., Li, B., Cao, G., Zhang, S., Zhou, S., Liu, H., Bogdan, A., Croitoru, C., Meliko, A. K., Short, C. A., & Li, B. (2022). Thermal comfort in hospital buildings – a literature review. *Journal of Building Engineering, 45*, 103463.
Zor, K., Çelik, Ö., Timur, O., & Teke, A. (2020). Short-term building electrical energy consumption forecasting by employing gene expression programming and GMDH networks. *Energies, 13*(5), 1102.

Chapter 22
Electrical Vehicles in the Context of Sustainability: An Analysis Through Social Media Platforms

İhsan Eken and Gabriela Oana Olaru

22.1 Introduction

Technology may be defined as a system based on change, transformation, and progress. Over time, the positive or negative effects of each innovation or discovery in the field of technology can be observed in humanity. Some researchers working on technology try to elucidate the effects of technology before they emerge, some in the process, and some with futuristic predictions based on the future scenarios. Therefore, the effects of technological developments are not limited to humanity but also express its responsibility to "everything" that exists on Earth. This study is designed to understand the level of knowledge and awareness of consumers in the automotive and renewable energy sectors regarding the environmental issue, which has been on the agenda more and more in recent years. The study used content analysis methods, which are one of the qualitative research approaches. The aim of the study is to identify the environmental awareness emphasis and potential consumer perception of electric vehicles in general and TOGG, the first domestic automotive brand of Turkey, by examining the video content and user comments published on the YouTube platform and included in the research. The main basis of the research is environmental problems, which have become one of the most pressing topics on the global agenda in recent years. In response to these problems, there has been a growing movement to develop and promote alternative practices that can reduce environmental damage. The issue of environmental pollution has been discussed at the global level since the 1970s. With globalization, the increasing population and motor vehicles are accepted as the main causes of the environmental problem, along with the fact that cities are becoming more crowded, and

İ. Eken (✉) · G. O. Olaru
The School of Communication, İstanbul Medipol University, Istanbul, Turkey
e-mail: ieken@medipol.edu.tr

H. Dinçer et al. (eds.), *Decision Making in Interdisciplinary Renewable Energy Projects*, Contributions to Management Science,
https://doi.org/10.1007/978-3-031-51532-3_22

urbanization is spreading in an irregular way. In the 1990s, environmental impact began to be investigated from a more comprehensive perspective, and the issue of climate change has begun to be discussed within an international framework. With the adoption of the United Nations Framework Convention on Climate Change, some concrete steps have also begun to be taken.

The environmental impact of the transportation sector, which is the focus of this study, has become increasingly evident in recent years, as the number of vehicles on the road has increased dramatically. Polluting emissions and carbon dioxide levels in the atmosphere have increased as a result, leading to the greenhouse effect and contributing negatively to climate change. To reduce the high CO2 emissions from the transportation sector, many countries are working to develop low-carbon methods using alternative fuel options. Electric vehicles, which have been on the agenda in recent years and for which studies have been conducted to reflect consumer preferences and attitudes toward vehicles, are thought to have a significant potential in reducing CO_2 emissions (Asamer et al., 2016; Bickert et al., 2015; Khoo et al., 2014; Mersky et al., 2016; Zhang & Yao, 2015).

The environmental and climate change issues that our world is facing, the negative effects of which are increasing day by day, the depletion of natural resources, and the awareness and sensitivity expectations of environmentally conscious consumers are also pushing the automotive industry to review vehicle production processes. With all these developments mentioned above, brands are trying to fulfill their duty by increasing the production of hybrid and electric vehicles. Electric vehicles are defined as "zero-emission vehicles" because they do not produce harmful emissions that are classified as environmental waste. In our country, various studies are being carried out on electric vehicles and environmental awareness, and efforts are being made to increase demand. The second part of our research will introduce electric vehicles and discuss their development over time. Again, in this section, the studies on electric vehicles in our country and the development of our domestic vehicle TOGG will be specifically discussed. The third section of our research will focus on the environmental impacts of gasoline vehicles, summarize climate and environmental goals considering the Sustainable Development Goals of the United Nations, and compile the studies on consumer acceptance of these vehicles in the literature. The fourth section of the study is the research section. The conclusion and evaluation section will include suggestions for future scenarios research considering the findings of the research.

22.2 The Market and Historical Development of Electric Vehicles

As Kocagöz and others have noted, innovative solutions have begun to emerge in recent years for a variety of reasons: the growing emphasis on environmental awareness, the changing behavior of consumers over time, and the innovations that technology has led to these solutions (Kocagöz et al., 2020). One of the most important issues at the heart of all these reasons is the environmental damage caused by the consumption of fossil fuels in large quantities by transportation vehicles. Hybrid and electric cars, in which the automotive sector concentrates its research and development activities, play an important place in addressing the environmental problems that are currently experienced.

According to a report published by the International Renewable Energy Agency (IRENA) in 2019, the global electric vehicle market, which reached 8 million units this year, is expected to exceed 1.1 billion units by 2050. The achievement of these increases in the use of electric vehicles predicted by IRENA will significantly help to reduce the environmental impacts of transportation. In this way, reduced greenhouse gas can help to reduce climate change caused by humans and to preserve the fossil fuels that are currently being used (Arce et al., 2014; Ghommem et al., 2012; Javid & Nejat, 2017).

To briefly describe the operating principle of electric vehicles, we can say that the engine in these vehicles is electric and the fuel tank is replaced by batteries (Andwari et al., 2017). The electric energy used by these vehicles is replenished by recharging the batteries. The charging operations required to increase energy efficiency, if obtained from alternative energy sources produced using solar and wind energy, has a positive environmental impact.

Electric vehicles also have some advantages and disadvantages for users. Although we are conducting a study that examines the awareness level of consumers in terms of sustainability, energy, and being friendly to nature, it is thought that it would be useful to generally mention the advantages and disadvantages that are among the factors that affect consumer preferences. The main advantages of electric vehicles include their independence from fossil fuels, which directly affects the environment, and therefore their reduction of greenhouse gas emissions (Bradley & Quinn, 2010; Andwari et al., 2017; Bühler et al., 2014). Other advantages of electric vehicles include lower noise, the ability to reach high speeds in a short period of time, high efficiency, and low cost (Skippon, 2014:15).

In addition to the advantages that influence consumer preferences, there are also disadvantages. Manufacturers are continuing their work to address the disadvantages of electric vehicles. It is important to eliminate these disadvantages to increase market share and achieve both commercial and environmental goals. Although research is ongoing on battery technologies, the fact that they have not yet been developed to the desired level limits consumer choice. The fact that electric vehicles are more expensive than internal combustion engine vehicles in terms of price is another disadvantage in consumer choice. One of the most important disadvantages is related to the range of the vehicles. The limited range of electric vehicles, the time it takes to charge these vehicles, and the lack of fast charging stations are important problems that need to be solved to turn consumer preferences toward these vehicles (Andwari et al., 2017:415; Ashique et al., 2017:1257; Kumar & Revankar, 2017; Hota et al., 2014:217).

The increase in advantages that shape consumer preferences and the completion of perceived disadvantages will help to accelerate the growth of the electric vehicle market, thus helping to reduce environmental impact. Summarizing the sales figures for 2022 and 2023 published by various institutions and organizations and seeing the level reached by the market will reflect the journey of electric vehicles to exist in the market.

According to the 2023 Global Outlook Report published by the International Energy Agency, global electric vehicle sales increased by 55% in 2022 compared to the previous year. This increase led to the use of over 10 million electric vehicles, and their share of total vehicle sales also increased to 14%. When sales distribution is examined by country, China is the country with the highest sales. Europe and the United States are the markets with the highest electric vehicle sales.

According to the report summary published by the Turkey Electric and Hybrid Vehicles Association (TEHAD), the Turkish market has shown significant growth. According to the data from the report, the number of electric vehicles sold in 2021 was 2849. In 2022, the market grew by 188%. Accordingly, a total of 8210 electric vehicles were sold in 2022.

Although the Turkish market is growing rapidly, it is important to increase its share in total vehicle sales to reduce environmental impacts. It is also predicted that the electric vehicle market in Turkey will show a faster growth trend with the entry of Turkey's Automobile Enterprise Group (TOGG), known as TOGG, into the market (Kocagöz et al., 2020). This prediction made by Kocagöz et al. is also mentioned in a separate note in the "2022 Electric Vehicle Sales Review" study published by Strategy & Part of PwC Network. The study defines Turkey as a "Focus Market" and expresses its goal of becoming a leader in electric vehicle production and market in the future scenarios.

The efforts to reduce the negative environmental impacts caused using fossil fuels in transportation, the development of electric vehicle models, and the short summary of the sales market, which is shaped by consumer preferences according to the advantages and disadvantages of these vehicles, are presented above. It is important to note that the issue of electric vehicles is being more in the spotlight today, but it is also worth discussing how these vehicles have developed over time and our country's journey to participate in these studies with TOGG's involvement in this section in terms of the overall topic.

Although electric vehicles are seen as a new topic that has been on the agenda in recent years, history shows that they have a history dating back to the 1800s. In fact, we see that they are based on the inventions of Aynos Jedlik in 1828 and that the first electric motor-powered and small-scale vehicle was produced by Professor Stratingh in the Netherlands in 1835. With the emergence and widespread use of internal combustion engines, electric vehicles lost their importance and withdrew from the market until the 1960s. The real interest began to emerge again in the 2000s, and especially the decreasing oil reserves, the damage it causes to the environment, and people's increasing awareness of these issues have brought electric vehicles to the agenda, and they have begun to take an important place in many strategies and policies. The development of battery and charging facilities, which are included in the technical elements of these vehicles, has led to an increase in the preference for electric vehicles, adding commercial value. In these years, the production of electric vehicles began in the United States and Germany with leadership goals. Especially the United States has implemented various policies to encourage the production and commercialization of electric vehicles. China is another important country that has launched an attack on the production of electric vehicles and related technologies and redesigned the automotive industry. In terms of the production of batteries used in electric vehicles, Poland and Hungary are also developing their production capacities together with Germany. It is foreseen that these countries will have an important role in global battery production soon. It is estimated that Sweden, Germany, Poland, and Hungary will have a significant production capacity in Europe by 2030 (Eddy et al., 2019).

In the race to produce electric vehicles and related technologies, Turkey is also working on production. Within these studies, which are carried out by different companies and are independent of each other, the SAPMAZ 3GEN, which was prototyped in 2013, is one of the first examples of electric vehicles. Renault Fluence ZE is the first electric vehicle that has been converted in Turkey (Kocabey, 2018:20). One of the most important steps taken in recent years in terms of electric vehicle production is the establishment of TOGG. This step, which is aimed at both the domestic and foreign markets, was established in 2018 by a partnership of Anadolu Group, BMC, Turkcell, Zorlu Holding, and the Turkish Chamber of Commerce and Industry (TOBB). In a message shared on the company's website by CEO Mehmet Gürcan Karakaş, TOGG is stated to have embarked on this journey with two strategic goals:

> When we set out for "more than a car", we had two strategic goals. First, to develop a mobility technology brand owned by Togg, Turkey, that can compete globally and has intellectual property. Second, to create the core of a mobility ecosystem that is accessible to everyone and can be developed by everyone, around our smart device, digital experience platform, and clean energy solutions, with the user at the center. (TOGG, 2021)

Among the goals highlighted by Karakaş, "clean energy solutions" stand out. In fact, TOGG has entered the automotive industry with a solution that reduces environmental impact, thanks to renewable energy sources. The following emphasis is particularly noteworthy in the continuation of these claims:

> We are a brand that is inherently green and sustainable. Our data-driven business model, our mobility and smart living solutions developed as a technology company, and our ecosystem built with an environmentally friendly and secure blockchain infrastructure, are rising on the basis of sustainability. (TOGG, 2021)

As can be seen, the CEO message on TOGG's website to consumers emphasizes the brand as green and sustainable, and details about clean energy solutions play an important role. In this regard, the TOGG initiative directly addresses the UN's Sustainable Development Goals, especially Goal 7, Goal 9, Goal 11, and Goal 13.

22.3 Methodology

According to a report published by the International Energy Agency (IEA), over 14% of greenhouse gas emissions that affect climate change originate from the transportation sector. With the increase in the number of vehicles, it is estimated that by 2030 this figure may reach approximately 22% (IEA Publications, 2023). The reduction of transportation-induced carbon emissions, which significantly impact climate change, emerges as a necessity (Tran et al., 2012). In light of the report released by the IEA, reducing carbon emissions may only be possible through joint declarations made by international organizations and states (Ebinger and Vandycke 2015). One of the most crucial solutions to decrease carbon emissions aligns with the "sustainable development goals" set by the United Nations, which can be diversified through transportation options with renewable energy sources (Ebinger and Vandycke 2015). Aspects necessary to maintain ecological balance, such as the use of clean energy sources and the sustainability of energy efficiency, can be achieved by increasing the production and use of electric vehicles. Particularly, increasing the use of electric vehicles is compatible with the 7th Goal (accessible, clean energy) emphasized in the Sustainable Development Goals by the United Nations, and the proliferation of these vehicles can impact the reduction of air pollution (Swain and Karimu 2019).

Sustainability is one of the most dominant and critical global issues of our time. Sustainability matters, such as climate change, depletion of natural resources, environmental pollution, and energy efficiency, are among the challenging problems humanity faces. In this context, the automotive industry plays a significant role in achieving sustainability goals. The TOGG brand, established for Turkey to hold a significant place globally, also holds an important place in terms of sustainable clean energy. TOGG can be seen as a crucial step in reducing environmental impacts during the transition phase from traditional internal combustion engine vehicles to electric and eco-friendly automobiles. When reviewing literature studies about TOGG, Yaprak (2023), in a study examining TOGG production using PESTLE analysis dimensions, emphasizes elements of lower carbon emissions and noise pollution in the environmental factor dimension. In another recent study related to TOGG, a quantitative research study has been conducted on the perceived image in the marketing field (Tilki et al., 2022). Another study conducted in the context of marketing, purchase intentions, and perceived brand image has been executed by Yılmaz et al. (2022) in the context of ethnocentrism, aiming to understand consumers' purchasing intentions. In the study conducted by Kocagöz et al. (2020), it is observed that environmental concerns are among the factors affecting consumers' purchasing intentions. The studies in the literature conducted by Uslu and Demirel (2022) and Yılmaz (2020) also focus on purchase intention, marketing, and promotion.

In the literature review conducted specifically regarding TOGG, no study directly addressing environmental elements has been encountered. Only in the research conducted by Yaprak (2023) and Kocagöz et al. (2020) are environmental impacts noted among the emerging elements. In this respect, this research is significant as it is one of the first studies directly targeting consumer awareness at a time when the environmental impact is important in the automotive sector. Similarly, this research is also important for understanding the environmentally friendly emphasis in TOGG's promotion, announcement, and marketing activities.

The aim of the study is to research the TOGG brand, defined as Turkey's first domestic electric car, through content analysis method, involving user comments along with video contents on social media platforms within the framework of sustainability.

Creswell (2016) defines data analysis within the qualitative research method under six main headings. These headings are document detection and classification, data detection, coding, detection of findings, presentation of results, and interpretation and inference. This study has been conducted with a case study, which is a qualitative method. Case study research (Creswell, 2018) is a method that requires the researcher to utilize multiple sources of information within real-life conditions over a specific period of time. These sources may include data collection methods such as observation, interview, audiovisual material, reports, or documents. This study has been conducted with the aim of situation description or revealing situation themes, which is a qualitative research approach (Yin, 2009).

The universe of the study is formed by YouTube channels emerging as a result of the "Top 10 Turkish YouTube Channels to Follow with Automobile Content" search conducted by Otokokpit and the shares made on TOGG's official YouTube channel (Otokokpit.com, n.d., tarih yok). In Otokokpit's research, "Doğan Kabak, Benzin Tv, otopark.com, Rhino Tube, and Carviser," along with TOGG's official YouTube page, constitute the universe of the research. Other channels that did not share about TOGG were not included in the research universe. The sampling method of the research is purposive sampling, which is one of the non-probabilistic sampling varieties. The study took place between December 27, 2019, and May 31, 2023. The universe of the study is constituted by 151 videos shared related to TOGG on YouTube pages. The obtained data have been analyzed using Maxqda 2022 software.

22.4 Findings

Within the scope of the research, video content and user comments on the most followed YouTube accounts in Turkey and TOGG's official YouTube account have been analyzed. 132 videos and user comments on TOGG's official account, 8 videos and user comments on Doğan Kabak's YouTube account, 3 videos and user comments on Benzin TV's YouTube account, 4 videos and user comments on Otopark.com's YouTube account, 2 videos and user comments on Rhino Tube Sertaç Kemiksiz's YouTube account, and 1 video and user comment on Carsiver's YouTube account have been analyzed within the scope of the study. In the scope of the study, 25,441 codes have emerged. Of the emerging codes, 10,630 belong to the Doğan Kabak YouTube channel, 10,463 to the TOGG official YouTube channel, 1826 to the otopark.com YouTube channel, 1374 to the Benzin TV YouTube channel, 1015 to the Carsiver YouTube channel, and 124 to the Rhino Tube Sertaç Kemiksiz YouTube channel.

When the analyses of video content and user comments published on the YouTube channel within the scope of the research are examined, it is seen that the name of the TOGG brand is coded by being commented 320 times with its name in a video on the Carsiver YouTube channel in terms of access and impact power. In this respect, the Doğan Kabak YouTube channel has been commented 2392 times with 8 videos and coded. When the table is examined numerically, although the TOGG official YouTube channel is seen to be the most coded YouTube channel, it is seen that TOGG has reached this coding number only with 132 videos and user comments. When the word "domestic" is examined in terms of access and impact power, it is seen that the word domestic is mentioned 3189 times in 8 videos on the Doğan Kabak YouTube channel. This number is far above the TOGG official YouTube account. It is seen that in the codings made with the word domestic in TOGG's YouTube account, this word is coded less than the other five YouTube channels. When the third most coded category, the word "price," is examined, it is seen that the YouTube channel that generated the most codes is Benzin Tv. Benzin Tv has been coded 378 times with 3 videos.

In the context of sustainability, which forms the main theme of the study, 2991 codes have been generated as a result of coding the headings of electric vehicle, future, energy, environment, fuel, hybrid, fossil fuel, environmental pollution, clean energy, and sustainability. This number of codes constitutes approximately 12% of the general code number. Among these concepts, the word that has been coded the most is electric vehicles. In the Doğan Kabak YouTube channel, this word has been coded 662 times in 8 videos. On the TOGG account, the word electric vehicle has been coded only 742 times in 132 videos and content comments.

Upon examining the table, 21 different themes have been coded in the context of video and user comment codings within the scope of the study. While the TOGG YouTube account is coded within the framework of all determined themes, the Rhino Tube Sertaç Kemiksiz account has been coded within the framework of 10 themes. The data is presented in Table 22.1.

22.5 Conclusion

TOGG, established as Turkey's first domestic car on a global scale, aims primarily to reduce carbon emissions and promote the use of electric cars, initially in Turkey and subsequently worldwide. Moreover, promoting the use of electric vehicles particularly aligns with the 7th Goal (affordable and clean energy) of the United Nations Sustainable Development Goals, and broader utilization of these vehicles could potentially contribute to reducing air pollution. TOGG vehicles emerge as an example of adopting and using electric vehicle technologies to contribute to sustainable development goals. These vehicles, operating based on clean energy sources,

Table 22.1 Data regarding the study

	TOGG (132)	Doğan Kabak (8)	Carsiver (1)	Benzin Tv (3)	Otopark (4)	Rhino (2)	Total
TOGG	3902	2392	320	134	416	35	7199
Local	1689	3189	246	230	571	26	5951
Price	1169	2088	111	378	258	20	4024
Electric vehicle[a]	742	662	54	201	153	10	1822
Technology	668	445	41	45	68	14	1281
Duration	483	276	84	45	101	5	994
Battery	675	845	58	142	144	8	1872
Media	270	240	24	51	40	2	627
Future[a]	235	193	28	25	41	3	525
Distance	49	51	18	42	5	1	166
Energy[a]	141	47	6	16	8		218
Environment[a]	94	23	3	9	7		136
Delivery time	93	10	2				105
Reputation	88	46	4	9	3		150
Fuel[a]	59	50	3	14	8		134
Charging Station	32	26	6	12	5		81
Hybrid[a]	31	33	5	16	17		102
Fossil fuel[a]	22	11	1	5			39
Environmental pollution[a]	5						5
Clean energy[a]	4	2					6
Sustainability[a]	2		1		1		4
Total	10,453	10,629	1015	1374	1846	124	25,441

[a]Themes formed within the sustainability theme are marked with superscript letter

support the United Nations' 7th Sustainable Development Goal by contributing both to environmental sustainability and accessible clean energy.

Highlighted items create awareness within the framework of the UN 7th Goal. However, these codings related to sustainability are found to be much less in other YouTube channels. It is thought that other content producers should emphasize these issues to increase consumer awareness and demand regarding environmental sustainability. In this respect, it will be beneficial for the TOGG brand to bring this issue to the forefront by collaborating with those who produce content on social media and can reach wide consumer masses.

Approximately 12% of the total codes emerged relate to the theme of sustainability. It can be seen that the official YouTube channel of the TOGG brand and other YouTube channels created by other influencers have not yet conducted sufficient work related to the issue of sustainability and that awareness regarding this issue has not yet been formed among users. If TOGG and influencers sharing about TOGG produce contents that provide more information about sustainability, it can be considered that users may behave sensitively in this regard by becoming more conscious within these contents.

When the shared video and content comments are analyzed, although the most coded YouTube channel is TOGG's official channel, it can be seen that TOGG reaches these numbers by creating many videos. Since influencers reach similar video and content comments with a few videos, it is thought that especially the TOGG brand can create more effective contents by working with influencers.

TOGG's sustainability-focused structure could serve as an inspiration source for other car manufacturers and car manufacturers wanting to invest in this area, by revealing the codes that constitute sustainability in the automotive industry, potentially contributing to the overall sustainability efforts of the sector. Therefore, this study examining the relationship between TOGG vehicles and sustainability has great significance in highlighting the role of the automotive industry in reaching sustainability goals and leaving a greener and more sustainable world to future generations.

References

Andwari, A., Pesiridis, A., Rajoo, S., Martinez-Botas, R., & Esfahaninan, V. (2017). A review of battery electric vehicle technology and readiness levels. *Renewable and Sustainable Energy Reviews, 78*, 414–430.

Arce, G., Carvalho, J., & Nascimento, L. (2014). A time series sequestration and storage model of atmospheric carbon dioxide. *Ecological Modelling, 272*, 59–67.

Asamer, J., Graser, A., Heilmann, B., & Ruthmair, M. (2016). Sensitivity analysis for energy demand estimation of electric vehicles. *Transportation Research Part D, 46*, 182–199.

Ashique, R., Salam, Z., Aziz, M., & Bhatti, A. (2017). Integrated photovoltaic-grid dc fast charging system for electric vehicle: A review of the architecture and control. *Renewable and Sustainable Energy Reviews, 69*, 1243–1257.

Bickert, S., Kampker, A., & Greger, D. (2015). Developments of CO2-emissions and costs for small electric and combustion engine vehicles in Germany. *Transportation Research Part D, 36*, 138–151.

Bradley, T., & Quinn, C. (2010). Analysis of plug-in hybrid electric vehicle utility factors. *Journal of Power Sources, 195*, 5399–5408.

Bühler, F., Cocron, P., Neumann, I., Franke, T., & Krems, J. (2014). Is EV experience related to EV acceptance? Results from a German field study. *Transportation Research Part F: Traffic Psychology and Behaviour, 25*, 34–49.

Creswell, J. W. (2016). *Araştırma Deseni Nitel, Nicel ve Karma Yöntem Yaklaşımları*. Siyasal Kitabevi.

Creswell, J. W. (2018). *Nitel Araştırma Yöntemleri Beş Yaklaşıma Göre Nitel Araştırma ve Araştırma Deseni*. Siyasal Kitabevi.

Ebinger, J., & Vandycke, O. (2015). *Moving toward climate-resilient transport: The World Bank's experience from building adaptation into programs*. World Bank.

Eddy, J., Pfeiffer, A., & van de Staaij, J. (2019, June). *Recharging economies: The EV-battery manufacturing outlook for Europe*. McKinsey & Company.

Ghommem, M., Hajj, M., & Puri, I. (2012). Influence of natural and anthropogenic carbon dioxide sequestration on global warming. *Ecological Modelling, 235–236*, 1–7.

Hota, A., Juvvanapudi, M., & Bajpai, P. (2014). Issues and solution approaches in PHEV integration to smart grid. *Renewable and Sustainable Energy Reviews, 30*, 217–229.

IEA Publications. (2023). *Global EV Outlook 2022*. International Energy Agency.

Javid, R. J., & Nejat, A. (2017, November 24). A comprehensive model of regional electric vehicle adoption and penetration. *Transport Policy, 54*, 40–42.

Khoo, Y. B., Wang, C. H., Paevere, P., & Higgins, A. (2014). Statistical modeling of electric vehicle electricity consumption in the Victorian EV trial, Australia. *Transportation Research Part D, 32*, 263–277.

Kocabey, S. (2018). Elektrikli Otomobillerin Dünü, Bugünü ve Geleceği. *Akıllı Ulaşım Sistemleri ve Uygulamaları Dergisi, 1*(1), 16–23.

Kocagöz, E., İğde, Ç. S., & Çetindağ, G. (2020). Elektrikli ve Akıllı, Yerli ve Milli: Türkiye'nin Otomobili Girişim Grubu'nun Tanıttığı Araçlara Yönelik Tüketicilerin İlk Değerlendirmeleri. *Erciyes Üniversitesi Sosyal Bilimler Enstitüsü Dergisi XLIX, 2*, 55–74.

Kumar, M., & Revankar, S. (2017). Development scheme and key technology of an electric vehicle: An overview. *Renewable and Sustainable Energy Reviews, 70*, 1266–1285.

Mersky, A. C., Sprei, F., Samaras, C., & Qian, Z. (2016). Effectiveness of incentives on electric vehicle adoption in Norway. *Transportation Research Part D, 46*, 56–68.

Otokokpit.com. (n.d.). *Otomobil İçerikli Takip Edilesi En İyi 10 Türkçe Youtube Kanalı.* Retrieved from Editörden Seçmeler https://www.otokokpit.com/otomobil-icerikli-takip-edilesi-en-iyi-10-turkce-youtube-kanali/

Skippon, S. (2014). How consumer drivers construe vehicle performance: Implications for electric vehicles. *Transportation Research Part F: Traffic Psychology and Behaviour, 23*, 15–31.

Swain, R. B., & Karimu, A. (2019). Renewable electricity and sustainable development goals in the EU. *World Development, 125*, 1–11.

Tilki, İ., Bozkurt, B., Aktan, M., & Koçak, A. (2022). Türkiye'nin Ülke İmajı ile Genel Ürün İmajının Yerli Otomobil TOGG'un Algılanan Ürün İmajına ve Satın Alma Niyetine Etkisi. *Bingöl Üniversitesi İktisadi ve İdari Bilimler Fakültesi Dergisi, 6*(2), 91–126. https://doi.org/10.33399/biibfad.1055642

TOGG. (2021). *Hakkımızda.* 07 30, 2023 tarihinde Türkiye'nin Otomobili Girişim Grubu. Retrieved from https://www.togg.com.tr/about

Tran, M., Banister, D., Bishop, J. D. K., & Mcculloch, M. D. (2012). Realizing the electric-vehicle revolution. *Nature Climate Change, 2*, 328–333. https://doi.org/10.1038/nclimate1429

Uslu, H., & Demirel, O. (2022). Elektrikli Otomobil Satın Alma İstekliliğini Etkileyen Faktörler: Konya İli Örneği. *Süleyman Demirel Üniversitesi Vizyoner Dergisi, 13*(35), 961–975. https://doi.org/10.21076/vizyoner.1024620

Yaprak, B. (2023). Togg'un pazarlama çevresinin değerlendirilmesi: PESTLE analizi. *Akıllı Ulaşım Sistemleri ve Uygulamaları Dergisi, 6*(1), 155–165. https://doi.org/10.51513/jitsa.1255172

Yılmaz, A. (2020). Reklam mecrası Instagram: TOGG ve Günsel elektrikli otomobil markaları üzerine ampirik bir araştırma. *Erciyes İletişim Dergisi, 7*(2), 895–915.

Yılmaz, A., Ulaş, M., & Yeşil, A. (2022). Etnosentrizm ve Tüketici İlişkiselliği: Yerli Otomobil Satın Alma Niyetinin Öngörüsü. *Selçuk İletişim, 15*(1), 110–141. https://doi.org/10.18094/josc.976314

Yin, R. K. (2009). *Case study research: Design and method.* Sage.

Zhang, R., & Yao, E. (2015). Electric vehicles' energy consumption estimation with real driving condition data. *Transportation Research Part D, 41*, 177–187.

Chapter 23
Unlocking the Benefits of Energy Efficiency: A Guide to Assessing and Improving Your Business's Energy Conservation

Solomon Eghosa Uhunamure and Abderrahmen Bouchenine

23.1 Introduction

The main tasks aimed at achieving the company's main goal in the field of energy efficiency are the following:

- Rational use through systematic monitoring and management of energy consumption
- Maximum utilization of secondary energy resources
- Compliance with technological discipline
- Improvement of management and technological processes and equipment
- Implementation of energy-saving measures

The proposed methodology for assessing the energy efficiency of a production complex contains the following main provisions:

- Calculation of energy efficiency indicators at the level of the production complex as a whole, at the level of manufacturing its products, and at the level of the technological process used to create a certain type of PC products
- Formation of blocks of indicators of efficiency of energy-consuming systems and energy saving, economic efficiency of energy consumption and efficiency of use of fixed assets, and environmental efficiency of consumption at each of the specified levels

S. E. Uhunamure (✉)
University of Cape Town, Cape Town, South Africa
e-mail: solomon.uhunamure@uct.ac.za

A. Bouchenine
College of Computing and Mathematics, King Fahd University of Petroleum and Minerals, Riyadh, Saudi Arabia

H. Dinçer et al. (eds.), *Decision Making in Interdisciplinary Renewable Energy Projects*, Contributions to Management Science,
https://doi.org/10.1007/978-3-031-51532-3_23

- Consideration of the following main processes for industrial production: material preparation, semifinished product production, additional processing, and final processing
- Use of indicators calculated using energy consumption and power consumption for both primary energy resources and secondary (final) energy resources
- Assessment of energy consumption growth prospects based on green economy indicators, which are based on indicators of greenhouse gas emissions, pollutant emissions, etc.
- Assessment of the significance of individual blocks (energy efficiency indicators) using the hierarchy analysis method

In conclusion, we note that the described methodology allows us to evaluate energy efficiency indicators in general and analyze each production process, starting with the preparation of fuel for the process and ending with the final processing of the product. Such an analysis is extremely necessary, since it allows you to assess the impact of each individual process on the final result while excluding the mutual influence of processes in the assessment (Xu et al., 2023).

Increasing the competitiveness of companies should be carried out in line with increasing energy efficiency and correspond to the energy-saving development of the industrial production economy within the framework of the new industrial policy. This should ensure sustainable socioeconomic development of industrial enterprises, industrial complexes, and the country as a whole (Han et al., 2023).

The main aims of the presented material are as follows:

1. Developing an energy strategy for the development of enterprises requires taking into account a whole range of factors and constantly reviewing them in connection with new technological achievements and initiatives of competitors. Without such a strategy, it is impossible to build a clear and precise program for improving energy efficiency and energy saving, without which it is impossible to claim the status of competitive production today (Aksoy et al., 2023).
2. The energy efficiency improvement program involves evaluating the energy efficiency of production as a whole, taking into account the interrelationships of various processes, and evaluating opportunities for optimizing individual processes, components, and system elements (Kou et al., 2023a, b). In this regard, it is proposed to determine energy efficiency using a system of indicators divided into three levels: the level of the production complex as a whole, the level of individual types of products produced by the production complex, and the level of the technological process for the production of products (Niu et al., 2023).
3. The scientific and methodological approach to research and competitiveness improvement based on their energy-efficient development is based on a comprehensive systematization of development factors from the standpoint of energy efficiency and energy conservation and assessment of their impact on its competitiveness. The approach implementation scheme includes a block of multifactor multilevel energy efficiency assessment, a block of measures and energy-efficient development projects, and a block of forming predictive

strategies for energy-efficient development aimed at ensuring competitiveness and achieving long-term competitive advantages (Ai et al., 2023).

4. The energy efficiency assessment characterizes the efficiency of energy-consuming systems and energy saving, the economic efficiency of energy consumption and the use of fixed assets, and the environmental efficiency of fuel and energy resources consumption (Wan et al., 2023).
5. The methodology of multi-criteria selection of priority energy-efficient projects is based on an indicative analysis, which makes it possible to make a final choice of the most optimal composition of priority energy-efficient development projects, the implementation of which is aimed at improving the efficiency of its production and economic activities and competitiveness (Yüksel & Dinçer, 2023).
6. Testing of the scientific and methodological approach to the study and improvement of PC competitiveness based on its energy-efficient development confirmed the validity of practical use of the developed methodological tools for assessing the energy efficiency of the production complex and selecting priority projects for its development aimed at improving competitiveness (Moiseev et al., 2023a, b).

23.2 Literature Review

It should also be noted that the technologies used for the production of copper products, almost all foreign competitors are using, are more modern and more efficient in terms of energy consumption and environmental impact (Bhuiyan et al., 2021; Dong et al., 2021; Mikhaylov, 2021; Barykin et al., 2022a; Liu et al., 2022a, b; Bhuiyan et al., 2022b; Danish et al., 2022a, b; Saqib et al., 2021; Mukhametov et al., 2021; Candila et al., 2021; Mikhaylov & Grilli, 2022; Li et al., 2022).

If we talk about the final assessment for the block of environmental efficiency of fuel and energy resources consumption, it should be noted that since the last two indicators turned out to be extremely low, this affected the overall cumulative assessment. Low energy efficiency indicators in the period under review were observed in the company for all units, which indicates the need to correct the current situation. Naturally, this state of affairs has an extremely negative impact on the company's competitiveness and urgently requires the development and implementation of appropriate measures to correct this situation (Khan et al., 2022; Dinçer et al., 2022a, b; Badr et al., 2022; Barykin et al., 2022b; Mehta et al., 2022; Kalinina et al. 2022; Shaikh et al., 2022; Mikhaylov et al., 2022a, b; Nyangarika et al., 2022).

At the same time, the changes in the situation for the company in recent years can generally be described as positive: there is a gradual increase in the company's energy efficiency. Of course, this is influenced by the fact that after a certain decline, there was an increase in supply in the markets of copper products (Eti et al., 2022). In addition, there is a positive effect from the implementation of measures related to improving the efficiency of consumption in the company (Denisova et al., 2019; Nyangarika et al., 2019a, b; Huang et al., 2021a, b; Mikhaylov, 2018, 2022;

Mikhaylov et al., 2019; Conteh et al., 2021; Sediqi et al., 2022, Khan et al., 2021; Bhuiyan et al., 2022a; Liu et al., 2021a, b; Daniali et al., 2021; Moiseev et al., 2023a, b).

Results of the energy efficiency assessment:

1. In terms of individual energy efficiency indicators (energy intensity and electrical intensity of gross revenue, coefficients of energy consumption and electricity consumption ahead of production growth, a number of indicators of environmental efficiency of consumption), the company looks even more preferable than its competitors (Dinçer et al., 2023a, b; Ivleva et al., 2023; Aissou et al., 2023; Mikhaylov, 2023).
2. It has a high level of consumption per unit of production in physical terms. This once again confirms the fact that these indicators exceed foreign ones by several times, which negatively affects competitors.
3. For the most part, the company's energy efficiency indicators are inferior to those of its competitors, falling short of the basic level.
4. Since 2016, there have been positive trends in changing the situation in most of the company's energy efficiency indicators, which is primarily due to rising prices for copper products and the gradual recovery of global copper markets in recent years (Dayong et al., 2020; Mikhaylov et al., 2018; Nyangarika et al., 2018; Danish et al., 2020, 2021; An et al., 2021; Uyeh et al., 2021; Tamashiro et al., 2021, 2023; Shaikh et al., 2021; Mikhaylov et al., 2023; Wang et al. 2023; He et al., 2023; Irshad et al., 2023; Iqbal et al., 2023; Aksoy et al., 2023).

Such conclusions largely do not correspond to the current energy strategy of Russia, which sets out new requirements for the energy efficiency of the country's economy and industrial complexes in the context of a changed economic and energy policy (Yüksel et al., 2023). It is clear that while maintaining the existing specific energy intensity of the economy in the future, achieving the projected pace of development of the country is problematic (Carayannis et al., 2023). In this regard, modern orders of the Government of the Russian Federation suggest a number of measures designed to change the current state of affairs: they approve requirements for the construction of industrial facilities and the operation of equipment in accordance with modern technologies and international standards, establish a system of payments for energy conservation, and develop the market for "white certificates" and energy service contracts (An et al., 2019a, b, 2020a, b, c, 2022; Mikhaylov & Tarakanov, 2020; Moiseev et al., 2020, 2021; Gura et al., 2022; Dooyum et al., 2020; Mikhaylov et al., 2020; Mikhaylov, 2020a, b, c).

23.3 Results

Energy conservation and energy efficiency are increasingly important topics for companies around the world. Concerns about the environment and the need to reduce costs are driving many companies to evaluate and improve their energy conservation practices. In this guide, I will introduce the benefits of energy conservation and energy efficiency for businesses, as well as tips on how to evaluate and improve your company's energy conservation (Dinçer et al., 2023a, b, c, d, e, f).

Energy conservation and energy efficiency are related but distinct concepts. Energy conservation refers to the responsible and careful use of energy, while energy efficiency refers to the efficient use of energy. Both are important for reducing energy consumption and associated costs. In addition, energy conservation and energy efficiency help reduce the carbon footprint of businesses, which is important for protecting the environment.

There are many benefits for businesses that adopt energy conservation and energy efficiency practices. First, energy conservation and energy efficiency can reduce a company's energy costs. This is especially important for companies that use large amounts of energy, such as manufacturing companies. In addition, energy conservation and energy efficiency can improve the company's image by showing that the company cares about the environment and is actively working to reduce its carbon footprint (Martínez et al., 2023).

The first step to improving your company's energy conservation is to understand how your company consumes energy. This can be done through an energy assessment, which is a comprehensive assessment of a company's energy consumption. During an energy assessment, an energy assessor will analyze a company's energy systems, including lighting, heating and cooling systems, appliances, and office equipment. Based on the results of the energy assessment, the company can identify areas where it can improve energy efficiency.

Once you understand how your business consumes energy, the next step is to conduct an energy audit. An energy audit is a detailed analysis of a company's energy consumption. During an energy audit, an energy assessor will analyze a company's energy consumption over a specific period of time. Based on the results of the energy audit, the company can identify areas where it can reduce energy consumption and improve energy efficiency.

Once you have identified areas where your business can improve energy efficiency, it is important to implement energy-efficient practices. This can include installing more efficient lighting and heating equipment, upgrading office equipment to more energy-efficient models, and implementing energy conservation practices such as turning off equipment when not in use.

Another way to improve your company's energy efficiency is to choose energy-efficient equipment and appliances. There are many equipment and appliances on the market that have been designed to be more energy efficient. By choosing energy-efficient equipment and appliances, your business can reduce its energy consumption and its associated costs (Eti et al., 2023).

In addition to evaluating and improving your company's energy efficiency, it is important for your company's employees to adopt energy conservation practices at home. This can include turning off lights when not in use, adjusting the thermostat to reduce energy consumption, and using energy-efficient appliances.

Technology plays an important role in energy conservation and energy efficiency. There are many technologies available that can help businesses reduce their energy consumption and improve energy efficiency. This may include implementing automation systems that turn off equipment when not in use, installing motion sensors to turn off lights when a room is not in use, and installing energy management systems to monitor and control energy consumption.

There are many programs and incentives available that can help businesses improve their energy efficiency. This can include government programs that offer subsidies or tax credits to businesses that implement energy conservation practices, as well as programs offered by utility companies that encourage businesses to reduce their energy consumption.

23.4 Conclusion

Finally, it is important to mention the energy efficiency certifications and ratings that are available. There are many organizations that offer energy efficiency certifications for buildings and equipment. By obtaining an energy efficiency certification, your company can demonstrate its commitment to energy conservation and energy efficiency.

Energy conservation and energy efficiency are important to reduce energy costs, improve company image, and protect the environment. By following the tips presented in this guide, your business can evaluate and improve its energy efficiency, reducing its energy consumption and its associated costs. In addition, by adopting energy conservation practices at home, your company's employees can help further reduce energy consumption.

References

Ai, R., Zheng, Y., Yüksel, S., & Dinçer, H. (2023). Investigating the components of fintech ecosystem for distributed energy investments with an integrated quantum spherical decision support system. *Financial Innovation, 9*(1), 27.

Aissou, A., Ahmed, H. S., Hussein, M. E., Hassan, M. R., Mikhaylov, A., & Mahmoud, E. (2023). Optimizing reliability of flow networks using RWGA subject to total capacity constraint. *Applied Mathematics and Information Sciences, 7*(4), 559–565. https://doi.org/10.18576/amis/170404

Aksoy, T., Yüksel, S., Dınçer, H., Hacioglu, U., Mikhaylov, A., & Muyeen, S. M. (2023). Assessing the electricity production capacities of emerging markets for the sustainable investment. *IEEE Access, 11*, 24574–24590.

An, J., Mikhaylov, A., & Moiseev, N. (2019a). Oil price predictors: Machine learning approach. *International Journal of Energy Economics and Policy, 9*(5), 1–6.

An, J., Mikhaylov, A., & Sokolinskaya, N. (2019b). Machine learning in economic planning: Ensembles of algorithms. *Journal of Physics: Conference Series, 1353*, 012126.

An, J., Mikhaylov, A., & Jung, S.-U. (2020a). The strategy of South Korea in the global oil market. *Energies, 13*(10), 2491.

An, J., Mikhaylov, A., & Kim, K. (2020b). Machine learning approach in heterogeneous group of algorithms for transport safety-critical system. *Applied Sciences, 10*(8), 2670.

An, J., Mikhaylov, A., & Richter, U. H. (2020c). Trade war effects: Evidence from sectors of energy and resources in Africa. *Heliyon, 6*, e05693. https://doi.org/10.1016/j.heliyon.2020.e05693

An, J., Mikhaylov, A., & Jung, S.-U. (2021). A linear programming approach for robust network revenue management in the airline industry. *Journal of Air Transport Management, 91*, 101979.

An, J., Mikhaylov, A., Dinçer, H., & Yüksel, S. (2022). Economic modelling of electricity generation: Long short-term memory and Q-rung orthopair fuzzy sets. *Heliyon, 8*, e12345.

Badr, A. M. N. A. A., Mohamed, T. I., Osman, N. A. Q., & Mikhaylov, A. (2022). A review of social media website users' interaction paths with governmental accounts during the COVID-19 pandemic. *Informatics, 9*, 50. https://doi.org/10.3390/informatics9030050

Barykin, S., Mikheev, A., Kiseleva, E., Putikhin, Y., Alekseeva, N., & Mikhaylov, A. (2022a). An empirical analysis of Russian regions' debt sustainability. *Economies, 10*(4), 296.

Barykin, S. E., Kapustina, I. V., Sergeev, S. M., Daniali, S. M., Kopteva, L. A., Semenova, G. N., Pryadko, I. P., Mikhaylov, A., Baboshkin, P., Datsyuk, P., & Senjyu, T. (2022b). Financial logistics models based on systematic approach improving management solutions. *F1000Research, 11*, 570. https://doi.org/10.12688/f1000research.111252.1

Bhuiyan, M. A., An, J., Mikhaylov, A., Moiseev, N., & Danish, M. S. S. (2021). Renewable energy deployment and COVID-19 measures for sustainable development. *Sustainability, 13*, 4418.

Bhuiyan, M. A., Dinçer, H., Yüksel, S., Mikhaylov, A., Danish, M. S. S., Pinter, G., Uyeh, D. D., & Stepanova, D. (2022a). Economic indicators and bioenergy supply in developed economies: QROF-DEMATEL and random forest models. *Energy Reports, 8*(2022), 561–570.

Bhuiyan, M. A., Zhang, Q., Khare, V., Mikhaylov, A., Pinter, G., & Huang, X. (2022b). Renewable energy consumption and economic growth nexus—A systematic literature review. *Frontiers in Environmental Science., 10*, 878394. https://doi.org/10.3389/fenvs.2022.878394

Candila, V., Maximov, D., Mikhaylov, A., Moiseev, N., Senjyu, T., & Tryndina, N. (2021). On the relationship between oil and exchange rates of oil-exporting and oil-importing countries: From the great recession period to the Covid-19 era. *Energies, 14*, 8065.

Carayannis, E., Kostis, P., Dinçer, H., & Yüksel, S. (2023). Quality function deployment-oriented strategic outlook to sustainable energy policies based on quintuple innovation helix. *Journal of the Knowledge Economy*, 1–19.

Conteh, F., Takahashi, H., Hemeida, A. M., Krishnan, N., Mikhaylov, A., & Senjyu, T. (2021). Hybrid grid-connected renewable power generation for sustainable electricity supply in Sierra Leone: Case study Lungi Town. *Sustainability, 13*(18).

Daniali, S. M., Barykin, S. E., Kapustina, I. V., Khortabi, F. M., Sergeev, S. M., Kalinina, O. V., Mikhaylov, A., Veynberg, R., Zasova, L., & Senjyu, T. (2021). Predicting volatility index according to technical index and economic indicators on the basis of deep learning algorithm. *Sustainability, 13*, 14011.

Danish, M. S. S., Bhattacharya, A., Stepanova, D., Mikhaylov, A., Grilli, M. L., Khosravy, M., & Senjyu, T. (2020). A systematic review of metal oxide applications for energy and environmental sustainability. *Metals, 10*(12), 1604.

Danish, M. S. S., Senjyu, T., Sabory, N. R., Khosravy, M., Grilli, M. L., Mikhaylov, A., & Majidi, H. (2021). A forefront framework for sustainable aquaponics modeling and design. *Sustainability, 13*, 9313.

Danish, M. S., Senjyu, T. S., Sabory, N. R., & Mikhaylov, A. (2022a). Lessons from COVID-19 conferring environmental re-engineering opportunity. In M. Danish & T. Senjyu (Eds.), *Eco-friendly and agile energy strategies and policy development* (pp. 99–118). IGI Global. https://doi.org/10.4018/978-1-7998-9502-2.ch005

Danish, M. S., Senjyu, T. S., Sabory, N. R., & Mikhaylov, A. (2022b). Interplay consequences of COVID-19 on global environmental sustainability. In M. Danish & T. Senjyu (Eds.), *Eco-friendly and agile energy strategies and policy development*. IGI Global. https://doi.org/10.4018/978-1-7998-9502-2.ch003

Dayong, N., Mikhaylov, A., Bratanovsky, S., Shaikh, Z. A., & Stepanova, D. (2020). Mathematical modeling of the technological processes of catering products production. *Journal of Food Process Engineering, 43*(2), 1–9.

Denisova, V., Mikhaylov, A., & Lopatin, E. (2019). Blockchain infrastructure and growth of global power consumption. *International Journal of Energy Economics and Policy, 9*(4), 22–29.
Dinçer, H., Yüksel, S., Mikhaylov, A., Pinter, G., & Shaikh, Z. A. (2022a). Analysis of renewable-friendly smart grid technologies for the distributed energy investment projects using a hybrid picture fuzzy rough decision-making approach. *Energy Reports, 8*, 11466–11477. https://doi.org/10.1016/j.egyr.2022.08.275
Dinçer, H., Yüksel, S., Mikhaylov, A., Barykin, S. E., Aksoy, T., & Hacioğlu, U. (2022b). Analysis of environmental priorities for green project investments using an integrated q-rung orthopair fuzzy modelling. *IEEE Access, 10*, 50996–51007. https://doi.org/10.1109/ACCESS.2022.3174058
Dinçer, H., Yüksel, S., Hacioglu, U., Yilmaz, M. K., & Delen, D. (2023a). Development of a sustainable corporate social responsibility index for performance evaluation of the energy industry: A hybrid decision-making methodology. *Resources Policy, 85*, 103940.
Dinçer, H., Eti, S., Gökalp, Y., Yüksel, S., & Çelebi, B. (2023b). The role of small-scale wind turbines in increasing the competitiveness of hospitals. In *Handbook of research on quality and competitiveness in the healthcare services sector* (pp. 336–352). IGI Global.
Dinçer, H., Yüksel, S., Mikhaylov, A., Muyeen, S. M., Chang, T., Barykin, S., & Kalinina, O. (2023c). CO2 emissions integrated fuzzy model: A case of seven emerging economies. *Energy Reports, 9*, 5741–5751.
Dinçer, H., Yüksel, S., Aksoy, T., Hacıoğlu, Ü., Mikhaylov, A., & Pinter, G. (2023d). Analysis of solar module alternatives for efficiency-based energy investments with hybrid 2-tuple IVIF modeling. *Energy Reports, 10*, 61–71.
Dinçer, H., Eti, S., Yüksel, S., Özdemir, S., Yflmaz, A. E., & Ergün, E. (2023e). Integrating data mining and fuzzy decision-making techniques for analyzing the key minimizing factors of carbon emissions. *Journal of Intelligent & Fuzzy Systems, 45*, 1–17.
Dinçer, H., Eti, S., Aksoy, T., Yüksel, S., Hacioglu, U., Mikhaylov, A., & Muyeen, S. M. (2023f). Analysis of environmental impact for material production investments using a novel soft computing methodology. *IEEE Access, 11*, 37987–38001. https://doi.org/10.1109/ACCESS.2023.3266524
Dong, B., Ikonnikova, I., Rogulin, R., Sakulyeva, T., & Mikhaylov, A. (2021). Environmental economic approach to optimization of transport communication in megacities. *Journal of Environmental Science and Health, Part A, 56*(6), 660–666.
Dooyum, U. D., Mikhaylov, A., & Varyash, I. (2020). Energy security concept in Russia and South Korea. *International Journal of Energy Economics and Policy, 10*(4), 102–107.
Eti, S., Dinçer, H., Yüksel, S., & Gökalp, Y. (2022). Analysis of the suitability of the solar panels for hospitals: A new fuzzy decision-making model proposal with the T-spherical TOP-DEMATEL method. *Journal of Intelligent & Fuzzy Systems, 44*, 1–13.
Eti, S., Dinçer, H., Gökalp, Y., Yüksel, S., & Kararoğlu, D. (2023). Identifying key issues to handle the inflation problem in the healthcare industry caused by energy prices: An evaluation with decision-making models. In *Managing inflation and supply chain disruptions in the global economy* (pp. 162–178). Springer.
Gura, D., Mikhaylov, A., Glushkov, S., Zaikov, M., & Shaikh, Z. A. (2022). Model for estimating power dissipation along the interconnect length in single on-chip topology. *Evolutionary Intelligence, 15*, 2369–2373. https://doi.org/10.1007/s12065-020-00407-7
Han, Z., Gong, L., Chen, H., & Yüksel, S. (2023). The role of mineral and oil resources in achieving a clean environment in emerging economies: An asymmetric analysis. *Resources Policy, 86*, 104099.
He, H., Huang, Y., Nakadomari, A., Masrur, H., Krishnan, N., Hemeida, A. M., Mikhaylov, A., Omine, E., & Senjyu, T. (2023). Potential and economic viability of green hydrogen production from seawater electrolysis using renewable energy in remote Japanese islands. *Renewable Energy, 202*, 1436–1447. https://doi.org/10.1016/j.renene.2022.12.046

Huang, Y., Masrur, H., Shigenobu, R., Hemeida, A. M., Mikhaylov, A., & Senjyu, T. (2021a). A comparative design of a campus microgrid considering a multi-scenario and multi-objective approach. *Energies, 14*, 2853.

Huang, Y., Yona, A., Takahashi, H., Hemeida, A. M., Mandal, P., Mikhaylov, A., Senjyu, T., & Lotfy, M. E. (2021b). Energy management system optimization of drug store operate electric vehicles charging station. *Sustainability, 14*, 1217836.

Iqbal, K. M. J., Khan, M. I., Mikhaylov, A., Shah, A. A., Yadykin, V., Filho, W. L., Tariq, M. A. U. R., & Ullah, W. (2023). Modeling principles, criteria and indicators to assess water sector governance for climate compatibility and sustainability. *Frontiers in Environmental Science, 11*, 989930.

Irshad, A. S., Ludin, G. A., Masrur, H., Ahmadi, M., Yona, A., Mikhaylov, A., Krishnan, N., & Senjyu, T. (2023). Optimization of grid-photovoltaic and battery hybrid system with most technically efficient PV technology after the performance analysis. *Renewable Energy, 207*, 714–730.

Ivleva, S., Hajimajeed, M., Kartashova, O., & Mikhaylov, A. (2023). A comparative study of regional variations in common visual impairments among school-age children. *Electronic Journal of General Medicine, 20*(5), em504. https://doi.org/10.29333/ejgm/13264

Kalinina, O. V., Barykin, S. E., Sergeev, S. M., Semenova, G. N., Fatkullina, A., Mikhaylov, A., & De La Poza, E. (2022). Smart city perspectives in post-pandemic governance: Externalities reduction policy. *F1000Research, 11*, 1032. https://doi.org/10.12688/f1000research.123195.1

Khan, A. A., Shaikh, Z. A., Baitenova, L., Mutaliyeva, L., Moiseev, N., Mikhaylov, A., Laghari, A. A., Idris, S. A., & Alshazly, H. (2021). QoS-ledger: Smart contracts and metaheuristic for secure quality-of-service and cost-efficient scheduling of medical-data processing. *Electronics, 10*, 3083.

Khan, A. A., Laghari, A. A., Gadekallu, T. R., Shaikh, Z. A., Javed, A. R., Rashid, M., Estrela, V. V., & Mikhaylov, A. (2022). A drone-based data management and optimization using metaheuristic algorithms and blockchain smart contracts in a secure fog environment. *Computers and Electrical Engineering, 102*, 108234. https://doi.org/10.1016/j.compeleceng.2022.108234

Kou, G., Pamucar, D., Dinçer, H., & Yüksel, S. (2023a). From risks to rewards: A comprehensive guide to sustainable investment decisions in renewable energy using a hybrid facial expression-based fuzzy decision-making approach. *Applied Soft Computing, 142*, 110365.

Kou, G., Yuksel, S., & Dincer, H. (2023b). A facial expression and expert recommendation fuzzy decision-making approach for sustainable business investments within the metaverse world. *Applied Soft Computing, 148*, 110849.

Li, J., Dinçer, H., Yüksel, S., Mikhaylov, A., & Barykin, S. E. (2022). Bipolar q-ROF hybrid decision making model with golden cut for analyzing the levelized cost of renewable energy alternatives. *IEEE Access, 10*, 35900. https://doi.org/10.1109/ACCESS.2022.3168315

Liu, L., Kato, T, Mandal, P., Mikhaylov, A, Hemeida, A. M., & Senjyu T. (2021a). Multi-area power system load frequency control by using model predictive control by introducing renewable energy sources. In *2021 International Conference on Science & Contemporary Technologies (ICSCT)*, pp. 1–6. https://doi.org/10.1109/ICSCT53883.2021.9642521

Liu, L., Kato, T., Mandal, P., Mikhaylov, A., Hemeida, A. M., Takahashi, H., & Senjyu, T. (2021b). Two cases studies of model predictive control approach for hybrid renewable energy systems. *AIMS Energy, 9*(6), 1241–1259.

Liu, Z.-J., Panfilova, E., Mikhaylov, A., & Kurilova, A. (2022a). COVID-19 crisis impact on the stability between parties in crowdfunding and crowdsourcing. *Wireless Personal Communications, 122*, 915–930.

Liu, Z., Panfilova, E., Mikhaylov, A., & Kurilova, A. (2022b). Assessing stability in the relationship between parties in crowdfunding and crowdsourcing projects during the COVID-19 crisis. *Journal of Global Information Management (JGIM), 30*(4), 1–18. https://doi.org/10.4018/JGIM.297905

Martínez, L., Dinçer, H., & Yüksel, S. (2023). A hybrid decision making approach for new service development process of renewable energy investment. *Applied Soft Computing, 133*, 109897.

Mehta, R., Verghese, J., Mahajan, S., Barykin, S., Bozhuk, S., Kozlova, N., Kapustina, I., Mikhaylov, A., Naumova, E., & Dedyukhina, N. (2022). Consumers' behavior in conversational commerce marketing based on messenger chatbots. *F1000Research, 11*(647), 10.12688/f1000research.122037.1.

Mikhaylov, A. (2018). The yield of the carry trade strategy. *Finance: Theory and Practice, 22*(3), 52–63. https://doi.org/10.26794/2587-5671-2018-22-3-52-63

Mikhaylov, A. (2020a). Cryptocurrency market analysis from the open innovation perspective. *Journal of Open Innovation: Technology, Market, and Complexity, 6*(4), 197.

Mikhaylov, A. (2020b). Geothermal energy development in Iceland. *International Journal of Energy Economics and Policy, 10*(4), 31–35.

Mikhaylov, A. (2020c). Cryptocurrency market development: Hurst method. *Finance: Theory and Practice, 24*(3), 81–91.

Mikhaylov, A. (2021). Lichens as indicators of atmospheric pollution in urban ecosystems. *Israel Journal of Ecology & Evolution, 67*(1–2), 60–68. https://doi.org/10.1163/22244662-bja10016

Mikhaylov, A. (2022). Sustainable development and renewable energy: A new view to a global problem. *Energies, 15*, 1397. https://doi.org/10.3390/en15041397

Mikhaylov, A. (2023). Understanding the risks associated with wallets, depository services, trading, lending, and borrowing in the crypto space. *Journal of Infrastructure, Policy and Development, 6*, 2.

Mikhaylov, A., & Grilli, M. L. (2022). Machine learning methods and sustainable development: Metal oxides and multilayer metal-oxides. *Metals, 12*(4).

Mikhaylov, A., & Tarakanov, S. (2020). Development of Levenberg-Marquardt theoretical approach for electric network. *Journal of Physics: Conference Series, 1515*, 052006.

Mikhaylov, A., Sokolinskaya, N., & Nyangarika, A. (2018). Optimal carry trade strategy based on currencies of energy and developed economies. *Journal of Reviews on Global Economics, 7*, 582–592.

Mikhaylov, A., Burakov, D., & Didenko, V. (2019). Relationship between oil price and macroeconomic indicators in Russia. *Finance: Theory and Practice, 23*(2), 105–116. https://doi.org/10.26794/2587-5671-2019-23-2-105-116

Mikhaylov, A., Moiseev, N., Aleshin, K., & Burkhardt, T. (2020). Global climate change and greenhouse effect. *Entrepreneurship and Sustainability Issues, 7*(4), 2897–2913. https://doi.org/10.9770/jesi.2020.7.4(21)

Mikhaylov, A., Bhatti, I. M., Dinçer, H., & Yüksel, S. (2022a). Integrated decision recommendation system using iteration-enhanced collaborative filtering, golden cut bipolar for analyzing the risk-based oil market spillovers. *Computational Economics*. https://doi.org/10.1007/s10614-022-10341-8

Mikhaylov, A., Yumashev, A., & Kolpak, E. (2022b). Quality of life, anxiety and depressive disorders in patients with extrasystolic arrhythmia. *Archives of Medical Science, 18*(2), 328. https://doi.org/10.5114/aoms.2020.101359

Mikhaylov, A., Dinçer, H., Yüksel, S., Pinter, G., & Shaikh, Z. A. (2023). Bitcoin mempool growth and trading volumes: Integrated approach based on QROF multi-SWARA and aggregation operators. *Journal of Innovation & Knowledge, 8*(3), 100378.

Moiseev, N., Mikhaylov, A., Varyash, I., & Saqib, A. (2020). Investigating the relation of GDP per capita and corruption index. *Entrepreneurship and Sustainability Issues, 8*(1), 780–794.

Moiseev, N., Sorokin, A., Zvezdina, N., Mikhaylov, A., Khomyakova, L., & Danish, M. S. S. (2021). Credit risk theoretical model on the base of DCC-GARCH in time-varying parameters framework. *Mathematics, 9*, 2423. https://doi.org/10.3390/math9192423

Moiseev, N., Mikhaylov, A., Dinçer, H., & Yüksel, S. (2023a). Effects of market capitalization shocks for open innovation models in e-commerce industry: Golden cut QROF MCDM analysis. *Financial Innovation, 9*(1), 32.

Moiseev, N., Mikhaylov, A., Dinçer, H., & Yüksel, S. (2023b). Market capitalization shock effects on open innovation models in e-commerce: Golden cut q-rung orthopair fuzzy multicriteria decision-making analysis. *Financial Innovation, 9*(1), 55.

Mukhametov, A., Bekhorashvili, A., Avdeenko, A., & Mikhaylov, A. (2021). The impact of growing legume plants under conditions of biologization and soil cultivation on Chernozem fertility and productivity of rotation crops. *Legume Research, 44*, 12.

Niu, X., Yüksel, S., & Dinçer, H. (2023). Emission strategy selection for the circular economy-based production investments with the enhanced decision support system. *Energy, 274*, 127446.

Nyangarika, A., Mikhaylov, A., & Tang, B.-J. (2018). Correlation of oil prices and gross domestic product in oil producing countries. *International Journal of Energy Economics and Policy, 8*(5), 42–48.

Nyangarika, A., Mikhaylov, A., & Richter, U. (2019a). Influence oil price towards economic indicators in Russia. *International Journal of Energy Economics and Policy, 9*(1), 123–129.

Nyangarika, A., Mikhaylov, A., & Richter, U. (2019b). Oil price factors: Forecasting on the base of modified auto-regressive integrated moving average model. *International Journal of Energy Economics and Policy, 9*(1), 149–159.

Nyangarika, A., Mikhaylov, A., Muyeen, S. M., Yadykin, V., Mottaeva, A., Pryadko, I., Barykin, S., Fomenko, N., Rykov, G., & Shvandar, K. (2022). Energy stability and decarbonization in developing countries: Random Forest approach for forecasting of crude oil trade flows and macro indicators. *Frontiers in Environmental Science, 10*, 1031343. https://doi.org/10.3389/fenvs.2022.1031343

Saqib, A., Chan, T.-H., Mikhaylov, A., & Lean, H. H. (2021). Are the responses of sectoral energy imports asymmetric to exchange rate volatilities in Pakistan? Evidence from recent foreign exchange regime. *Frontiers in Energy Research, 9*, 614463. https://doi.org/10.3389/fenrg.2021.614463

Sediqi, M. M., Nakadomari, A., Mikhaylov, A., Krishnan, N., Lotfy, M. E., Yona, A., & Senjyu, T. (2022). Impact of time-of-use demand response program on optimal operation of Afghanistan real power system. *Energies, 15*, 296.

Shaikh, Z. A., Moiseev, N., Mikhaylov, A., & Yüksel, S. (2021). Facile synthesis of copper oxide-cobalt oxide/nitrogen-doped carbon (Cu2O-Co3O4/CN) composite for efficient water splitting. *Applied Sciences, 11*, 9974. https://doi.org/10.3390/app11219974

Shaikh, Z. A., Kraikin, A., Mikhaylov, A., & Pinter, G. (2022). Forecasting stock prices of companies producing solar panels using machine learning methods. *Complexity, 2022*, 9186265. https://doi.org/10.1155/2022/9186265

Tamashiro, K., Alharbi, T., Mikhaylov, A., Hemeida, A. M., Krishnan, N., Lotfy, M. E., & Senjyu, T. (2021). Investigation of home energy management with advanced direct load control and optimal scheduling of controllable loads. *Energies, 14*, 7314. https://doi.org/10.3390/en14217314

Tamashiro, K., Omine, E., Krishnan, N., Mikhaylov, A., Hemeida, A. M., & Senjyu, T. (2023). Optimal components capacity based multi-objective optimization and optimal scheduling based MPC-optimization algorithm in smart apartment buildings. *Energy & Buildings, 278*, 112616. https://doi.org/10.1016/j.enbuild.2022.112616

Uyeh, D. D., Asem-Hiablie, S., Park, T., Kim, K.-M., Mikhaylov, A., Woo, S., & Ha, Y. (2021). Could Japonica rice be an alternative variety for increased global food security and climate change mitigation? *Food, 10*, 1869. https://doi.org/10.3390/foods10081869

Wan, Q., Miao, X., Wang, C., Dinçer, H., & Yüksel, S. (2023). A hybrid decision support system with golden cut and bipolar q-ROFSs for evaluating the risk-based strategic priorities of fintech lending for clean energy projects. *Financial Innovation, 9*(1), 1–25.

Wang, Q., Guo, J., Li, R., Mikhaylov, A., & Moiseev, N. (2023). Does technical assistance alleviate energy poverty in sub-Saharan African countries? A new perspective on spatial spillover effects of technical assistance. *Energy Strategy Reviews, 45*, 101047. https://doi.org/10.1016/j.esr.2022.101047

Xu, X., Yüksel, S., & Dinçer, H. (2023). An integrated decision-making approach with golden cut and bipolar q-ROFSs to renewable energy storage investments. *International Journal of Fuzzy Systems, 25*(1), 168–181.

Yüksel, S., & Dinçer, H. (2023). Sustainability analysis of digital transformation and circular industrialization with quantum spherical fuzzy modeling and golden cuts. *Applied Soft Computing, 138*, 110192.

Yüksel, S., Dinçer, H., & Uluer, G. S. (2023). Socio-economic indicators of nuclear energy investments: A study on countries. In *Inclusive developments through socio-economic indicators: New theoretical and empirical insights* (pp. 33–44). Emerald Publishing.

SPRINGER NATURE

GPSR Compliance

The European Union's (EU) General Product Safety Regulation (GPSR) is a set of rules that requires consumer products to be safe and our obligations to ensure this.

If you have any concerns about our products, you can contact us on ProductSafety@springernature.com

In case Publisher is established outside the EU, the EU authorized representative is:

Springer Nature Customer Service Center GmbH
Europaplatz 3
69115 Heidelberg, Germany

The manufacturer's authorised representative in the EU is Springer
Nature Customer Service Centre GmbH, Europaplatz 3, 69115 Heidelberg,
Germany. If you have any concerns regarding our products, please
contact ProductSafety@springernature.com

Printed and bound by CPI Group (UK) Ltd, Croydon, CR0 4YY
28/09/2026
02224070-0006